Taking SketchUp Pro to the Next Level

Go beyond the basics and develop custom 3D modeling workflows to become a SketchUp ninja

Aaron Dietzen aka 'The SketchUp Guy'

BIRMINGHAM—MUMBAI

Taking SketchUp Pro to the Next Level

Group Product Manager: Rohit Rajkumar
Publishing Product Manager: Kaustubh Manglurkar
Senior Editor: Hayden Edwards
Senior Content Development Editor: Rashi Dubey
Technical Editor: Shubham Sharma
Copy Editor: Safis Editing
Project Coordinator: Sonam Pandey
Proofreader: Safis Editing
Indexer: Subalakshmi Govindhan
Production Designer: Aparna Bhagat
Marketing Coordinator: Elizabeth Varghese

First published: September 2022

Production reference: 1250822

Published by Packt Publishing Ltd.
Livery Place
35 Livery Street
Birmingham
B3 2PB, UK.

ISBN 978-1-80324-269-9

www.packt.com

To my amazing wife, Katie. There is no way I could tackle something as huge as writing a book without you there to support me. Thank you for always believing in me and sticking with me. I love you!

– Aaron Dietzen aka 'The SketchUp Guy'

Contributors

About the author

Aaron Dietzen aka 'The SketchUp Guy' has been modeling in 3D since high school when he started at his father's software company. Over the next few decades, he worked in technical support, training, and product management, learning about dozens of modeling and CAD software packages along the way.

One of the software packages Aaron learned about was SketchUp, which he taught himself how to use. He currently works for Trimble, creating video content about SketchUp. Often referred to as "The SketchUp Guy," Aaron focuses on helping SketchUp users to develop their own workflows and become more efficient modelers and does his best to make learning enjoyable. Aaron lives in Colorado with his wife, three kids, and a couple of rescued puppies.

I want to thank everyone who has ever given me an opportunity to create, teach, and just be me. To my parents, who supported me to create, thank you! To my bosses and managers, who encouraged me to learn and teach, thank you! To my friends and family, who have stuck with me and supported my creative efforts, thank you! To the amazing team at Packt and Duane, thank you for helping me to write this book. Finally, to YOU, who are reading this book. Thank you for allowing me to be a part of your effort to grow. It is an honor to be a part of your learning journey.

About the reviewer

Duane Kemp, a 3D reconstruction specialist and family man living in Switzerland, started working with Google SketchUp 6 in 2007. He used SU8 as a project manager to reproduce an old Swiss mill and build execution plans for its 2008 *Moulin de Rivendell* ongoing renovation project. His company, Kemp Productions' animations were presented at *SketchUp Basecamp 2012* and have featured in books and articles: *CatchUp Edition 15*, Daniel Tal's *Rendering in SketchUp*, Matt Donley's *SketchUp to LayOut* article *Mason Farm Renders*, and more.

The winner of various 3D competitions, he admins Facebook's *Trimble SketchUp Group* and has promoted the use of SketchUp since 2012 on social media platforms. Duane likes to test plugins/rendering software and enjoys using SketchUp.

This is a wonderful introductory, step-by-step overview of SketchUp Pro's basic features, usage, available tools, and companion apps... important aspects of the software to get a user up and running! It will be a great accompaniment for new and old users as they explore updated and newly introduced features of SketchUp Pro and LayOut as well.

To Rashika and the Packt team, thanks for the opportunity to consult on Aaron's recommendable contribution to the SketchUp community.

Table of Contents

3

Modifying Native Commands 59

4

Taking Inferencing to the Next Level 79

5

Creating Beautiful Custom Materials 97

Part 2: Customizing SketchUp and Making It Your Own

6

Knowing What You Need Out of SketchUp 131

7

Creating Custom Shortcuts 147

8

Customizing Your User Interface 165

9

Taking Advantage of Templates 195

14

Introduction to LayOut 345

15

Leveraging the SketchUp Ecosystem 385

Preface

SketchUp is the quickest and easiest way to create 3D models. Not only is SketchUp super easy to learn with its approachable interface, but it is an extremely capable piece of software that can take you far beyond what you may have initially thought possible. With this book, you will be able to quickly level up from a basic user and get started on the path of becoming a SketchUp ninja!

In each chapter, you will walk through the capabilities of SketchUp, challenging you to use tools in ways you may not have thought of in the past. This includes thoughts on organizing your model, modifying native commands, customizing your interface, utilizing inferencing, and more.

Additionally, you will learn about extensions that can be added to SketchUp to supplement the tools you have been using, allowing you to make your 3D modeling process quicker, easier, and more powerful. By the end of this book, you will have a better understanding of how to use SketchUp tools and be on your way to customizing SketchUp so that it is tailored to your own one-of-a-kind workflow.

Who this book is for

This book is for designers, architects, and professional modelers who have used SketchUp before, are perhaps self-taught or have completed software training, and find themselves needing more than just the basics from SketchUp. We assume that you have spent some time in SketchUp before and are capable of basic modeling.

What this book covers

Chapter 1, Reviewing the Basics, gives a quick recap of SketchUp functionality.

Chapter 2, Organizing Your 3D Model, includes tips that will help keep your SketchUp model organized.

Chapter 3, Modifying Native Commands, discusses getting more out of the native commands using modifier keys.

Chapter 4, Taking Inferencing to the Next Level, explains how to save yourself time and energy by making inferencing work for you.

Chapter 5, Creating Beautiful Custom Materials, discusses making the outside of your model look amazing!

Chapter 6, Knowing What You Need Out of SketchUp, explains that before you can develop your own custom SketchUp workflows, you need to nail down what you need.

Chapter 7, Creating Custom Shortcuts, explains how to speed up your modeling time using shortcut keys.

Chapter 8, Customizing Your User Interface, discusses how you can tailor your on-screen space for your specific workflow.

Chapter 9, Taking Advantage of Templates, explains how you can get quicker with templates designed specifically for what you plan to model.

Chapter 10, Hardware to Make You a More Efficient Modeler, discusses some hardware upgrades to consider as you develop your SketchUp skills.

Chapter 11, What Are Extensions?, explains the extensions that help you to take SketchUp far beyond what is possible with just native commands.

Chapter 12, Using 3D Warehouse and Extension Warehouse, discusses SketchUp's integrated repositories of extensions and 3D entourage.

Chapter 13, Must-Have Extensions for Any Workflow, explains the extensions that will fit into almost any workflow.

Chapter 14, Introduction to LayOut, is a primer on SketchUp's backend software.

Chapter 15, Leveraging the SketchUp Ecosystem, explains that there is so much more to SketchUp than just SketchUp Pro.

To get the most out of this book

You will need a copy of SketchUp Pro installed on your Windows or macOS computer. All examples in this book were created using SketchUp 2022, but future versions should work as well. Later chapters mention LayOut, SketchUp Viewer, SketchUp for Web, and SketchUp for iPad, but they are all a part of a SketchUp Pro subscription (`https://sketchup.com/plans-and-pricing`).

Software/hardware covered in the book	Operating system requirements
SketchUp Pro	Windows or macOS
LayOut	Windows or macOS

There are also references throughout the book to files hosted on SketchUp's online repositories. You will need an internet connection and a Trimble ID in order to take advantage of these resources.

The assumption of this book is that you have done basic learning about SketchUp (either through online training sessions, videos, or other books) and are ready to move from being a new user to an intermediate/advanced user.

Download the color images

We also provide a PDF file that has color images of the screenshots and diagrams used in this book. You can download it here: `https://packt.link/aUcVr`.

Conventions used

There are a number of text conventions used throughout this book.

`Code in text`: Indicates code words in text, database table names, folder names, filenames, file extensions, pathnames, dummy URLs, user input, and Twitter handles. Here is an example: "When you import an image into SketchUp, it brings the entire image into the `.skp` file and saves it there."

Bold: Indicates a new term, an important word, or words that you see onscreen. For instance, words in menus or dialog boxes appear in **bold**. Here is an example: "Once you have drawn the profile geometry, erase the extra edges and double-click the profile, then right-click, and choose **Make Group** from the context menu."

> **Tips or Important Notes**
> Appear like this.

Get in touch

Feedback from our readers is always welcome.

General feedback: If you have questions about any aspect of this book, email us at `customercare@packtpub.com` and mention the book title in the subject of your message.

Errata: Although we have taken every care to ensure the accuracy of our content, mistakes do happen. If you have found a mistake in this book, we would be grateful if you would report this to us. Please visit `www.packtpub.com/support/errata` and fill in the form.

Piracy: If you come across any illegal copies of our works in any form on the internet, we would be grateful if you would provide us with the location address or website name. Please contact us at `copyright@packt.com` with a link to the material.

If you are interested in becoming an author: If there is a topic that you have expertise in and you are interested in either writing or contributing to a book, please visit `authors.packtpub.com`.

Part 1: Getting More Out of Native Tools

The default tools in SketchUp are even more powerful than you realize. Let's start by reviewing the fundamentals of SketchUp and see how we can take advantage of native functionality to make you a better 3D modeler.

This part contains the following chapters:

- *Chapter 1, Reviewing the Basics*
- *Chapter 2, Organizing Your 3D Model*
- *Chapter 3, Modifying Native Commands*
- *Chapter 4, Taking Inferencing to the Next Level*
- *Chapter 5, Creating Beautiful Custom Materials*

1

Reviewing the Basics

SketchUp is known for being an easy software to pick up and start using. In fact, it is probably safe to say that most users are self-taught! One of the issues that many users run into when they teach themselves, be it through online videos or via third-party books, is that they take away only what they need at that moment. This means that there is more to learn when it comes to the basics.

In this chapter, we will take a look at some of the basic functionality in SketchUp and dive into how to get the most out of a few commands. Yes, things such as extensions and creating custom shortcuts are important, and we will be covering those items in future chapters. While it would be great to do a comprehensive review of every way that every command can be used, we will have to settle for a few specific use cases. This chapter should serve as a spark to get you to think about how to get more out of the native input and editing commands. This knowledge will be the base upon which you will grow your SketchUp expertise!

In this chapter, we will cover these main topics:

- Navigating in SketchUp
- Exploring edges and faces
- Getting more out of Move
- Deforming geometry
- Diving deep into Follow Me

Technical requirements

This chapter will assume that you are using SketchUp Pro and have access to either a three-button mouse or a touchpad.

Navigating in SketchUp

One of the main things that makes SketchUp so easy to work in is the fact that you are always moving through your 3D model. Unlike other 3D modeling programs, SketchUp allows you to navigate in 3D space without the use of special viewports or widgets to change how you are viewing your model in 3D space.

Basic navigating can be thought of as three distinct pieces:

- **Orbit**
- **Zoom**
- **Pan**

The great thing about the way the SketchUp navigation is set up is that you can use the Orbit and Zoom commands without leaving the input or modification command you are currently using! In fact, depending on the hardware you are using, you can change the views of your model while in the middle of creating or modifying geometry.

The navigation commands can be used in one of three ways, depending on your hardware. It is generally accepted that the best way to navigate in SketchUp is with a three-button mouse, but it is possible to do it with a standard two-button mouse or with a trackpad:

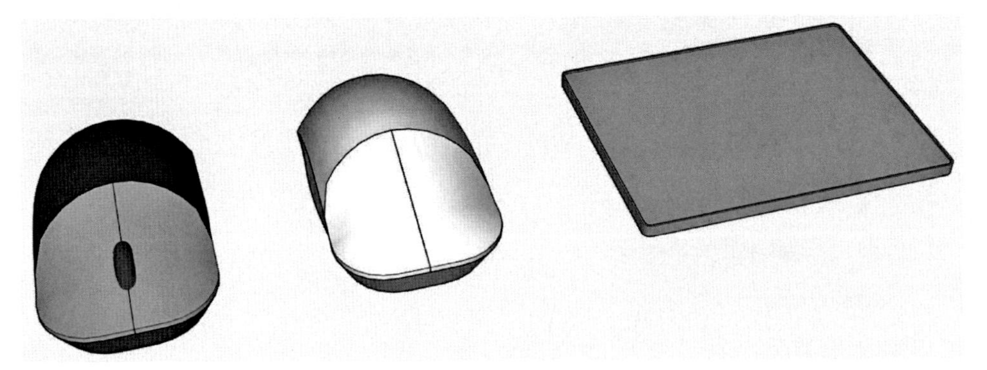

Figure 1.1 – Left to right, a three-button mouse, a two-button mouse, and a touchpad

When we talk about using a three-button mouse, we are talking about a mouse with a trackwheel in the middle of the left and right mouse buttons. While not actually a third button, the track wheel allows quick access to both zooming and orbiting commands with one finger, which can be a huge time-saver while designing. In the next sections, we will talk about how to use the commands with any of the hardware mentioned and which ones might be preferable.

It may seem odd to learn how to use any navigation technique other than the primary one (if you have a three-button mouse, why learn to use a trackpad for navigation?), but any SketchUp master knows that you cannot depend on hardware alone. I have many stories about needing to open or modify a model while on the go with no place to use my mouse, having to navigate with a touchpad, or using SketchUp at a customer site where only an old two-button mouse was available. Knowing how to use any of the hardware options will ensure your ability to move through a SketchUp model, regardless of your input device.

Speaking of Hardware Options...

We will be discussing using other hardware such as tablets or 3D mice in *Chapter 10, Hardware to Make You a More Efficient Modeler*.

Navigating with the Orbit, Zoom, and Pan icons

Obviously, your primary focus when using SketchUp will be to create and edit 3D geometry. While this will be the primary focus of most of this book, it is important to stop and look at the commands you will use more often than any input command or tool – the navigation commands:

Figure 1.2 – The Orbit, Zoom, and Pan (hand) icons are part of the default toolbar

Regardless of your input hardware, you can always use the **Orbit** and **Zoom** icons to move through your model. The best part about using these icons is that you can use them in the middle of any command. Once you are done, you will be returned to where you were before you clicked the icon. Let's give it a try:

1. Use the **Rectangle** command to draw a rectangle on the ground.
2. Click the **Push/Pull** icon on the **Standard** or **Large** toolbar.

3. Click and release on the rectangle.

   The face of the rectangle should move up and down on your screen as you move your cursor. Now, let's see what happens when you start using the Orbit command.

4. Click the **Orbit** icon on the toolbar.

   Notice how your Push/Pull state freezes? This will allow you to spin around your model without having to finish the Push/Pull command. The Push/Pull is paused until you complete your Orbit.

Click and Release or Click and Drag

Unless otherwise noted, input and editing should be done by click and release, rather than clicking and dragging. While SketchUp will allow you to do either in many cases, click and release will allow you to do things such as change your views or add modifiers to a command without having to hold down the mouse button the whole time.

While in the Orbit command, you can click anywhere on your model and use that point as a handle to spin your entire model in 3D space.

Centering Your Orbit

Notice that the camera spins around the center of your view. How you orbit around your model will depend on how your model is situated in your current view. This may mean using a combination of Zoom, Pan, and Orbit commands to get to the ideal view of your model.

Remember, while it can seem as though you are moving a piece of your model, it is important to remember that you are actually spinning the entire 3D workspace. You are effectively moving your camera, or your view of your 3D model, and not the geometry in the model.

Now that we have finished orbiting, let's get back to Push/Pull.

5. Click on the **Push/Pull** icon again.

   Notice how you are back to Pushing/Pulling the face of the original rectangle? Remember, entering the Orbit command will temporarily suspend the current command and return you to it once the orbit is complete.

   Let's give Zoom a quick try! While still dragging the face of the rectangle around, activate the Zoom command.

6. Click on the **Zoom** icon.

   Notice how the Push/Pull freezes again and you click and drag the magnifying glass up and down your model to zoom in and out? Once again, you can click back on the **Push/Pull** icon to return to your command.

7. Click on the **Push/Pull** icon again.

Another command that you may want to learn to use via the toolbar is **Zoom Extents**. This command will blow the model up as large as it can get on your screen. This can also be clicked at any point during input or editing. Unlike the Orbit and Zoom commands, you cannot access this command via hardware; the toolbar icon or menu command is the only way to use Zoom Extents.

Finally, the **Pan** icon (the one with the hand) can be used to grab your model by a point and move it around without turning it. Pan moves the entire model normal to your view, so instead of spinning around your model, you will be moving straight up, down, and side to side. Just like Zoom and Orbit, the Pan command will suspend your current action until you return to the command you were originally in.

Navigating with a touchpad

Anyone designing on a laptop has had that experience when they have had to whip out SketchUp on the go and take a look at a model using just the hardware on the laptop. While many laptop users rely on an external mouse as their primary means of input, there are those special few who actually prefer to use a touchpad when they can!

Using a touchpad to navigate is a process of combining modifier keys or buttons with gestures on the touchpad. These keys will vary depending on your operating system:

- To orbit your model on a Windows computer, hold down the middle button.
- On a Mac, hold down the *Control* and *Command* keys.

Just like orbiting with the buttons, the active command suspends until you are done moving. The advantage of using this process is that you don't have to click back to the command to resume. As soon as you release the middle button/modifier keys, you are just back in the command!

Zoom works in very much the same way, except you do not need to use any keys. Regardless of the operating system, you can zoom in or out by sliding two fingers up or down the touchpad.

To pan while using a touchpad, you will add a modifier key to Orbit. Just as with Orbit, the modifier key will change depending on your operating system:

- To pan your model on a Windows computer, press *Shift* while orbiting (middle button and *Shift*).
- On Mac, press the *Shift* key while orbiting (*Control* + *Command*, and *Shift*).

I know that this can seem like a lot of keys to hold down, especially on a Mac, but with a little practice, it really can become second nature.

Navigating with a three-button mouse

It has been said that the best hardware is the hardware that you have available to you. This is mostly true, but many SketchUp users will say that the best input device to use is a three-button mouse. The reason is the third "button." With the scroll wheel, you can instantly start orbiting or zooming without the need to press any keys or click any toolbar icons.

To zoom, you only need to roll the scroll wheel up and down. The point that is under your cursor at the time that you start scrolling will be the center of your zoom. You will zoom in or out of that single point.

Zooming Direction

Zooming direction is controlled by the operating system. You may need to modify your system settings to toggle between Natural and Reverse scrolling. This will change whether scrolling up is a zoom-in or a zoom-out. Neither one is inherently better for SketchUp. Use the one that feels the most natural for you.

To orbit with a three-button mouse, you simply have to press down on the scroll wheel. While holding the scroll wheel, you will move the entire model in 3D space. The location that your cursor was at when you pressed down on the scroll wheel will act as a handle to spin the model around.

Middle Mouse Button

System settings and hardware drivers may give you the option to *program* the middle button. This can be a great way to add functionality to your mouse, but for SketchUp, it is important to have the scroll wheel click set to middle button. Without it, you will not be able to orbit using the scroll wheel.

Finally, panning is performed by pressing the *Shift* key while in orbit. Double-clicking the scroll wheel over any point in the model will pan the model to center that point on your screen.

The best way to navigate in SketchUp

With practice, any one of the input devices mentioned previously can become efficient tools for navigating SketchUp. Far and away though, most users find that a three-button mouse allows them to spend less time getting to the view they need and more time actually modeling. If you have been using a touchpad or a two-button mouse (or a one-button mouse, yikes!), you should consider getting hold of a three-button mouse. The addition of a button dedicated to moving your model in 3D space is a huge time-saver and will help navigating to become second nature.

Now that we have covered how to navigate through a SketchUp model, let's dive into what actually makes up a SketchUp model.

Exploring edges and faces

When it comes to what a model is made of in SketchUp, there are really only two pieces: faces and edges. I know that, eventually, you will end up with a model with groups, curves, polygons, and components, but those are just containers for edges and faces, which are building blocks that make up anything and everything that can be created in SketchUp:

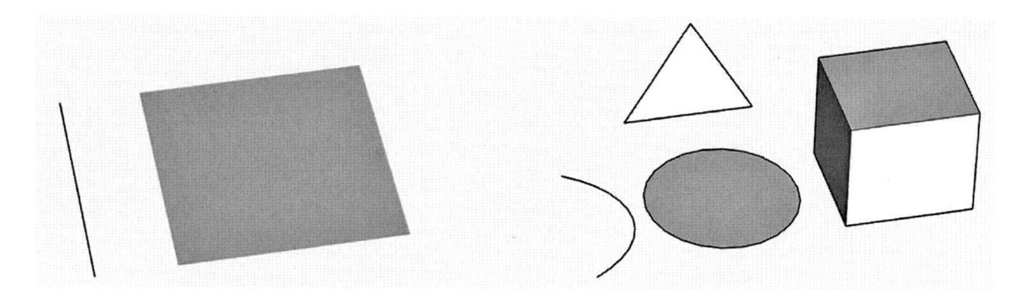

Figure 1.3 – Edges and faces (left) make up everything on the right

For this reason, it is very important to understand how these items are created and work together in order to gain mastery of the 3D modeling process.

Edges are lines, right?

Technically, this is a true statement. **Edges** are a connection between two points in 3D space. In SketchUp, we call them edges because they generally define one or more faces, and the term **line** is reserved for a specific command.

Edges are created when you use any of the input commands but are most simply created using the Line command. The Line command, of course, connects two points in 3D space with a single edge. This can be useful for creating boundaries of faces or breaking them into smaller pieces.

Edges also make up curves, arcs, and circles. While you do not have to draw each individual edge when you create one of these entities, it is very important to note that they are all made up of a series of edges.

Edges in arcs and circles

When you use the **Circle** command or any of the **Arc** commands to draw a curved line in a model, you are actually creating a string of connected edges. In fact, any of these commands will allow you to define the number of edges before or after a circle or arc is created. The measurement box in the lower right of your screen will allow you to enter the number of sides before you start drawing. Once an arc or circle has been created, you can edit the number of segments in the **Entity Info** window. Changing this number will increase or decrease the number of edges drawn while keeping the geometry in the same place with the same curve.

Welding edges

In SketchUp, you can select any number of edges and **weld** them using the right-click menu. Once one or more edges are welded, SketchUp will recognize them as a curve. A curve is a series of connected edges that are selected together and recognized as a single entity. This is different from an arc or circle though, as you cannot edit the number of segments in a curve, nor can you modify how many edges make up the curve, as shown here:

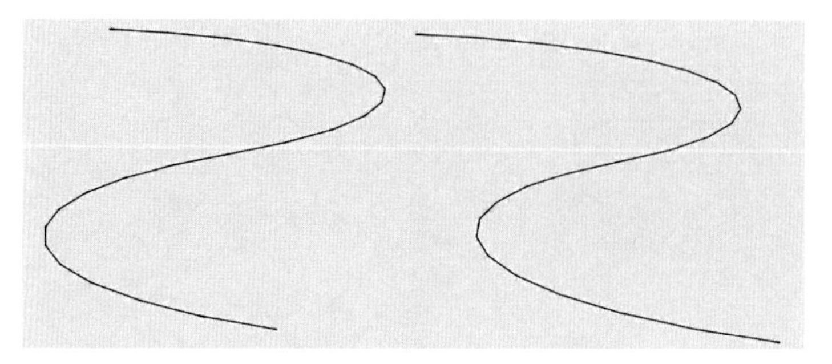

Figure 1.4 – With endpoints visible, you can see the difference between the segments on the left and the single-piece curve on the right

Let's do an experiment:

1. Use **2-Point Arc** to draw an arc with the default 12 edges of any size.
2. Select the arc and look at **Entity Info**. See how it calls it an arc and you have an editable **Segments** field?
3. Now, right-click on the arc and select **Explode Curve** from the context menu.

   Look at the **Entity Info** window now. You now have 12 edges selected. The option to modify the number of segments is gone.
4. Finally, right-click on the highlighted edges again and choose **Weld Edges** from the context menu.

The **Entity Info** window now shows that you have a curve selected. It also tells you that the curve has 12 segments, but the field is disabled. This means that while you get to know how many edges are in the curve, you cannot modify that number.

So, why would you ever want to explode curves or weld them back together? The truth is, that sometimes, the modeling process will cause curves to explode. Intersecting geometry can break curves and circles and leave you will a bunch of disconnected edges. Using **Weld Edges** will get them to stay connected, making them easier to work with. This will also affect how these edges work with a command such as **Push/Pull** or **Follow Me**. There may also be times that it serves your model to have those edges broken apart rather than stay connected. The important thing to remember is that all this geometry, arcs, circles, and curves, is simply made of a series of edges that can be broken apart or put back together as needed.

Face or a surface?

A **face** is created any time three or more edges connect in a single plane. Unlike some other file formats, SketchUp faces can be made of any number of sides and may contain holes. As soon as an edge breaks a face into more than one piece, it becomes multiple faces.

If one or more faces are connected at the edge and that edge is smoothed (either using the Eraser with the Smooth modifier key or using the **Soften/Smooth** window) they will merge together, forming a **surface**. A surface is any number of faces that have been connected by smoothed edges. If you select a surface, the entire surface will be highlighted. If you choose to show hidden edges (from the **View** menu), you will see the hidden edges as dashed lines. In this view, you will be able to select individual faces, despite them being part of a surface.

It is important to know the difference between the two as certain commands, such as Push/Pull, cannot be used on surfaces. However, you *can* use Push/Pull on individual faces when hidden edges are visible!

Crossing faces

When you draw an edge across another edge, they break at the intersection. One edge crossing a second edge will result in four total edges, meeting at one central point. The same is not true for faces. Try this example:

1. Draw a rectangle on the ground.
2. Pan over and draw a vertical rectangle. Use the *Right Arrow* key to constrain the rectangle to the red plane.
3. Select the vertical rectangle by the midpoint of a vertical side.
4. Move it so that it crosses the face of the first rectangle.
5. Place the vertical rectangle so that the edges do not touch:

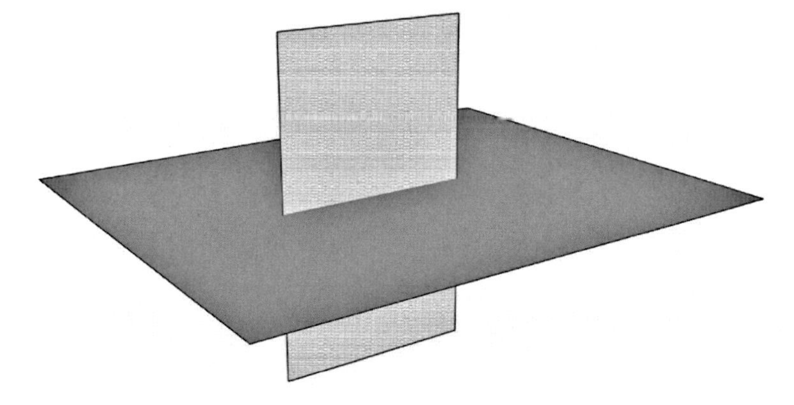

Figure 1.5 – Faces can cross without merging geometry

See how the rectangles cross over each other without *merging* together. At this point, you can select the vertical rectangle again by any point and move it out of the original rectangle. If you place the vertical rectangle so that the edge intersects any edge of the first rectangle, then the edges will become connected, and the faces themselves will still simply pass through each other.

To cause a face to intersect another face, you must tell it to intersect.

6. With both rectangles crossing through one another, select both faces (use *Shift* + **Select** to highlight both).
7. Right-click on either plane.
8. Select **Intersect Faces** from the context menu.
9. Select **With Selection**.

Notice how both faces are now broken where they cross each other. Edges will automatically connect to other edges, but faces will not.

As you advance with your 3D modeling skills, you will become more and more efficient using different native commands and extensions and will develop your own workflows for different types of models. The thing to remember though is that every single model is made of edges and faces. Knowing how these basic pieces work and interact is key to understanding how SketchUp works. Another key is in knowing how to use commands beyond their initial function. In the next section, we will see how to use a tool such as Move to perform multiple actions.

Getting more out of Move

Don't worry, this is not the part of the book where the author encourages you to stand up and stretch before proceeding (but if you feel the need, go ahead, I will be right here when you are done). This is the part of the chapter where we do a deeper dive into using a few of the Modify tools.

More to Move than you remember

One of the biggest parts of moving from a beginner user to an intermediate user is learning to use the basic commands beyond their basic functions. Many new users only scratch the surface of what they can do with tools such as Move, Select, or Rotate.

The first thing you will need is a couple of pieces of geometry:

1. Draw a rectangle on the ground.
2. Use **Push/Pull** to create a box.
3. Triple-click on the box to select all geometry.
4. Right-click and select **Make Group**.

This will be a good piece of geometry to work with, but you will need to create a different group to work with as well. Next to the box, let's create a cylinder.

5. Draw a circle on the ground.
6. Use **Push/Pull** to create a cylinder.
7. Triple-click on the box to select all geometry.
8. Right-click and select **Make Group**.

   Now, you have some things to move! Let's start by exploring the Move and Rotate options of a group.

9. Use **Select** to highlight the new group.
10. Select the **Move** command and move the cursor over the selected group.

At this point, you have a couple of options for moving this box. You can click anywhere on the highlighted group and use that point as a *handle* to move by. Clicking and releasing on any point will connect the group to the cursor by that point. Clicking and releasing a second time will place the group at the new cursor location. You can use any geometry in the group as an arbitrary move point, but if you pay attention, you can be much more exact in your selection.

Move your cursor over the cylinder group and look at the corners of the group. Notice how there is a circle on every corner? Clicking on one of these points will allow you to use the corner as a point by which to move the group. If you orbit to a three-quarter or isometric view of the group, you should see a point on the seven corners of the group that you can see. You should also notice that there is one more point that represents the corner of the group that you cannot see. If you move your cursor over that point, the geometry in the group will temporarily become transparent, allowing you to see and select the back corner.

This is great for geometry such as this cylinder that does not have geometry filling in the corner of the group as the box does:

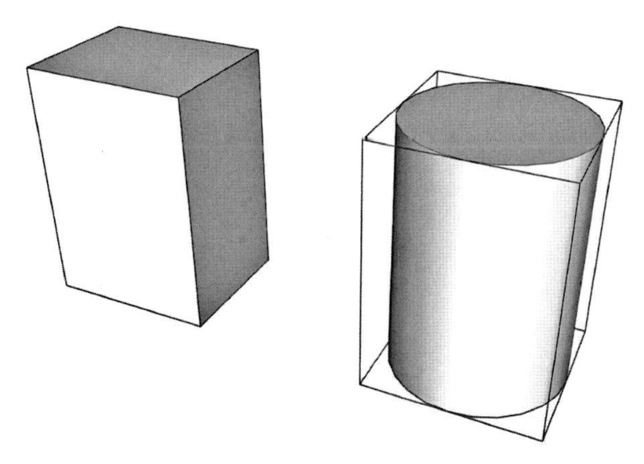

Figure 1.6 – The bounding box wraps tightly around the box while the cylinder does not reach the corners

But wait, there's more! Look at the modifier keys listed in the status bar. If you press the key to toggle through grip types (*Alt* on Windows or *Command* on macOS), you can change the points visible on your group. Tapping the modifier key will toggle between the following point sets:

- Corners of group
- Midpoints of group edge
- Middle of group sides
- Center of group

> **Modifier Keys**
>
> We will be talking a lot more about using modifier keys in *Chapter 3, Modifying Native Commands*, which includes a comprehensive list of every modifier key for all commands!

Careful selection of the point you use to move items can save you a lot of time and energy. Clicking the correct start point when you are moving can allow you to complete the move in one go, rather than performing multiple moves to get your selected geometry exactly where it needs to be.

Copies and arrays

Move, like many SketchUp tools, can be used to perform more than just one action. For example, Move can be used to create an array of the selected item. Odds are good that you have learned to use the Modify command to copy before, but just to make sure, we will cover it quickly here:

1. Use the **Move** tool to select the box.
2. Move the box to the side (away from your cylinder).
3. Tap the Copy modifier key (*Ctrl* on Windows or *Option* on macOS).
4. Click on the new location and receive a copy of your box at this location.
5. Now, before you click anything else, type `5X` and hit *Enter*.

   Now, you have a total of five copies of your box. Each box is spaced the exact same distance as your first copy. This is referred to as an **array** in SketchUp.

   Now, let's say that you were hoping to have a total of five boxes. Typing `5X` gave you five copies plus the original for a total of six. The good news is that as long as you do not start a different command or select different geometry, you are still in the array-making process! Let's get rid of that extra copy.

6. Type `4x` and press the *Enter* key.

   The extra copy is now gone! You can increase or decrease the number of copies by typing `NUMBERX` and hitting *Enter* as many times as you like, as long as you stay in the command.

This is the process of creating an external array. That is, the copies are placed one after another, external to the original move. Next, let's create an internal array.

7. Select the cylinder group and use the Copy modifier to move a copy a long way from the original in the opposite direction of the box group.
8. Immediately after placing the copy, type /4 and press the *Enter* key.

   In this case, the number of copies specified is placed, evenly spaced between the original and the copy. Just like an external array, you can change the number of copies by simply typing a new number.
9. Type /8 and press *Enter*.
10. Type /3 and press *Enter*.

Using the native Move command with a few simple modifier keys makes it easy to quickly create a series of geometry. Just imagine how much more work it would be to place each copy in an array or, worse yet, recreate the geometry over and over! Understanding how and when to modify commands will help you become a faster, more efficient SketchUp modeler.

Radial arrays

This same process can be used with the **Rotate** command as well. Start a new model and let's give this a try:

1. Create a box and put it in a group.
2. Use **Move** with the grip toggle to move the box so that its front bottom edge is at the origin.
3. Then, move the box up the green axis.
4. With the box still highlighted, select the **Rotate** tool.
5. Select the **origin** as the center point of rotation.

   As you start to rotate the box around the origin, press the Copy modifier key (*Ctrl* on Windows or *Option* on macOS).
6. Rotate the copy around to the green axis on the far side of the origin (180 degrees) and click to place the copy.
7. Type /6 and press the *Enter* key.

   With that, you have created an internal radial array. The same rules apply here, as you can change the number of copies between the original and copy by simply entering a new number at any point.

   The process of making an external radial array is just as simple.
8. Use **Undo** to remove the copies from the previous step.

9. Use **Rotate** to make your first copy 30 degrees from the original.
10. Type `11x` and press *Enter*.

Now, you know how to make a clock face!

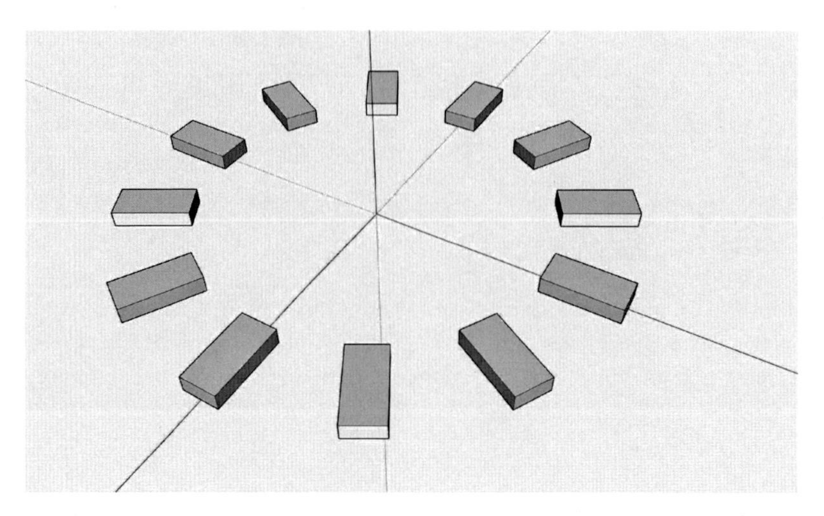

Figure 1.7 – A dozen boxes in an array around a central point is exactly how you would make a clock

The important thing to take away here is that the basic commands can be used to do more than their name will imply. Creating things such as copies of arrays using the Move or Rotate commands are simple ways to get more out of a single command. Before worrying about what extensions are out there to expand how you model, remember to explore everything that the native commands are capable of. We will be diving much deeper into this idea in *Chapter 3, Modifying Native Commands*. Up next, let's take a look at how we can use Move and a few other native commands to create some organic geometry.

Deforming geometry

SketchUp is well known for allowing users to draw basic geometry (polygons, lines, and arcs, for example) quickly and easily. One of the biggest misconceptions about SketchUp is that if you want to go beyond drawing basic shapes or straight lines, you need to depend on extensions. While extensions are amazing (and we will be covering them in *Chapter 11, What Are Extensions?*) any SketchUp ninja knows that you can lean on the basic input and modification tools to generate advanced geometry quickly. In this section, we will see how to use native tools to deform basic geometry into any shape you need.

For this example, let's get hands-on and draw a tree, something like the following:

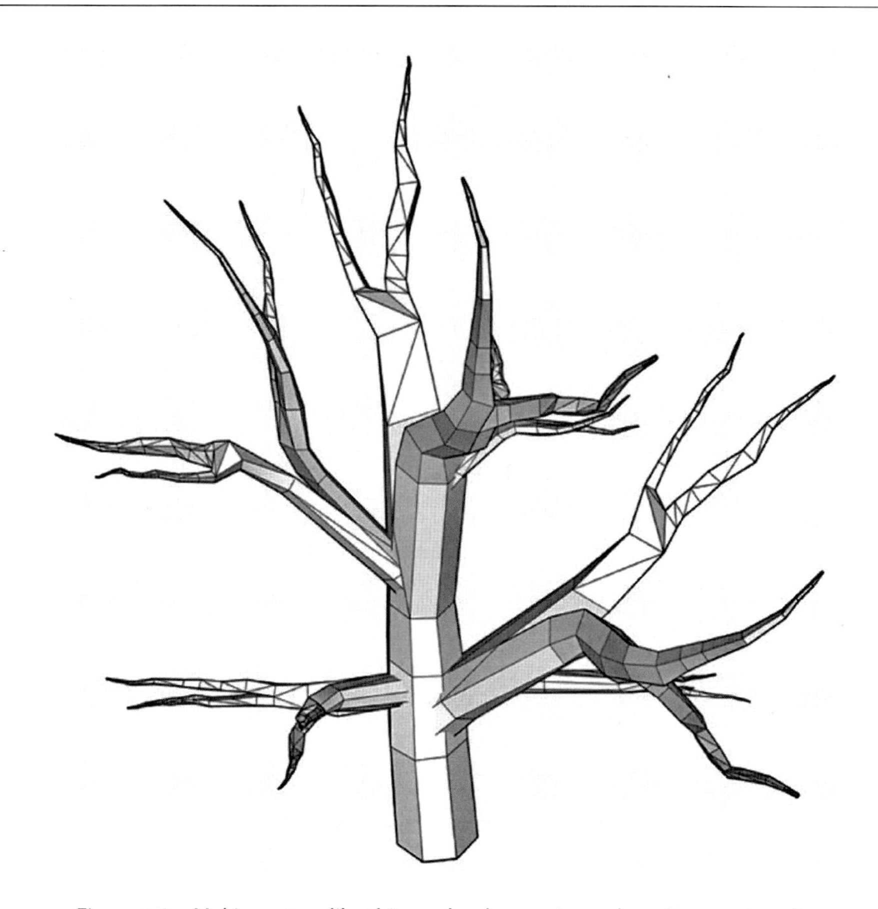

Figure 1.8 – Making a tree like this can be done using only native commands

Now, there will be a little bit of sculpting and using your creative abilities in this, but we will cover the basic steps that will allow you to create something like this using only native SketchUp commands.

Let's open a brand new SketchUp model and dive into some tree modeling:

1. Use the **Polygon** tool to draw an eight-sided shape on the ground (fairly large, as this will be the base of the tree).
2. Use **Push/Pull** to pull the shape up.
3. Select the top of your extruded octagon.
4. Use the **Scale** command to shrink the polygon to around 90% of its original size.
5. Select the **Rotate** tool to spin the top polygon 10 to 20 degrees.

At this point, you should have a basic looking tree trunk, something like this:

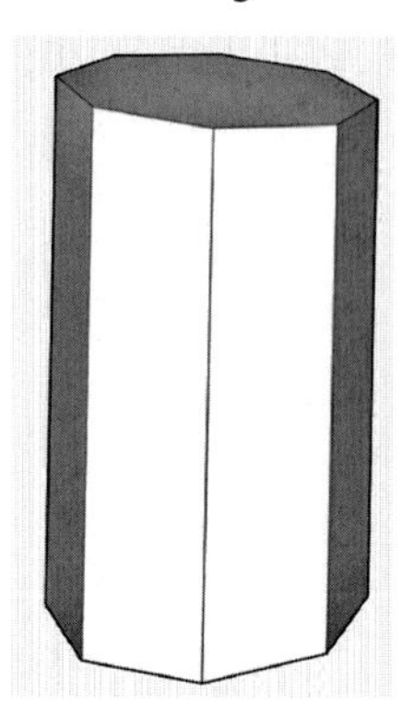

Figure 1.9 – Tree trunk

From here, we will do the same steps, but add another Rotate into the mix. In order to get a curve in the tree's trunk, we will rotate the top shape out of the plane.

6. Use **Push/Pull** to pull up the top of your trunk, then scale it just a bit (maybe to 95%).
7. Now, use **Rotate** to angle the top polygon on the red axis (tap the *Right Arrow* key to lock in the red axis). You don't need to rotate it a lot; 5 to 10% should look good.

More Inferencing

If you want more information on inferencing, check out *Chapter 4, Taking Inferencing to the Next Level*, where we will dive deep into everything that inferencing has to offer.

8. Use **Push/Pull** again to continue the shape.

   See where we are going with this? Just by using this workflow, we can create a shape that gets smaller and smaller as it gets longer, and using the Rotate command from the sides will allow you to have the trunk twist and turn as much as you like!

9. Use this workflow to add some more taper to the trunk and a few more turns. Remember, you don't have to limit your angle to the blue, red, or green axis.

   Let's hop in and split the trunk to allow multiple branches to come out!

10. Zoom in on the face of your top polygon.
11. Draw a line to split the shape in half.
12. Use **Push/Pull** to pull up one half of the shape.
13. Use the Move command to move the shape away from its original location, ideally, in the opposite direction of the second half that is about to be pulled up. You will likely need to toggle **Auto-Fold** to move in the direction you want (toggle with *Alt* in Windows or *Command* on macOS).

14. Now, pull up the second half.

    At this point, you should have something that looks like this:

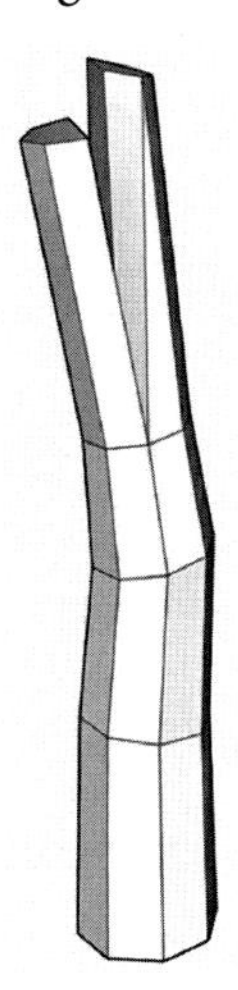

Figure 1.10 – Your tree is growing

It is looking more and more like a tree! Let's keep repeating this process on both branches.

15. Use **Push/Pull**, **Scale**, **Rotate**, and **Move** to continue these branches. If you like, go ahead and split the ends of one of the smaller branches.
16. Continue these branches until they get too small to extend out.

    At this point, you should have something like this:

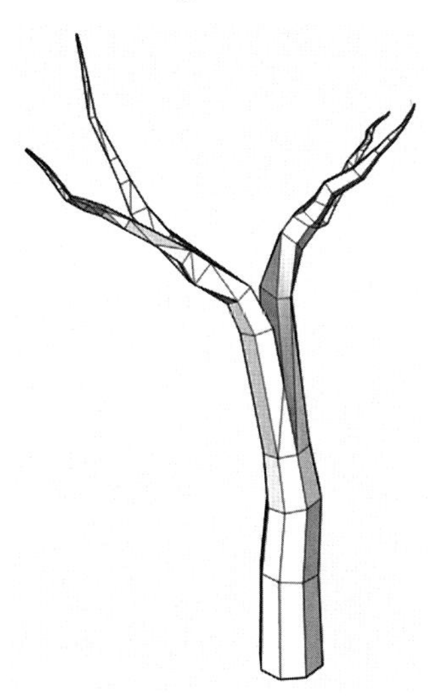

Figure 1.11 – Most of these branches were created using only the Scale and Push/Pull tools

Not a bad-looking tree, but honestly, a little bit light, as far as branches go. Fortunately, we can multiply the number of branches quickly! What we can do is grab a portion of the original model and copy it. Let's give it a try.

17. Use the **Select** tool to select the top two-thirds of the tree.
18. Use **Move** with the Copy modifier to move a copy of the selection off to the side. Make sure to move it far enough that none of the copied geometry crosses the original.
19. Right-click on the selected copied geometry and choose **Make Group** from the context menu.
20. Now, select the **Move** tool and use the Rotate handles on the group to tilt and spin the group. The goal here is to put this set of branches in a different orientation than the original.
21. Now, move this group so that the base overlaps the original geometry.

At this point, you should end up with something like this:

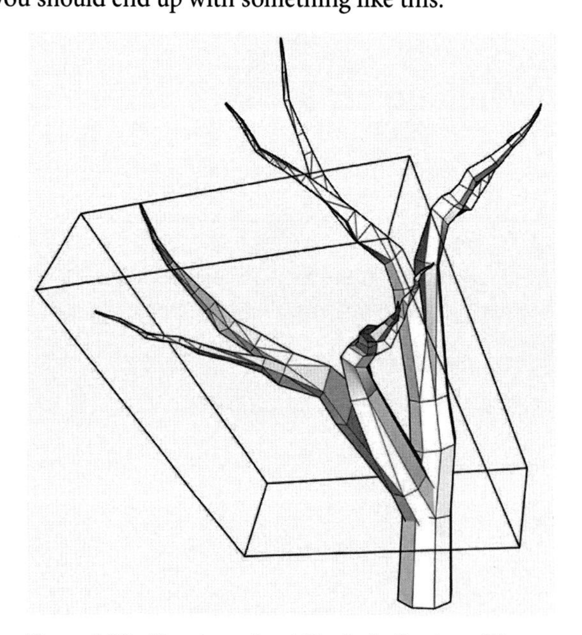

Figure 1.12 – Your tree should look similar, but different

From here, you can continue to use these same steps to copy and group sections of the tree to make more and more branches. You may want to do a little scaling here and there to mix up the size of the branches. Another option is to double-click to edit the groups and use Move and Rotate on the geometry to add even more variation!

In the end, you will have a twisting, turning shape created with only native tools. Note that we did start with an octagon, just to keep the model as light as possible, but you could start your tree with a shape with more sides. For that matter, you could, if you wanted, draw a totally original shape to start your trunk.

One of my personal favorite things about SketchUp is that it is simply a tool for drawing shapes. Great SketchUp modelers can model almost anything using these basic tools. Remember, it is you, the designer, that makes amazing models. The software you use is just the tool to help you realize your creation. Another basic tool that can do far more than users give it credit for is the Follow Me tool. In the next section, we will spend some time learning how to push Follow Me beyond its basic functionality.

Diving deep into Follow Me

Another command that people only scratch the surface of when they learn SketchUp is **Follow Me**. Generally, we see how Follow Me can be used to put crown molding in a room or can be used with a circle to make a bottle or pawn chess piece. Follow Me is made to extrude faces along paths, so these sorts of things are a perfect use of the tool.

Seems simple enough, right? There are, however, a handful of tips that will move Follow Me from a command you use in a few specific instances to a tool that you will keep on hand for daily use! In this section, you will see how Follow Me can become your go-to tool for creating fillets, rounding over corners, or even generating geometry quicker than any other SketchUp tool.

Follow Me and groups

The first thing to think about with Follow Me is how you set up your initial geometry. Something that many people don't think about is using a group to separate the Follow Me geometry from base geometry or preselecting paths. Let's create an example:

1. Draw a large rectangle on the ground.
2. Push/Pull the rectangle up so you create a box.
3. Now use the **Line** and **Arc** tools to create a quarter circle in the middle of one of the sides of the top of the rectangle, as shown in this figure:

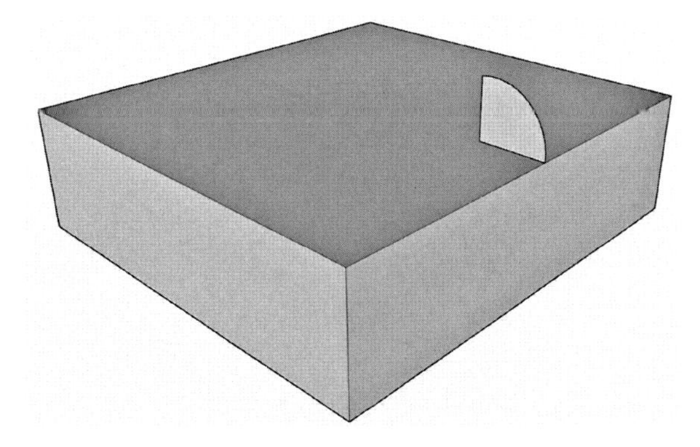

Figure 1.13 – Ready for Follow Me

4. Select the **Follow Me** tool.
5. Select the face you created with the Line and Arc tool.
6. Move your mouse along the top edge of the rectangle.

   Now, unless you are very steady-handed, there is a chance that your cursor may wander just a bit off the line as you move it around the model. When you do this, the profile you are pulling may jump all over the place, creating a crazy extrusion. Sometimes, it will jump down a side of the rectangle; other times, the profile will disappear completely.

   Many first-time, self-taught users think that this is the way to use this tool, but there is a better way. The key to a quick and easy Follow Me is to preselect your path!

7. Use **Select** with the *Shift* modifier key to select the top edge of the rectangle.
8. Choose the **Follow Me** tool.
9. Select the same face as last time.

   Boom! With that, we have a nice, curved arc that follows the top of the rectangle!

   However, if you select any portion of the new geometry, you will see that it is connected to the rectangle. Now, in many instances, there may be nothing wrong with this, as you may want a rim or surface detail to merge with the part geometry. In the case of something such as a chair rail or baseboard, you may want to keep them separate so you can edit them easier in the future or use a different material for them. Let's redo that Follow Me command with one change.

10. Use the **Undo** command to undo Follow Me.
11. Double-click on our arced face (double-clicking will select the face and all the edges that define it).
12. Right-click on the face and choose **Make Group** from the context menu.
13. Now, use **Select** to select the edges around the top of the rectangle. Your selection should look like this:

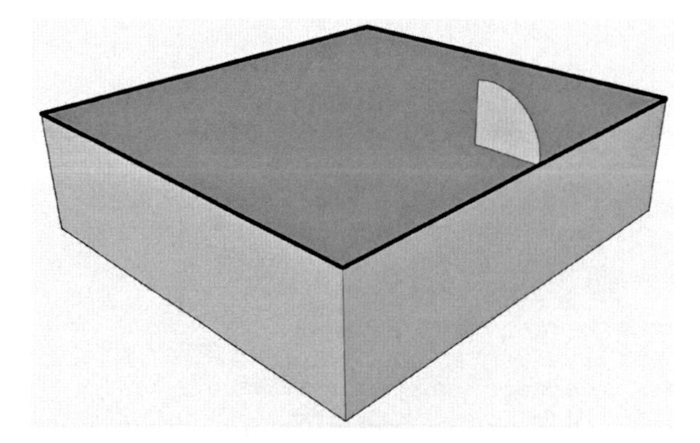

Figure 1.14 – Pre-selected Follow Me path

14. Choose the **Follow Me** command.

    The next step is to choose the face that should be used for Follow Me, but if you hover over the group that you created, you will notice that it is not a valid selection. At this point, we need to get inside the group temporarily to select the face.

15. Right-click on the group and choose **Edit Group** from the context menu.
16. Select the face.
17. Use **Select** to click outside of the current group.

Notice that if you select the geometry that Follow Me created, it is in its own group. This means you can apply a different tag to this group than the base geometry or easily select all the geometry to apply a specific material. Having it in a separate group makes it much easier to work with!

One of the issues that many users run into with Follow Me is problems with the ends of the extrusion. Remember, when you create a face to be used in Follow Me, it should be perpendicular to the path. Let's take a look at two examples, as follows:

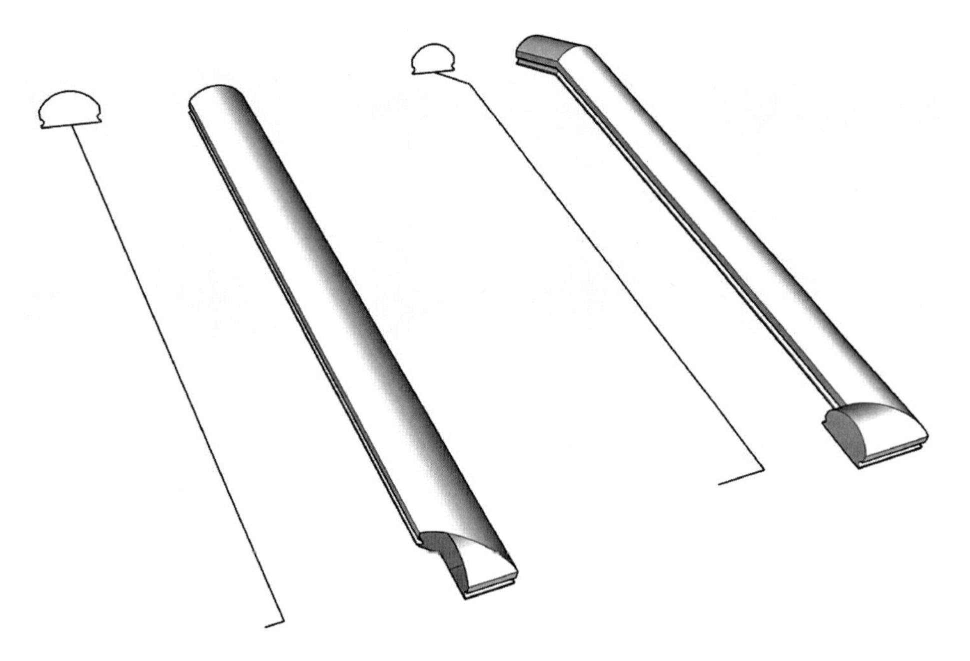

Figure 1.15 – These two examples show the face and path before and after Follow Me

In these examples, a simple profile was created for a handrail. In the example on the left, the profile is not perpendicular to the path and ends up squashed in the final extrusion. To make things worse, the end looks odd because the path is shorter than it should be.

In the example on the right, the path was extended out at both ends, and the profile was placed perpendicular to the end of the path. There may be cases like this where you may want to add geometry to the path or have your extrusion run long of where they will be in your final model. Remember, if you group the face, then the geometry will be easy to trim back after you have created a good-looking extrusion.

Let's see another example of this with a half-torus. In this example, we will make two shapes and see how important it is to spend a little extra time on your path before using Follow Me. Start a new model and follow these steps:

1. Draw a large circle on the ground with its center at the origin.
2. Draw a line along the green access from one edge of the circle to the other, cutting the circle in half.
3. Erase half of the circle and the line through the middle.

   This gives you a perfect half-circle, right? In some cases, this would be a perfect half-circle, but for Follow Me, it might not be ideal. Let's add another circle to this example and see what happens when we try to create a half-torus.
4. Use the **Circle** command and the *Left Arrow* key to draw a circle at one end of the half-circle path.
5. Put the circle in its own group.
6. Run **Follow Me** with the half-circle as the path and the grouped circle as the face:

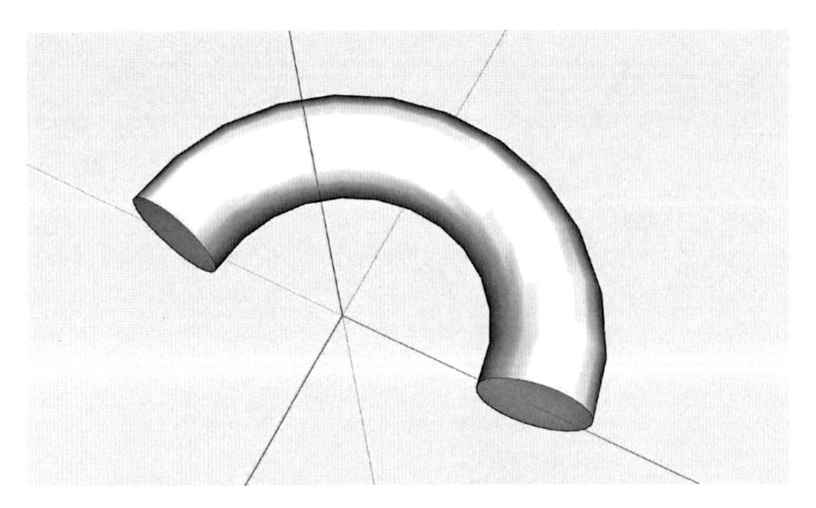

Figure 1.16 – This process did not create a perfect half-torus

Look at the geometry that has been created. Specifically, look at the ends. What is important to notice is how the faces of the extrusion are tilted out from the green axis. Also, while the shapes at the end may seem like perfect circles, check their dimensions using the Tape Measure. If you measure the end vertically and then horizontally you will find that they are not the same.

The issue with this Follow Me is that the shapes were not perpendicular to the ends of the paths, so Follow Me pushed the shape along the path at a little bit of an angle. Let's try this again with one small change. Just so we don't forget what this practice produced, group-select the whole thing and then move it off to the side.

7. Draw a new circle at the origin about the same size as the very first one.

   Look at the circle. See how the axis hits the circle where two edges meet. What we want to do is rotate the circle just a little bit so that the green axis crosses the circle in the middle of an edge, rather than at an endpoint.

8. Select the **Rotate** command.
9. Use the origin as the center of rotation.
10. Use the center of an edge as the start point for the rotation.
11. Use the green axis as the second point.
12. Draw a line across the circle like last time.
13. Erase the bottom half of the circle and the line on the green axis.

    Look at the edges of the circle you have created. The difference is subtle but important. We will see a big difference when we perform the same Follow Me steps with this geometry as a path.

14. Use the **Circle** command and the *Left Arrow* key to draw a circle at one end of the half-circle path.
15. Put the circle in its own group.
16. Run **Follow Me** with the half-circle as the path and the grouped circle as the face:

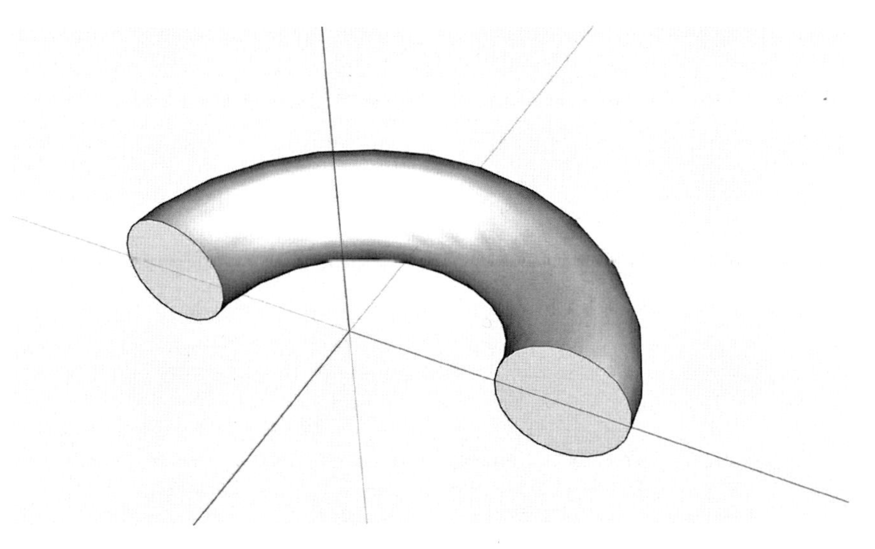

Figure 1.17 – The perfect half-torus

Look at this extrusion compared to the previous effort. The ends line up with the green axis and if you measure the ends, they are still perfect circles! Remember this example next time you use Follow Me and you will avoid issues with the extrusion coming up short or having weird end shapes!

Turning shapes with Follow Me

As mentioned before, Follow Me is often used to make shapes similar to a woodworker turning a piece of wood on a lathe. A profile is aligned with the center of a circle and Follow Me makes a shape based on the profile spinning around the circle. A few examples of this are shown here:

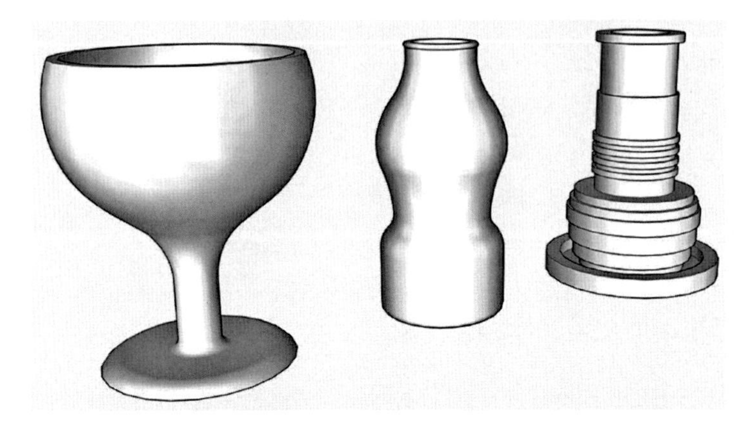

Image 1.18 – Turned shapes using Follow Me

Since the intention of this book is to introduce intermediate concepts and push you past the basic use of the SketchUp toolset, let's consider a slightly different way to use this setup. Aligning a profile with the center of a shape to follow is a great concept and can be used to quickly create some great shapes, but people often don't consider that the shape that is being followed does not have to be a circle. Let's start a new file and model a fancy pedestal in just a few steps:

1. Draw a square centered at the origin (use the Toggle Select Center modifier to start a rectangle on origin (*Ctrl* on Windows or *Option* on macOS) and watch for the dotted line to indicate square geometry).

   While it is not strictly required that you do a Follow Me like this at the origin, it does make it easier on you. By centering the path geometry on the origin, you have a visual indicator of the center of the *spin* of the geometry.

   The next step will be to create the profile. I find that the easiest way to do this is to create a working plane that I can draw my shapes onto. As long as one edge of the plane is along the origin, I know that Follow Me will create the properly *turned* geometry.

2. Draw a rectangle on the red axis (*Right Arrow*) from the origin up from the square on the ground. You should end up with geometry similar to this figure:

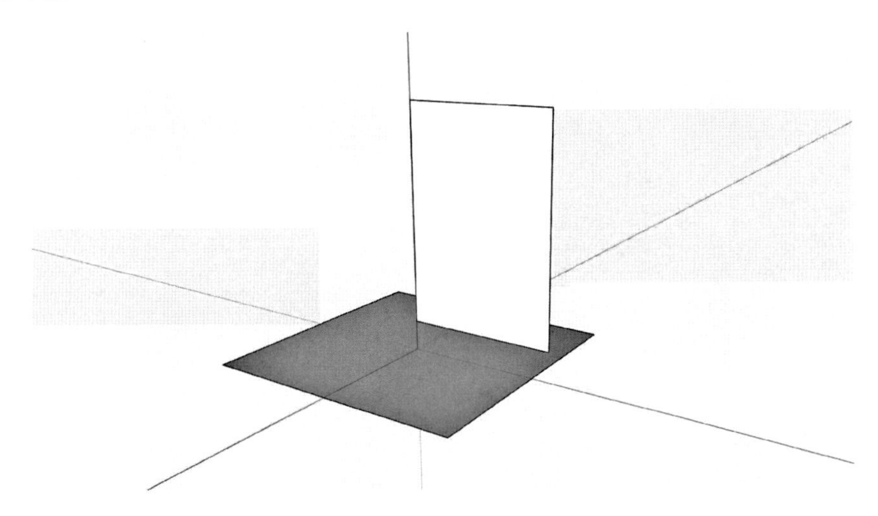

Figure 1.19 – Ready to start creating a profile

3. The next step is to create the geometry of the profile. For this example, create something similar to what I have drawn here:

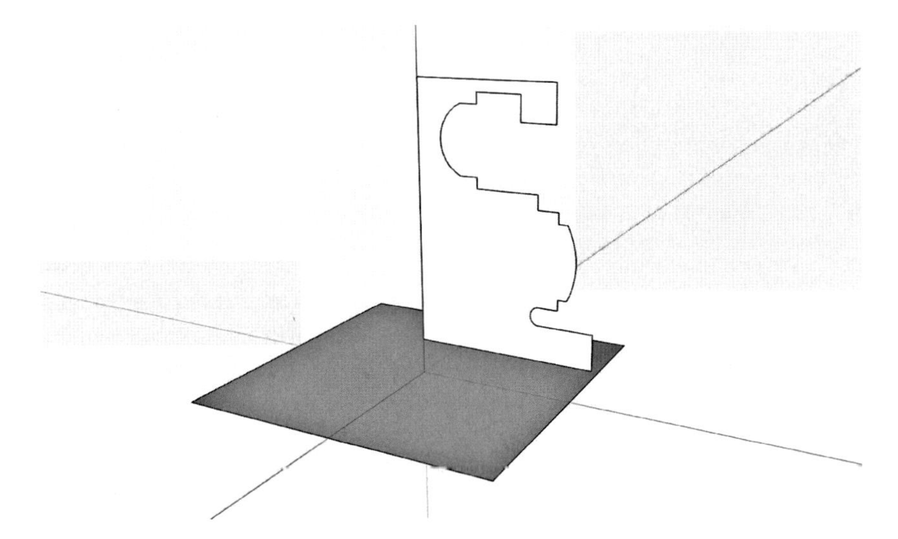

Figure 1.20 – Your profile may vary, but try to create something with a few corners and arcs

4. Once you have the profile geometry drawn, erase the extra edges and double-click the profile, then right-click, and choose **Make Group** from the context menu.
5. Select the square (you can choose the edges or simply double-click the face of the square to highlight all of the edges), choose **Follow Me**, then click the profile:

Figure 1.21 – Final geometry

That's it! Your fancy pedestal is created. Any time you have geometry that follows along a shape, whether it is a linear extrusion similar to what we created in the first example, or if the geometry wraps around an enclosed shape like this pedestal, keep Follow Me in mind as you generate geometry!

Summary

The goal of this chapter was to show you, as a SketchUp user, that the native commands can be pushed beyond their basic functions. You should now have a better understanding of what makes a model, and how you can use native commands in more than just one way.

While this was not a comprehensive list of every way that input or modification commands can be used, it should prompt you to look at the commands in a new way. The hope is that you will now question how these tools function and how you might use the native tools to create geometry beyond how you currently use them. The ability to use the native toolset to its fullest potential will allow you to model quicker and create more detailed geometry without relying on extensions.

In *Chapter 2, Organizing Your 3D Model*, we will discuss how to use groups, components, tags, and scenes to organize your model, making it easier to work with and move through.

2

Organizing Your 3D Model

Hopefully, the last chapter prompted you to rethink how you might use the SketchUp commands when you create a 3D model. With this chapter, I am hoping you will rethink how you structure your 3D models.

Generally, it is very apparent when someone with experience creates a 3D model versus someone just starting off. A well-organized model demonstrates a modeler's experience and ability to think through their model as they go. A model that is put together in a structured manner is not just easier to look through, but easier to work with, as well! Whether this model will be passed to a co-worker or collaborator at some point in the future, or it is you who will be coming back to make revisions six months after your final save, if a model is well laid out, it will save time and energy!

In this chapter, we will construct a model of a dining table complete with place setting and chairs. Each piece will be added to the model in an organized manner, making it quick and easy to return and make changes as needed. This will serve as a template for how your models should be put together in the future.

In this chapter, we will cover these main topics:

- Mastering Groups and Components
- Getting more out of tags
- Understanding the Outliner
- Using scenes for more than appearances
- Removing model excess
- Using visibility commands

Technical requirements

This chapter will assume that you are using the newest version of SketchUp Pro. You will also need to download the model `Taking SketchUp Pro to the Next Level - Chapter 2.skp` from 3D Warehouse:

`https://3dwarehouse.sketchup.com/model/31e77d33-ac9b-4663-915e-07d3916259b0/Taking-SketchUp-Pro-to-the-Next-Level-Chapter-2`

Mastering Groups and Components

By this point, we have already reviewed how important it is to use **Groups** and **Components** to separate geometry. I would say that one of the very worst things a SketchUp user can do when they create a 3D model is to ignore Groups while modeling. I have downloaded many models from 3D Warehouse only to find that they are monolithic blobs of geometry, nearly impossible to edit or manipulate due to all the geometry being merged. Using Groups to separate the "pieces" of your model is not an intermediate or advanced process, but simply how you should be doing things from the start.

Starting with Groups

Before diving into when and where to use Groups and Components, let's touch on a single command that may help you organize every model you work on from this point forward: **Make Group**.

I know, you have used the **Make Group** command in the past and it's not a "new" command, but I would like to suggest you think about using it in a new way. To simplify your modeling process and to help you keep things organized, use **Make Group** before you start modeling.

Try this:

1. Open a brand new SketchUp model.
2. Right-click on the blank screen.
3. Click on **Make Group**.

   At this point, it may seem as if nothing has happened, but you are currently looking at an empty Group. Anything you draw at this point will be a part of this new Group.

4. Draw a rectangle on the ground plane.

   See how the Group indicator appeared around your new rectangle as soon as it was drawn? Creating a Group before you start drawing ensures that this geometry will not only be separate from edges and faces drawn in the future but will allow you to quickly place this Group into the proper tag or convert it into a Component as needed.

   Make Group can be used any time you create geometry in SketchUp, not just when you are starting a new model. Let's add another Group to this Group.

5. Use **Push/Pull** to pull this rectangle up into a box.
6. Use **Select** to click outside of the Group.

Closing a Group

Selecting outside of a Group will close the Group. If you don't want to change commands to Select to close a Group, you can right-click in the empty space inside the Group boundary and select **Close Group** from the context menu.

7. Right-click anywhere other than the box you just created and choose **Make Group**.
8. Draw a circle on the side of the box in the previous Group and use **Push/Pull** to pull it out into a cylinder.
9. Close this new Group.

You now have geometry in two separate Groups. Using **Make Group** at the beginning of the modeling process means that you don't have to mess with any selection acrobatics to highlight the edges and surfaces you want in each Group if you had done so after the geometry was created.

You probably noticed that when you right-clicked, you had the option of choosing **Make Component...** as well. So why did we use **Make Group** in the previous example? This was one of those times where we used the simplified command for the sake of instruction (**Make Component...** brings up a dialog that I did not want to cover, at this point, so **Make Group** was used). If you know that you want to create a Component, then do so from the start! If you are not sure when to use Groups or Components, then read on!

Groups versus Components

Groups and Components are collectively referred to as **Objects** in SketchUp and both serve the basic functionality of separating geometry from other geometry. Additionally, Components have some extra functionality that makes them special (sorry, Groups, but it's true). In addition to being a container for geometry, Components also have additional functionality:

- Component copies are connected; change one, you change them all.
- Components have the ability to store additional user-defined information or ICF file data.
- Components can be used to create **Dynamic Components** used to add additional behavior or data.

This list contains all the reasons you should consider using a Component when you create an Object. Generally, my thought process on when to use a Component looks something like this:

- Will this Object repeat through my model?

  If yes, use a Component.

- Do I need to report upon and information of this Object (things such as material, length, price, and so on)?

  If yes, use a Component.

- Am I making a Dynamic Component?

  If yes, use a Component.

If you answered *no* to all of these questions, you are probably safe to make a Group. The good news is, if you decide later on that you want a Component, you are only one right-click away from converting your Group into a new Component.

It should be noted that there are those who make *everything* a Component. While there is an argument to be made for this, as copies of Components will end up creating a smaller overall file size than the same number of Groups in a file, one of the nice parts about using a Group is that it involves just a single click or shortcut key and you are done. Creating a Component will always require that you fill out information in a dialog.

> **Naming Components**
>
> Skipping the Component information and just clicking the **Create** button is tempting, but we are in a chapter on keeping our work organized, so let's commit to a simple rule: *All Components get a descriptive name*. No exceptions! For that matter, it is not a bad idea to name Groups, either.

I tend to lean toward my own simplified set of rules for Component creation. In this setup, every "thing" is a Component. Any time I want to keep a set of Components together, they go into Groups.

Nesting Groups and Components

Nesting is simply the act of putting a Group or Component inside of another. This is a great way to keep your model organized. Any time you create a set of Objects, placing them inside of another container will make your model much easier to work with and make the process of copying or editing geometry much easier.

As we discuss nesting, we should address the previous question in this new context: when nesting, should you use Groups or Components? Once again, a lot of this comes down to personal preference, but I will offer a simple suggestion: if this collection of Objects needs to be reported upon or if it will be copied and you want the copies to be the same, use Components. Here is a visual example of combining.

Let's get hands-on with an example:

1. Open the model `Taking SketchUp Pro to the Next Level - Chapter 2.skp`.
2. Open the **Component** window.

> **The Component Browser**
>
> Another benefit of modeling with Components is that they show up in the **Component Browser** window. From here, you can simply drag and drop Components saved in the file into your modeling window. Unlike Groups, Components can exist in your model without being on the screen. This is another way to keep your model organized.

3. Drag the **Salt Shaker** Component onto the table.
4. Drag the **Pepper Shaker** Component onto the table, next to the **Salt Shaker** Component.

   Right now, you have two Components sitting on a third (the table). At this point, you could say you are done with the model and move on, but let's say you want to ensure that the two shakers stay next to each other. To do that, we will put them into a Group.
5. Select both the **Salt Shaker** and **Pepper Shaker** Components.
6. Right-click and choose **Make Group** from the context menu:

Figure 2.1 – These two Components are now held together in a single Group

   In this case, it makes sense to use a Group to keep these Components together since that is the only function of the container.

   There are of course exceptions to every rule, and this one is no different. There are cases where you may want to place Components into other Components.
7. Arrange the **Fork**, **Knife**, **Spoon**, and **Plate** Components on the tabletop.
8. Select all five Components and right-click.

9. Choose **Make Component...** from the context menu.
10. Name this Component `Place Setting`, then click **Create**:

Figure 2.2 – Table for one

Now, why did we make this a Component rather than a Group? In this case, we want to make a few copies of the Place Setting around the table.

11. Use the **Rotate** command to make a radial array of the four **Place Settings** Components around the table.

    Great! Now our table is set, but as often happens when designing in the real world, the requirements for this model have changed. Let's say the client has requested that you add a salad fork to the Place Setting. Since we made the Place Setting a Component, we can add the salad fork to all four instances of Place Setting easily!

12. Double-click on one of the **Place Setting** Components to open the Component.
13. Drag the **Salad Fork** Component next to **Fork**.
14. Close the **Place Setting** Component.

    Notice that all four settings get updated. This is using the power of Components along with nesting to simplify the process of working on your model and keeping it nice and organized, as shown here:

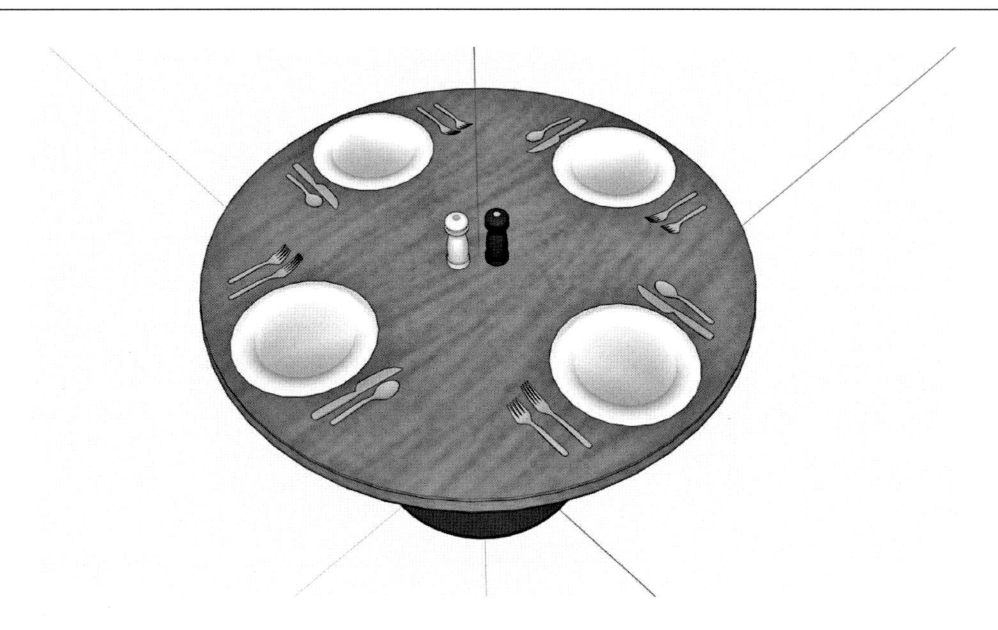

Figure 2.3 – A perfect table for four

Let's finish this table up with some chairs.

15. Drag a **Chair** Component from the **Components** window and place it in front of one **Place Setting**, then use **Rotate** to make an array four chairs in total around the table.
16. Select everything (the table, chairs, shakers and place settings) and place them into a new Component called `Four Top`.
17. Now make a copy of that Component along the red axis.
18. Select both Components and copy them along the green axis.
19. Now select all four Components and make a Group of them.

> **Naming Groups**
>
> While Groups will not prompt you for any information, you can give them names. Select any Group and enter a name for it in the **Instance** field in **Entity Info**.

With this, we have created a small seating section of a dining room:

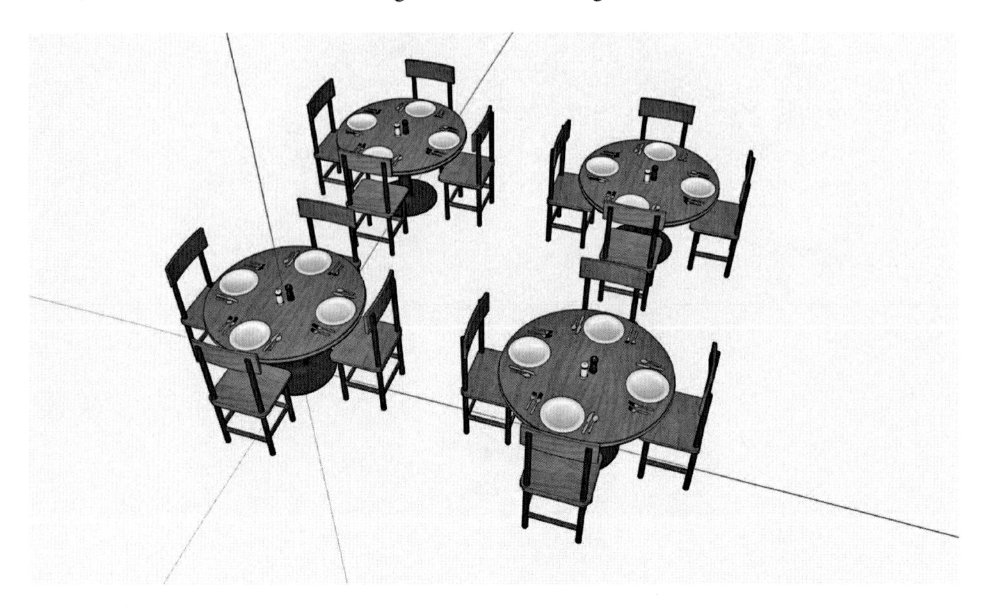

Figure 2.4 – Components in Groups in Components with Groups

If we needed to move all of these tables together, it would be as simple as selecting and editing the Component!

Remember, creating Groups and Components not only make your future job of working in this model easier by keeping things organized, but the use of Components mean you only need to edit a single instance in order to change everything. Now that we have everything organized, let's talk about controlling the visibility of our model.

Getting more out of tags

The next level of organization you should be thinking about is using **tags**. Tags allow you to control the visibility of a set of Objects with a single click. One of the things that many beginners struggle with in SketchUp is limiting what is on the screen at any given time. Being able to control what you see on the screen ensures you can focus on what you are working on. Tags make it simple to control what you see and how you see it, making it easier to work with your model.

> **More Than Just Visibility**
>
> While, for many, the primary use of tags is to toggle visibility, tags can also be used to set line styles. Remember, if you use tags for this purpose, every edge to which you assign the tag will receive the given line style.

Let's organize our table with some tags:

1. Let's create some new tags. Open the **Tags** window and click the plus icon at the top to create the following tags:

    - `Settings`
    - `Condiments`
    - `Tables`
    - `Chairs`

    At this point, your **Tags** window should look like this:

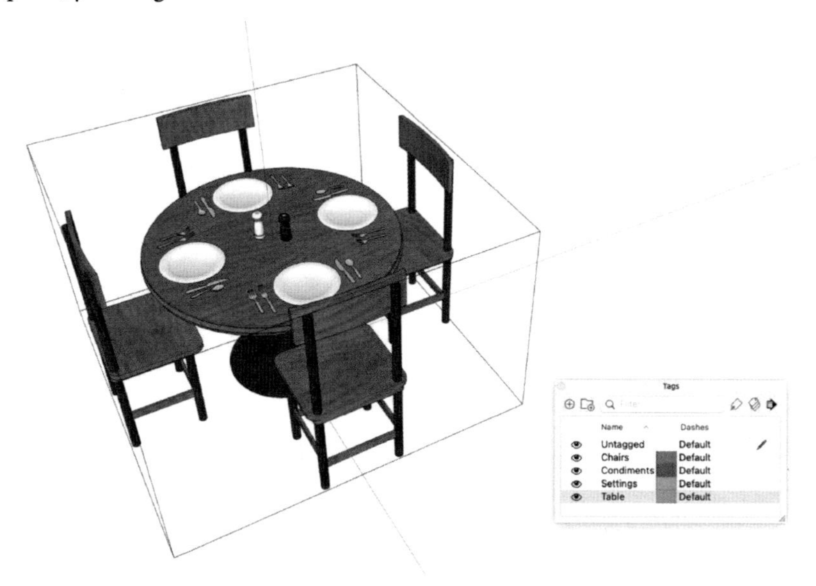

Figure 2.5 – The Tags window with tags ready to be applied

Now that you have the needed tags created, let's assign them to the proper Objects.

2. Double click the **Four Top** Component, choose the **Tag** command, then highlight the **Condiments** tag in the **Tags** window.
3. Click on the Group that contains the **Salt Shaker** and **Pepper Shaper** Components.
4. Next, highlight the **Settings** tag in the **Tags** window and click on all four **Place Settings**.
5. Assign the **Chairs** tag to the chairs.
6. Finally, assign the **Table** tag to the **Table** Component.

At this point, everything has its own tag and visibility control. Turn the **Table** tag on and off and you will see all the **Table** Components disappear. This is, again, the power of Components. Since the entire Object, which you named `Four Top` in the previous section, is a Component (and not a Group), you only have to apply tags to one instance and all copies inherit the tag assignments.

Only Tag Objects

Tags are great for organizing your model but be sure to only apply tags to Groups and Components. Raw geometry (all of those edges and faces) should stay untagged. Applying tags to raw geometry can create a mess of a model.

Having these tags is a great help, but if we want to turn off all the parts of our model, we need to click each tag individually. How can we make that easier? Just like Objects, tags can be nested as well. Let's grab all the tags we just created and put them into another tag. In this case, we will create a tag folder to hold the tags we created.

7. Use *Shift* to select all four tags.
8. Click the **Tag Folder** button at the top of the **Tags** window.
9. Name the new tag folder `Four Top`.

Now you can toggle visibility of the entire tag folder, or navigate into the folder to toggle individual tags as needed.

Downloaded Tags

Tags are saved as a part of all SKP files. This means if you import a file into your model, you are importing its tags, as well. Importing a piece of furniture or entourage from 3D Warehouse is a huge time-saver, but make sure you keep an eye on the tags that they add to your model. Get more information on working with 3D Warehouse models in *Chapter 12, Using 3D Warehouse and Extension Warehouse.*

Taking the time to create and assign tags to your model as you go will help to keep your model organized and make it so much easier to work with. This is true not just for you, but for others who may need to use your model as well. One of the hallmarks of a usable SketchUp model is clearly named, properly applied tags. Now let's dive into one of the most powerful tools at your disposal for model organization, the Outliner.

Understanding the Outliner

One of the most overlooked tools in SketchUp is the **Outliner** window. Many users rarely, if ever, open it, much less use it in their regular modeling sessions. In case you are one of those who have not spent time with the Outliner, it is basically a list showing every Object in your model along with the full nesting hierarchy. Not only is it a display of all the pieces that make up your model; it's also a way for you to select and toggle the visibility of everything!

Figure 2.6 – The Outliner shows every Component and Group in the entire model

We have done a good job of creating Groups and Components and assigning tags to them, so our simple dining room is easy to work with so far, but the Outliner will let you see the state of your model as well as giving you access to items that are not individually tagged. For example, let's say you want to turn that Salad Fork Component off. Let's do that using the Outliner:

1. Click the arrow to the left of one of the **Four Top** Components to expand it.

   You will see each of the **Place Settings** Components, the **Table** Component, the **Chair** Components, and a Group that contains the **Salt Shaker** and **Pepper Shaker** Components.

2. Expand one of the **Place Setting** Components.
3. Click on the **Salad Fork** Component.

   When you select anything in the Outliner, you are selecting that item (and in the case of nested Components, all instances of that Component) in your model. This has already saved you numerous double-clicks and orbits to get to a view of the item you want to work with. Let's go further!

4. Right-click on the selected **Salad Fork** Component in the Outliner.

   At this point, while still in the Outliner, you have access to all of the commands you would see in the modeling window had you right-clicked on the actual Object. From here you can erase, open, or hide the Component. Let's see what happens when you hide the Salad Fork.

5. Select **Hide** from the context menu.

   It's gone! Well, it's not gone, just hidden. **Hide** is a great way to temporarily take Objects off your model screen, but can be tricky to use because once an Object is hidden, you cannot interact with it. This causes a headache for many users because they will "lose" items in their model by hiding them and never getting them back. With the Outliner, you can see everything, even hidden items.

   Notice that the **Salad Fork** Component is still in the Outliner, but the eye icon to its left is toggled off (and the text is light gray). This makes it very easy to right-click and select **Unhide**. Unlike tags, visibility in the Outliner allows you to control individual Objects and not everything as a collection. This gives you a lot more control of what you are seeing in your model.

6. Right-click and select **Unhide** for the **Salad Fork** Component.

   The Outliner can be used for so much more than just toggling visibility. For example, notice that the Group that contains our shaker Components is named **Group**. Another thing that will make your model easier for future users of this file (including future you) is giving everything a name. When Groups are created, they simply get the name Group (not having to enter information makes Groups quick and easy to use). Let's change the name of this Group using the Outliner.

7. Select the **Group**.
8. Click again on the name field.
9. Type `Shakers` and press *Enter*.

Entering Objects through the Outliner

Double-clicking on an Object in the Outliner will open that Object in the modeling window. This makes it very easy to open Groups or Components for editing without having to burrow down into your nesting hierarchy.

This chapter only scratched the surface of what you can do with the Outliner. If it is not a tool that you currently use, I would highly recommend that you start keeping an eye on it or consider using it as your model get more complicated. Not only will it help you to navigate a complicated model, but it will also help remind you to keep things organized! Speaking of organization, let's hop into the **Scenes** window and see how we can use scenes to make it easier to work with our SketchUp model.

Using scenes for more than appearances

Most users depend on **scenes** to position the camera, or to prepare a view to be used for output from LayOut. While this is a primary function of scenes, there are other ways that you can use scenes to help you model. For this next example, let's make a few scenes that will allow us to quickly change what is visible in your model:

1. Open the **Scenes** window if it is not already open.
2. Click the **Show Details** button to show all the options for scenes.

   In this portion of the window, you can do some basic functions including naming the scene or specifying that scene should be in an animation, and it also includes all of the properties that will be saved with the scene. By default, all the listed properties will be saved with the scene and restored when a scene is activated. By choosing to turn on only some of the properties, we can create scenes that help during the modeling process.

   For example, let's create a scene that hides all of the Place Setting and Shaker Objects on the table. This is something that you may want to do to optimize your file as you model. Remember, the number of things on the screen at any given time will affect how SketchUp performs. Making a scene with detailed items turned off that you can quickly switch to lets you speed up SketchUp when needed.
3. Turn off all checkboxes except for **Visible Tags**.
4. Verify that all tags are currently visible, then click the **Add Scene** icon (the plus at the upper left of the **Scenes** window).

   Since we created this scene with everything visible, this is our "all on" scene. Enter the name `All On` and press *Enter*.
5. Now turn off the **Settings** and **Condiments** tags.
6. Click the **Add Scene** icon again and rename the new scene to `Empty Tables`.

You now have two scenes that you can use to display or hide all the stuff on the table. Click back and forth between the two scenes and see how easy it is to make the change to what's visible in your model. Clicking the **All On** scene will show everything in the model:

Figure 2.7 – All On scene

If at any point you want to see just the tables and chairs, you can hide everything on top of the tables by clicking the **Empty Tables** scene:

Figure 2.8 – Empty Tables scene

Note that you can orbit around the model and still switch these scenes. Since there is no connection between the camera location and the scenes, you can jump between them without having to worry about your point of view changing.

So, we know that we can use scenes to move the camera and create views for modeling or output, and we can also tie scenes to tag visibility, but let's try making a scene with another property. We will now create a couple of scenes that change how the Objects appear on screen:

7. Click the **Add Scene** icon.
8. With this new scene highlighted in the **Scenes** window, turn **Visible Tags** off and turn on the **Style and Fog** property and the **Shadow Settings** property.
9. Click on the **Update Scene(s)** icon (just left of the **Add Scene** icon). Click **Update**.
10. Name this scene `Standard Style`.
11. In the **Style** window, choose the **Monochrome** style.
12. In the **Shadow Settings** window, turn shadows on.
13. Click the **Add Scene** icon.
14. Name this scene `Mono Shadow`.

    You now have four scenes that allow you to control what is visible and how it looks. At this point there are four states you can create by combining these four scenes:

Figure 2.9 – Four scenes create four possible views of your model

We are using our scenes as controls to quickly toggle properties of our model. Those of you reading this book who are good at math may be looking at this thinking, "I have four scenes that allow me to view my model four ways, so how is this an advantage of creating four scenes?"

That is a great point. The real advantage happens when we add another control using scenes. Let's make a couple of scenes that allow us to toggle off the chairs:

15. Click the **Add Scene** icon.
16. With this new scene highlighted in the **Scenes** window, turn the **Style and Fog** and **Shadow Settings** properties off and turn on the **Hidden Objects** property.
17. Click on the **Update Scene(s)** icon (just to the left of the **Add Scene** icon). Click **Update**.
18. Name this scene `With Chairs`.
19. In the **Outliner**, open the **Four Top** Component and toggle visibility off for the four **Chair** Components.
20. Click the **Add Scene** icon.
21. Name this scene `No Chairs`.

You now have a set of controls to turn the chairs on and off. When you combine this control with the previous setup, you can now create twice as many views of your model:

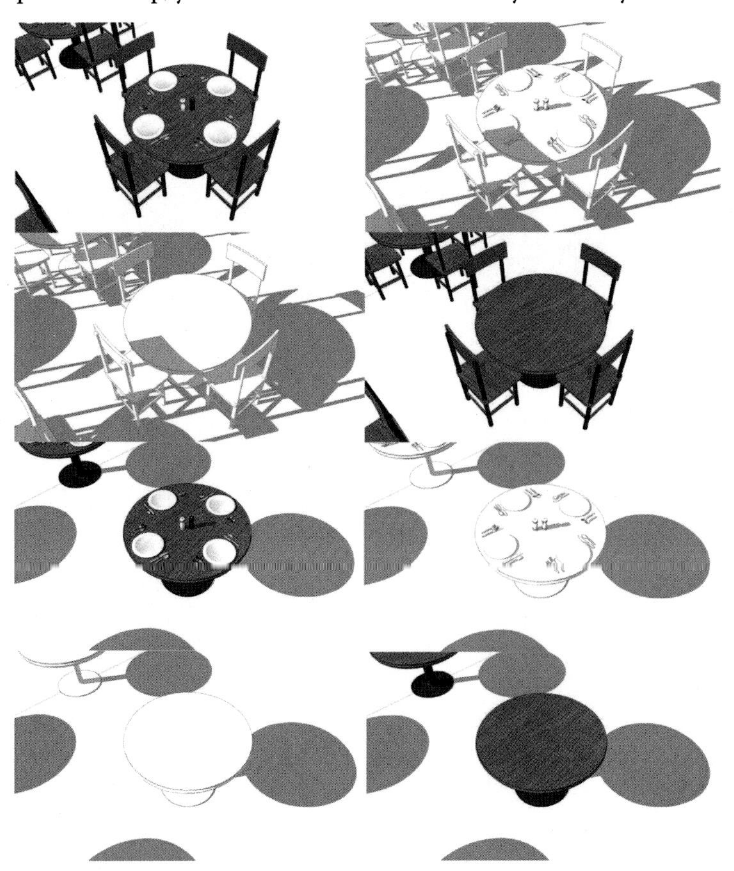

Figure 2.10 – Six Scenes create eight possible views of your model

See how the ways to view your model grows? And we are only using a few properties! With more variations of the properties (such as multiple scenes with different tags visible, or more style scenes), you can create dozens of views with just a few scenes.

Remember, scenes do not have to be binary. When using the default scenes (with all properties enabled) you create scenes that force everything when they are enabled. By choosing to only enable some of the properties in a scene, you can create view controls that can be used together!

You have now learned a lot about organizing and controlling the view of your model, which is an essential skill for creating an optimum SketchUp file. Another important ability of a great modeler is to get rid of the excess.

Removing model excess

Many new SketchUp users have started playing around with a model, making it bigger and bigger, trying out commands and just seeing what they can make, when the model starts to slow down. This can become apparent when saving or opening the model (long load times) or when orbiting the model (jerky or lagging movement through the model). While hardware can impact this (a brand-new desktop computer with 64 GB of RAM will handle a larger model than a 10-year-old laptop with a quarter of the RAM), a model that is free from excess data will behave much better than a model with extra, unneeded data.

Up to here, we have talked about keeping your geometry organized with Objects, controlling visibility with tags, using scenes to create quick toggles to change how your model appears on the screen, and finding Objects in your nesting structure with the Outliner. These are great ways to create and organize your model; they do not, however, help you to prevent your models from becoming bloated and hard to work with.

While steps have been taken in recent releases to minimize file size and speed up the modeling process, a simple fact remains: an organized, optimized model will perform better than a messy, bloated one. Using what was just covered will help to keep your model organized, but this section will look at how to optimize your model.

Something that is very important to remember is that a SketchUp file (`.skp`) is made up of more than just what is on your screen. An `.skp` file includes the geometry and colors that you see on the screen, and so much more! Every file you create includes the following:

- Geometry (faces and edges, hidden or visible)
- Materials (including imported imagery)
- Group and Component information
- Tags and scenes
- Styles

- Model info (such as units, text default, and the author)
- Geolocation information
- And all the other stuff you may add during modeling (guides, sections, dimensions, text, and so on)

In this section, we will look at how to clean up a model if it has gotten out of hand. We will start, however, by thinking about creating models with the least amount of geometry necessary.

Optimizing geometry

While we have established that there is a lot of data that goes into a SketchUp file, let's start with what you can see. Whenever a SketchUp file is saved, it contains all of the geometry you have drawn on the screen. Less geometry in your model yields a smaller file size, which will open and save quicker and allow you to orbit around without slowing down.

Remember, SketchUp needs to remember what and where every piece of geometry is, so only including what you need in your model will make for better modeling performance. While that seems obvious ("I'm not going to model things that I don't want in my model!"), there are things that people do that causes their models to bloat with unnecessary geometry. Take a look at the two models here:

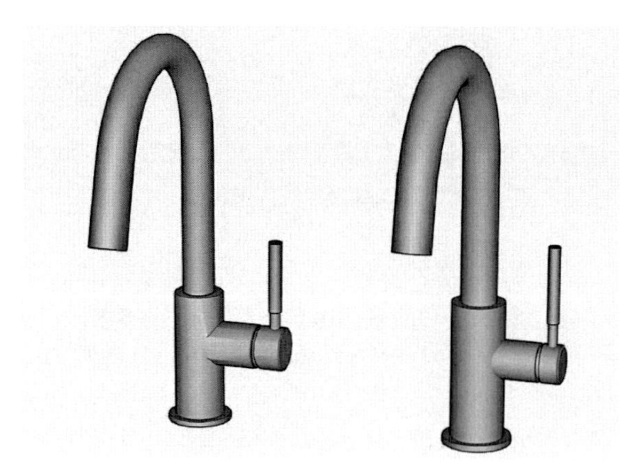

Figure 2.11 – These seem the same, but one has three times as much geometry as the other

In this example, the model on the right has a third of the number of edges and planes as the one on the left side. Despite this, they look the same! If we toggle on **Hidden Geometry**, you can really see the excess geometry:

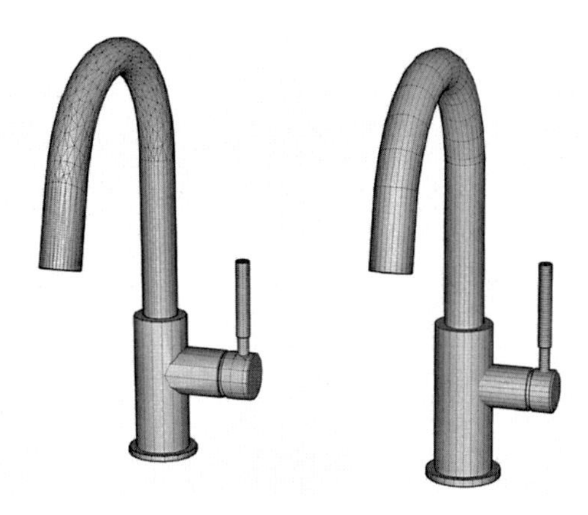

Figure 2.12 – Hidden Geometry reveals the excess geometry

In this example, we are looking at a few thousand extra edges or faces, but if this was continued through an entire model, it could add up to hundreds of thousands of extra pieces of geometry.

> **Model Statistics**
> To see what makes up your model, check out the **Model Statistics** tab in the **Model Info** window. It will allow you to see everything in your file and may help troubleshoot those that perform less satisfactorily.

One of the best parts of SketchUp can lead to the downfall of the indifferent modeler. SketchUp allows you to control how many edges are used in your curves and circles. Many new modelers assume, "The more, the better!" and crank circles up to the max (999 sides). The problem that this causes, of course, is unneeded geometry. How much geometry *should* you have in a model? That all depends on what the model is being used for.

Let's take another look at the example from above. With something like a faucet, like many items you may model, it will probably be presented in one of three ways:

- It will be seen as a small detail in 2D output (construction drawing).
- It will sit in the background of the scene (design render or vignette).
- It will be the hero of a render (product image).

Each of these uses requires the faucet at a different level of detail. This means that there are actually three different answers to the question, "How much geometry is the right amount?"

Here are three models that could be used in each of the preceding examples, along with statistics for each model. **Hidden Geometry** has been toggled to visible to illustrate the differences between models. The first one would be perfect for dropping into something like a construction drawing or 2D construction detail:

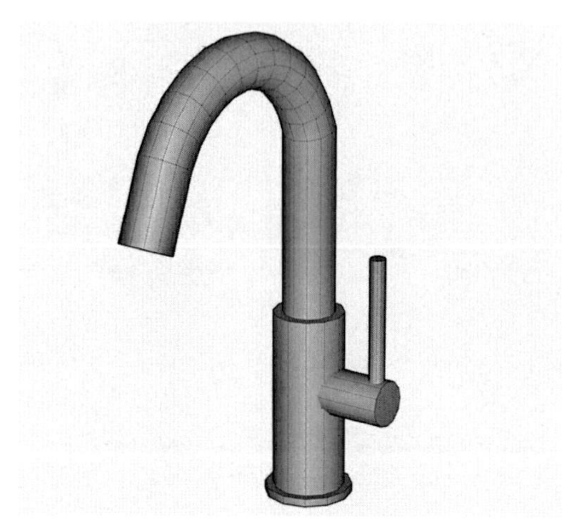

Figure 2.13 – Low-detail model perfect for a construction drawing (213 faces and 492 edges)

This next version of the faucet would be great for something like a design render showing the whole kitchen, or even a counter with a sink. There is enough detail here that it could be used in a render and stand up to fairly close scrutiny as part of a larger scene:

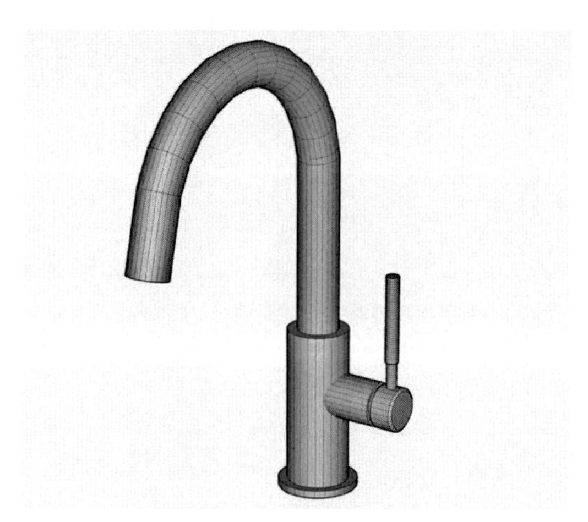

Figure 2.14 – Medium-detail model good for a general purpose (559 faces and 1,323 edges)

This model was purpose built for a product render. There is a lot of detail and a complex mesh creating the curve of the faucet that would not be noticed in most applications, making the extra geometry it brings to your model excessive and unneeded, unless the purpose of your model is to show off this faucet:

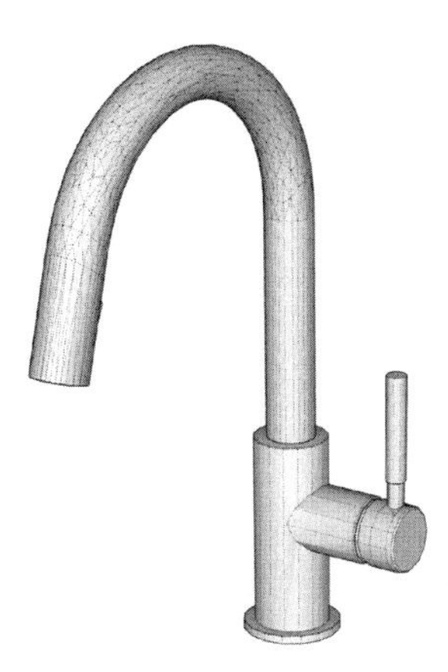

Figure 2.15 – High-detail model ideal for a hero render (2,190 faces and 3,831 edges)

Being conscious of how a piece of a model will be used and modeling appropriately for that specific use is something that sets a SketchUp ninja apart from the novices.

Optimizing materials

Much like geometry, image data imported into SketchUp can cause a file to become much larger than it really needs to be. Look at the two images in the model here:

Figure 2.16 – (left) A 4k image that has a file size of 16.8 MB; (right) the same image in SketchUp but only 627 KB

To understand why these files are so different but look so similar in SketchUp, you need to understand how SketchUp uses imagery. When you import an image into SketchUp, it brings the entire image into the `.skp` file and saves it there. However, SketchUp downsamples the file to 1,024 pixels when it displays it in your model.

> **Maximum Texture Size**
>
> To have your imported images look as detailed as possible, make sure to turn on **Use maximum texture size** in the **OpenGL** tab of the SketchUp **Preferences** window.

This means that importing a huge image into SketchUp causes your model to be much larger without making it look any better. To keep your model optimized, you may want to consider downsizing images before importing them into the model. You do not have to import to exactly 1,024 pixels, as SketchUp will resize to the exact size as needed but resizing an image to approximately the size that will be used will help to keep your model compact.

Something else to note is how much performance was immediately impacted. The smaller image imported into SketchUp quickly and was easy to move around the model. When the larger image was imported, however, it jumped and lagged as it was moved around.

Another thing that creators of oversized model files do is import images of solid colors. Check out this example:

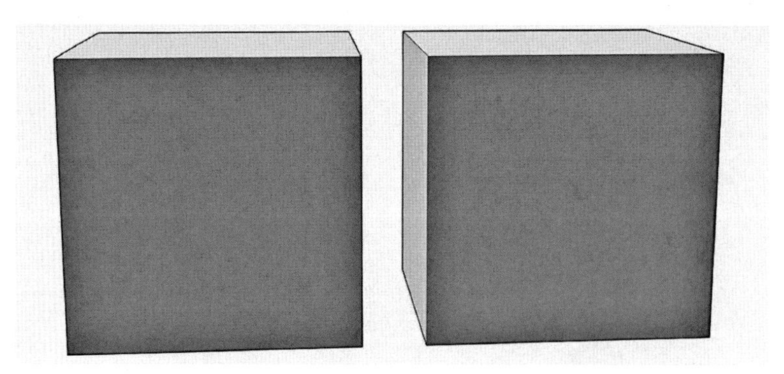

Figure 2.17 – (left) The cube has an imported image; (right) the cube has a color

In this example, the square on the left is an imported image adding 53 KB to the model, while the image on the right is a color added using a color from the color wheel in the **Color** window. This color adds virtually no additional data to the file. If this is done through your model, you can end up with a bloated, slow model with no improvement to the look of the model. On top of that, it takes time to import color swatch files into SketchUp while color samples are just a click away.

Extra Components, styles, and materials

Any time you import or create a Component, it is saved as a part of your model, even if you delete it from your screen. Any style you "try out" as you build your model is saved in your model. Every material you use, if it is picked from a library or imported, even if you don't apply it to anything, is also saved.

This data is a part of your model even if you are not using it in your model. This extra data, especially unused Components, can lead to files that are several times larger than they need to be. The good news is you can clear up much of this unused data with a single command. That command is called **Purge unused**.

To run Purge unused, pull up the **Preferences** window and click on the **Statistics** tab.

Figure 2.18 – The Statistics tab of the Model Info window (where Purge unused lives) displays all the information about your model

The **Purge unused** option is in the Statistics tab of the Model Info window. Clicking it will remove Components, styles, and materials that are saved in your model but not actively being used. In some cases, running this simple command can cut the size of a file in half!

A word of warning, though. If you have Components in your Component browser that you want to keep around, even though they are not currently in the model, or if you have colors that you plan to use in the future, but have not used yet, they will be lost when you run Purge unused. This command does not know what you plan to use in the future. It only knows what is on the screen, in the modeling area right now, and if it is not there, it is gone.

Purge unused is a command that everyone should run on their model at least once. If you ever plan to share your model, post it on 3D Warehouse, or just optimize the use of your computer's storage, clicking this command will help dump the junk and leave you with a better SketchUp model.

Using visibility commands

While everything covered in this chapter so far will help you organize your models and make them easier to navigate and work with, there are situations where you will need to change something in your model as you work on it. You might want to temporarily hide one Object temporarily, or show what has been hidden in your model just for the duration of one modify command. These situations do not require a modification to your workflow but can be addressed using a few Edit and View commands.

Visibility while editing Objects

Any time you edit a Group or Component, you can temporarily hide everything else in the model with a single command. In the **View** menu is a sub-menu called **Component Edit**. In this sub-menu there are two toggles: **Hide Rest of Model** and **Hide Similar Components**. Clicking on one of these options will place a checkmark next to them. If they are checked, they will automatically activate anytime you edit an Object (yes, the name of the sub menu is **Component Edit**, but it works when editing Groups as well as Components).

Hiding geometry outside of the current context is a great way to simplify what you have on your screen and can make working in your model much easier. These toggles can make a big difference as far as what is on your screen at a given time:

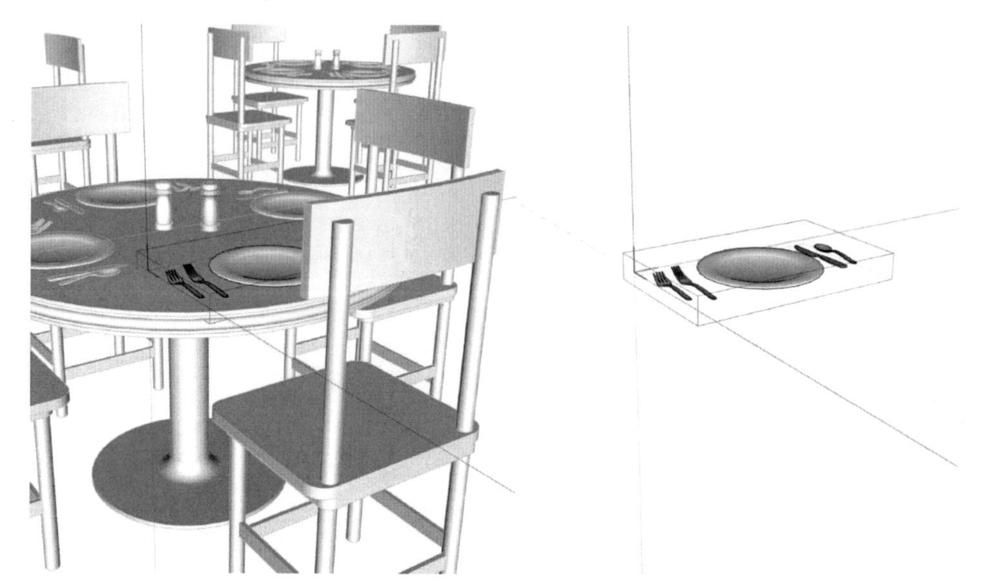

Figure 2.19 – Editing the Place Setting with Component Edit toggled off and on

Let's take a look at an example. With both options toggled off (no checkmarks), do this:

1. Open one of the **Four Top** Components by double-clicking on it.
2. Now double-click on the **Shaker** Group.

   Notice that everything that is outside of the Group is now a hazy, faded color, but still visible. This does a good job of calling attention to the contents of the Group but does not necessarily make the Group 100% accessible. If you orbit under the table, you will see that, even though the table is faded, it still prevents you from seeing the contents of the Group you are currently editing.

> **Controlling the Fade**
> You can control how much the items outside of the current Object are faded. In **Model Info**, under the **Components** tab, there are two sliders. These sliders allow you to control how faded the geometry is displayed. The two sliders allow you to control the fade Components that are copies of the current Component, or the entire rest of the model separately.

   Let's see how the model changes when we change those toggles.

3. Choose **Component Edit** from the **View** menu then click on **Hide Rest of Model**.

   Now, all you can see are the two shakers, because that is all that is visible inside the Group. The other **Shaker** Groups are still visible on your screen because you are editing a Group that is inside of a Component. **Hide Rest of Model** hides everything that is outside of the current Object but keeps it visible if it is a copy of an open Component. Let's see how to hide that geometry.

4. Click outside of the **Shaker** Group to get back to the **Four Top** Component.
5. Choose **Component Edit** from the **View** menu and click on **Hide Similar Components**.

Now everything that is not inside the currently selected Component is hidden. These two commands are amazing tools to have as you work on models with lots of Objects.

> **Quick Toggles**
> Many expert modelers will assign shortcut keys to one or both commands making it easy to toggle them while in the middle of their modeling workflow. We will cover setting up custom shortcuts for these commands in *Chapter 7, Creating Custom Shortcuts*.

Hiding things from yourself

While tags are definitely the best way to control visibility, there are times where you may want to hide a single Object or piece of geometry. This is most easily done with the **Hide** command. **Hide** can be found under the **Edit** menu or, more easily, in the right-click context menu.

Let's look at how to use, and more importantly how not to use, the **Hide** command:

1. Click outside of any open Objects to close them.
2. Right-click on one of the **Four Top** Components.
3. Choose **Hide** from the context menu.

   And the table is gone! Seems simple enough, right? Yes, making things disappear is easy. The real trick is in getting them back. Since you cannot see the hidden table, you cannot right-click it to choose **Unhide** from the context menu. Many a novice modeler has lost pieces of their model using the **Hide** command without understanding where hidden things go.

   Before hiding something, you should know that hidden does not mean erased. Hidden items are in your model. Other geometry can and will interact with them. They still add to the geometry and data in your file. You cannot see hidden geometry or Objects, but they are still there.

   So, how do you get hidden geometry back? There are two options. The first depends heavily on you not having hidden more than one item at a time.

4. Open the **Edit** menu and click on **Unhide**.
5. Choose **Last** from the sub-menu.

At this point, the hidden **Four Top** should be back. The **Unhide** command has the option of unhiding the last Object or all Objects (or just those selected, but we will get to that in a moment). These are not very granular controls, so you need to be very careful using the **Hide** command. Personally, I recommend never hiding more than one face or Object using this command and, even then, only leave it hidden long enough to complete a set of commands before unhiding it again.

It is also important to know that hiding is context sensitive. This means that you can hide an Object that is inside of another Object but cannot unhide it if you leave the context. That sounded a little confusing, so let's do a quick hands-on example:

1. Double-click to enter one of the Four Top Components.
2. Right-click on any of the **Place Setting** Components and choose **Hide** from the context menu.
3. Click outside of the Four Top Component to close it.
4. Open the **Edit** menu, click **Unhide** then choose **All** from the sub-menu.

   Nothing happened! Does this mean that the hidden Component is gone forever? How do you get it back? The **Unhide All** command is context sensitive meaning it will unhide all hidden Objects in the current context. Since no Objects are currently open, it is unhiding all hidden items that are not in a Group or Component.

5. Double-click to enter the same **Four Top** Component.
6. Open the **Edit** menu, click **Unhide** then choose **All** from the sub-menu.

Another option is to use the Unhide Last command. Unhide Last will unhide the last thing hidden, regardless of context. This is great for getting back something inside of a Group or Component. There is, however, a limitation to Unhide Last. If you run Unhide All, you will lose the ability to run Unhide Last, even if Unhide All did not unhide anything. This is another illustration of why you should be careful and sparing with the Hide command.

Seeing the invisible

Unhiding is great, but there has to be a way to see things when they are hidden, right? Of course there is – so let's dive into that right now. First, let's make a mess in our model then use the visibility toggle to allow us to clean things up:

1. Right-click and use **Hide** to hide one of the **Four Top** Components.
2. Double-click to enter another **Four Top** Component and hide the **Shakers** and a couple of **Place Settings** instances.
3. Double-click to enter the **Table** Component.
4. Select the face that creates the table's top.
5. Right-click and hide the face.
6. Click outside of the **Table** Component to close it.
7. Then click outside of the **Four Top** Component to close it.

   At this point we have several things missing. And **Unhide Last** or **Unhide All** will only get us one piece back, at best. Fortunately, we have a simple tool to get it all back. In the **View** menu there are two toggles that we are going to use: **Hidden Geometry** and **Hidden Objects**. Let's start with the second.

8. Open the **View** menu and toggle on **Hidden Objects**.

   As soon as you do this, all of the Groups and Components we hid are shown with dashed lines and a faint hatched fill. These hidden Objects can be interacted with the same as visible items, making it possible to modify, delete, or unhide any of them! Let's get our hidden Objects back.

9. Right-click on the hidden **Four Top** Component.
10. Choose **Unhide** from the context menu.

    The whole thing is back! Let's try to get the rest back. Since the **Shakers** and **Place Setting** are inside of a Component, you cannot right-click and select **Unhide** (**Unhide** will not be in the context menu) until you open the **Four Top** Component.

11. Double-click the **Four Top** Component with the hidden Objects inside.

12. Right-click and unhide the hidden Objects.

    That brings back all of our hidden Groups and Components! Our hidden table top, however, is still missing. Not only that, but toggling on **Hidden Objects** did not even show the face that we hid. This is because the top of the table was not an Object but a single face. To see our hidden face we need to toggle on **Hidden Geometry**.

13. Open the **View** menu and toggle on **Hidden Geometry**.

    There's the missing table top, and so much more! Notice the dashed lines on the plate, the silverware, and the shakers. **Hidden Geometry** will not only toggle geometry that was hidden by an intentional **Hide** command, but also geometry that was hidden in order to smooth the geometry. Let's take care of our table, then look at some of those hidden edges.

14. Enter the table Component and unhide the hidden table face.

    Whenever you run certain commands such as **Soften/Smooth**, or if Push/Pull or Follow Me are used to extrude a curve, you will have hidden geometry automatically created in your model. In *Chapter 1, Reviewing the Basics*, we talked about how surfaces are created by hiding the edges between faces. When a surface is selected the whole thing is highlighted. If Hidden Geometry is turned on, however, you can interact with any of the faces or hidden edges as if they were normal geometry.

15. Open the **Four Top** Component, then the **Place Setting** Group, and then the **Plate** Component.

16. Click on a face.

    When hidden geometry is visible you can interact with the pieces between hidden edges. This means you do things that you cannot do with the full surface, such as use **Push/Pull** on a single face or paint a section of a smoothed surface. Let's use these hidden faces to add detail to our plate.

17. Use *Shift* and **Select** to select a ring around the edge of the plate. Your selection should look like this:

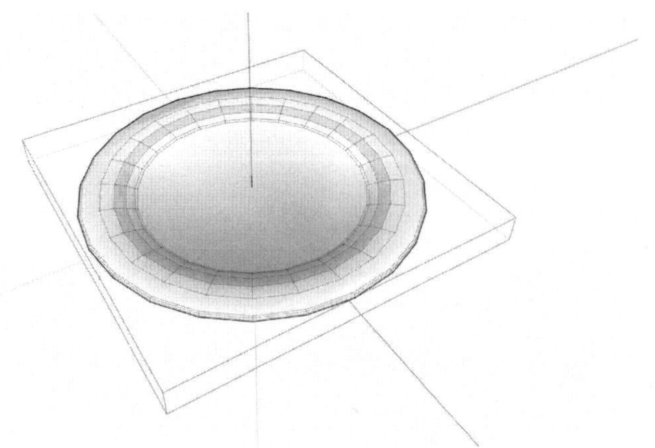

Figure 2.20 – Selected hidden geometry

18. Now use the **Paint Bucket** tool to fill the selected faces with a blue color.
19. Use **Select** to click outside of the Objects until you are completely out of all Components.
20. Open the **View** menu and toggle off **Hidden Objects**.
21. Open the **View** menu and toggle off **Hidden Geometry**.

More Toggles

Just like Component visibility, many users will set shortcuts for one or both visibility toggles. Custom shortcuts will be covered in *Chapter 7, Creating Custom Shortcuts*.

With that, you not only have all of your model back, but you have a smooth, seamless line painted around your plate:

Figure 2.21 – Final model

Understanding how to use visibility commands is something that will not only make it easier to work with, but will also improve the performance of more complex models.

Summary

In this chapter, we started by diving into ways that you can customize your modeling workflow to make models that are more organized and performant. We saw how to use Groups, Components, and the Outliner to organize your model and learned when to purge the unneeded data that can cause model bloat. We also saw how to use scenes and visibility toggles to allow you to see exactly what you need while working on your model.

In *Chapter 3, Modifying Native Commands*, we will dive deep into native input and modification commands and how you can get more out of them than you thought possible using modifier keys.

3
Modifying Native Commands

Back in *Chapter 1, Reviewing the Basics*, we briefly touched on the idea of using **modifier keys** to get more out of native commands. In this chapter, we review how modifiers work, then review every command and how it uses the modifier keys. If you are impatient, you can jump to the end of the chapter and use the complete modifier list for your **operating system** (**OS**). For now, let's hop in and get modifying!

In this chapter, we will cover these main topics:

- Using modifiers
- Modifying Select
- Using drawing modifiers
- Modifying tools
- Modifying SketchUp – quick reference

Technical requirements

This chapter will assume that you are using SketchUp Pro and the `Taking SketchUp Pro to the Next Level - Chapter 3.skp` file from 3D Warehouse:

`https://3dwarehouse.sketchup.com/model/c783842e-1642-4c9d-b9ba-d5d525f93b5e/Taking-SketchUp-Pro-to-the-Next-Level-Practice-File-for-Chapter3`

Using modifiers

One of the things that makes SketchUp such an appealing software to new users is the fact that commands do exactly what they say: the **Line** tool draws lines, and the **Move** tool moves things from one location to another, for example. When you're first learning SketchUp, this makes for a short learning curve and helps the software feel intuitive.

Using modifier keys alongside native commands allows SketchUp to add a layer of functionality on top of the simplest of basic commands. Tapping or holding a modifier key will allow a command that you already know to do more than the simple command name suggests possible. With a few exceptions, most modifiers are listed in the status bar at the bottom of the screen when a command or tool is selected. So, why include a chapter on this information? In many cases, the name of the modifier does not make it clear exactly what it does. The bulk of this chapter will explain exactly how to use the modifier for each command or tool.

The modifier keys

It may seem a little crazy, but when you get down to it, this entire chapter is dedicated to three keyboard keys. Depending on your OS, the modifier keys you will use will be different.

On Windows, you will use these keys:

- *Shift*
- *Ctrl*
- *Alt*

While on macOS, you will use these keys:

- *Shift*
- *Option*
- *Command*

If you are one of those people who have multiple OSs, or if you work with others and regularly have to know how to work on Windows or macOS, the mapping of these keys is one-to-one:

- *Shift* works the same, regardless of your OS.
- *Ctrl* modifiers on Windows are the same as *Option* modifiers on macOS.
- *Alt* modifiers on Windows are the same as *Command* modifiers on macOS.

If you look at your keyboard, you will see that all modifier keys are in the lower left or right corner. This means that the standard way to interact with your keyboard is to rest your non-mouse hand near the modifier keys. Once you start a command, you can quickly tap or hold modifier keys to change the function of a command. A big part of becoming a better and faster SketchUp modeler is to be able to quickly use and modify commands. Keeping your hands near the modifier keys allows you to quickly multiply the functions you can perform with minimal effort.

Press or Hold

Most modifier keys function as toggles, allowing you to tap the key once to change the functionality of the command. This means that changing what a command does is as simple as *tapping* the key once (or in some cases twice). There are, however, a few modifier keys that must be *held down* to function. In these cases, you will need to keep a finger on the modifier key until you complete the command.

Now that you know what the modifier keys are and where they are, let's look at how to use them. First, let's see how modifiers can be used with the **Select** command.

Modifying Select

The **Select** command is often thought of as the default or null state in SketchUp. It is the command that you are in when you are not drawing or using one of the tools. The **Select** shortcut is the biggest key on the keyboard by default, allowing you to quickly and easily activate it. **Select** is unlike any other command in SketchUp, and as such, has its own way of using modifier keys.

When modifying a selection, it is important to remember that modifiers function differently with **Select** than they do with other commands. With **Select** modifiers, you must hold down the key for it to work. This means keeping your non-mouse hand on the keyboard, and holding down a key until your selection is complete.

Normal **Select** will activate whatever item (for example, edge, face, object, dimensions, or section plane) is directly under the very tip of your cursor. If you click on one item, it lights up in the highlight color. When you click on a second, it highlights while the previous item returns to normal (that is, deselects).

The modifier keys for **Select** are as follows:

Modifier	**Windows**	**Mac**
Toggle Selection	*Shift*	*Shift*
Add to Selection	*Ctrl*	*Option*
Remove from Selection	*Shift + Ctrl*	*Shift + Option*
Invert Selection	*Shift + Ctrl + I*	*Shift + Command + I*
Select All	*Ctrl + A*	*Command + A*
Deselect All	*Ctrl + T*	*Shift + Command + A*

Figure 3.1 – Select modifiers

Add to Selection and **Remove from Selection** are fairly obvious in their functions. Using these modifier keys will add items to your selection, or remove them. One thing to remember about modifiers is that they do not only affect single-click selections but dragging the **Select** or **Lasso Select** cursor over your model. This means that if you are holding down *Ctrl* (or *Option*) when you click and drag a selection window across your model, anything inside that selection window is added to your current selection.

The *Shift* key behaves slightly differently as it functions as both **Add to Selection** and **Remove from Selection**. This is simple to keep track of when you are using a single click on items in your model. Hold down *Shift* and click on an item and it highlights. Oops! You did not mean to select that item? Not a big deal, still holding down *Shift*, just click it again and it de-highlights.

The complexity of using the *Shift* key comes when using drag-select. Since *Shift* + **Select** toggles an item's selected status, dragging a window over a combination of selected and non-selected items will toggle them all from selected to deselected, and vice versa.

The second half of the table in *Figure 3.1* is not technically modifier keys, but good shortcuts to know when using **Select**. These shortcuts allow you to quickly and easily make big changes to your selection. Use these shortcuts to invert your selection, select everything, or deselect any highlighted items. It is important to note that these shortcuts all run *in context*. This means that **Select All** will actually only highlight the items that are in the group that is currently open.

Knowing how to use the modifier keys alongside the **Select** command will help to change you from a *swipe a selection window across the model and pray you get all the pieces you actually need highlighted* type of user into a selection master creating complex selections with just a few clicks (and keystrokes). Now that we know how modifiers can multiply the functionality of a simple command such as **Select**, just imagine what it can do for the drawing commands!

Using drawing modifiers

Every single native command used to draw something on the screen has at least one modifier connected to it. While some of these are simple modifications such as locking to an axis while defining geometry, some will actually change where or how geometry is created. In this section, we will list out all the drawing commands and give examples of how and when you might use each of the modifier keys.

> **Inference Locking Uses Keys, Too**
>
> Input commands do lean heavily on inferencing for establishing where geometry will be in 3D space and allow the user to lock to inferences using certain keys. Inferencing will be covered in fine detail in the very next chapter, *Chapter 4, Taking Inferencing to the Next Level.*

In this section, each drawing modifier is listed along with any modifier keys that can be used. The description will focus on how the modifiers change what the command does, and worry less about the command without modifiers (I am assuming that you have a grasp of how the basic commands work, at this point).

Line

The **Line** command is arguably the simplest of commands to use in SketchUp: two mouse clicks give you a line between them. Modification for this command is simple. If you hover over an existing face and hold down the *Shift* key before clicking the first point of your line, the cursor will lock to the plane of that face. If you pick the first point, and then move your cursor over any face (the same one or a new one), then the endpoint will lock the plane of that face.

This description can seem a little confusing, so let's look at an example:

1. Open the `Taking SketchUp Pro to the Next Level - Chapter 3.skp` practice file.
2. Click the scene named **LINE**.
3. With the **Line** command active, hover the cursor over the green face and hold down the *Shift* key.
4. Still holding *Shift*, move your cursor off the green face.

   Notice that the cursor moves, but there is a dotted line tracing back to the green face. If you start your line anywhere on the screen, that first point of the line will be in the same plane as the green surface.

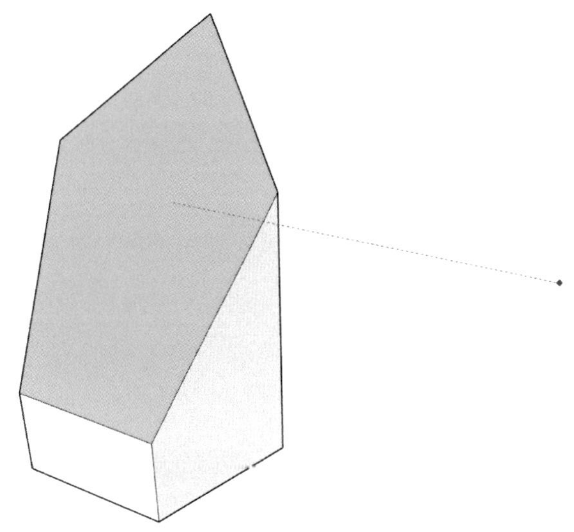

Figure 3.2 – Inferencing a plane

5. Still holding *Shift*, click once to the right of the green face.
6. Still holding *Shift*, move your cursor to the left side of the green face and click again.

   If you orbit around what you have just drawn, you will notice that the line crosses the green surface, even though it is not *in plane* with any of the axis. If you use **Select** to pick a part of the green surface, you will see that the line splits the surface in two!

This is the power of locking to a face when drawing in 3D space.

Freehand

Freehand cannot be modified until after the line is completely drawn. After drawing it, but before starting a new line or changing commands, press *Alt/Command* or *Ctrl/Option* to increase or decrease the number of sides in the curve, respectively. Endpoints will temporarily show up on the curve to show you how many edges are currently being used. Once you have the number you want, you do not need to do anything to accept the change, just click to start a new freehand line or change commands. Here is an example of the same curve with a different number of sides:

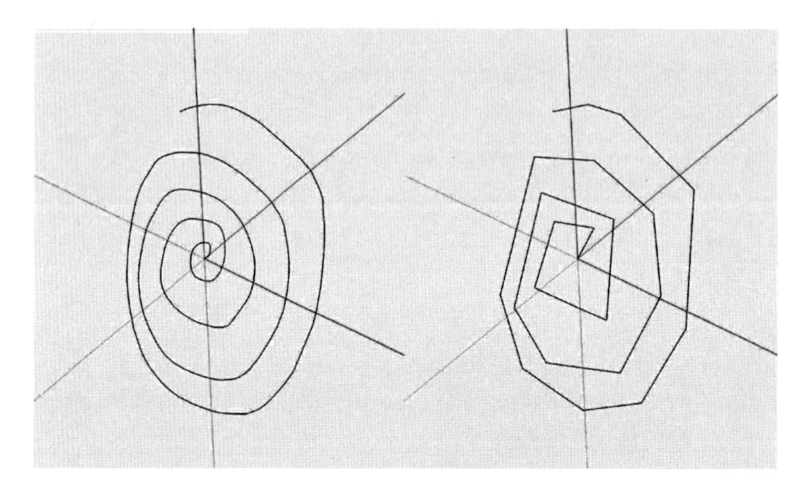

Figure 3.3 – The same Freehand curve with a different number of sides

Arc/2-Point Arc/3-Point Arc/Pie

SketchUp provides you with four different commands to draw arcs:

- **Arc** will allow you to draw a section of a circle around a center point.
- **2-Point Arc** will allow you to set the start and endpoints, then define the bulge.
- **3-Point Arc** will have you define the beginning and endpoints and angle of the arc.
- **Pie** will draw an arc like the **Arc** command but will fill it in, creating a face.

Regardless of which tool you prefer to use to draw an arc, you will be using the same modifier keys.

Just like **Line**, holding down the *Shift* key while you hover over a face will lock you into the plane of that surface.

You can also change the number of edges in an arc as you are drawing it. Holding down the *Ctrl/Option* key and pressing the - key will reduce the number of edges in the arc by one at every press. Each press of the + key will increase the edges by one. This process can be performed at any point from the start of the command until the final arc is placed.

Rectangle

You are probably seeing a pattern here with the *Shift* key as, just like with the **Line** or **Arc** commands, holding down *Shift* while hovering over a face will lock onto the plane of that face.

Tapping the *Ctrl* or *Option* key will change whether you are drawing from a corner or from the center. This is a toggled state, so pressing *Ctrl/Option* multiple times will switch back and forth. Center or corner input can be changed at any time before you select the second point of the rectangle.

Rotated Rectangle

The **Rotated Rectangle** command allows you to define a sloped plane, and then draw a rectangle on that plane. The *Alt/Command* modifier will allow you to lock the baseline of the protractor on input.

Let's hop in and use **Rotated Rectangle** with the modifier key. Start with these steps:

1. Click the start point of the first side of the rectangle. Once you have established the start point, your cursor is connected to the compass. This line will define the first side of the plane of the rectangle.
2. Click the end point of the first side of the rectangle. Once you click the second point, the compass appears perpendicular to the end of the newly created line. Now, the rectangle you are creating can be previewed as it rotates and stretches around the line you have created.
3. Move your cursor away and around the baseline and click a third time to create a rectangle.

If you did not get exactly what you were expecting, just pan over and give it another try. **Rotated Rectangle** can take some getting used to, but a little practice can help you to understand exactly how it works.

Now, let's do that again, but with the modifier key:

1. Click the start point of the first side of the rectangle. At this point, when you move the cursor around the model, the compass is jumping to different planes, depending on what your cursor is snapping to. This can make it tricky to get to a specific length on a specific plane. This is where the modifier key comes in.
2. Press the *Alt/Command* key and move your cursor some more.

   See how you are locked into a plane but have not yet set your second point? This is what the modifier key does.
3. Click to set your second and third points to create the rectangle.

Using the modifier with **Rotated Rectangle** helps you to lock into a specific plane at the beginning of the input.

Circle

The **Circle** command always draws the circle from the center and holding down the *Shift* key will lock to the plane that your cursor is over. Holding down the *Ctrl/Option* key and tapping the - or + key will decrease or increase the number of sides in the circle, respectively.

Polygon

When using **Polygon**, holding the *Shift* key while picking the center point will lock to the plane that your cursor is hovering over.

Once you have selected the center point and started pulling out the polygon outline, you will see that the polygon is sitting inside of a dashed circle. The dashed line represents the distance from the center of the polygon to the corner of the polygon (the polygon starts as inscribed; the corners of the polygon touch the circle). Tapping the *Ctrl/Option* key will toggle the polygon to circumscribed (the centers of the edges of the polygon touch the circle):

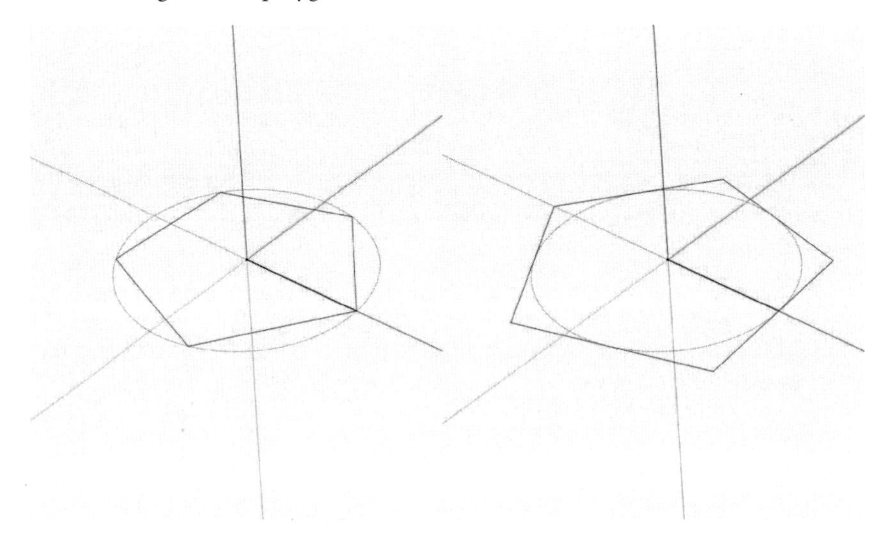

Figure 3.4 – Inscribed and circumscribed pentagons

If you hold down the *Ctrl/Option* key and press the - or + keys, you will decrease or increase the number of sides of the polygon, respectively.

As you have seen, the modifier keys for the drawing commands help you to fine-tune the input process. In many cases, using a modifier saves you from having to modify a shape once it is drawn or allows you to change the properties of an item as you draw it. In the next section, we will see how using modifiers with some of the editing tools can transform the tool into something completely new!

Modifying tools

Modifying a tool can completely change the functionality. If the native tools count for 16 items in your *SketchUp Toolkit*, then learning to add modifiers adds another two dozen tools! A handful of these tools will allow you to modify geometry in ways many new users do not even think is possible. In this section, we will review every tool and see how and why to use each modifier.

Unlike modifiers for **Select** or most of the drawing commands, almost all modifier keys for tools function as toggles. This means you tap the modifier key once to change the state of what the tool is doing. Unless otherwise noted, you can simply tap the key once to activate/change the modifier.

Eraser

The **Eraser** command will allow you to do so much more than just delete items from your model. With the modifier key, you can change the way that geometry looks and behaves by hiding, softening, and smoothing geometry.

> **I Have a Small Confession to Make**
>
> Back when I was first teaching myself to use SketchUp, I had no idea how to soften geometry. If I did something that broke an arc and caused the surface pulled from the arc to break up, I would erase the whole thing and start over. If only I had this list of modifiers back then!

The modifier keys for **Eraser** will change the function of what the tool is doing. Without any modifier, **Eraser** will remove anything the cursor drags over. Tapping any of the modifier keys will make the modifier hide, soften and smooth, or smooth and hide edges. That's right! When using modifiers, the **Eraser** tool can only be used on edges. This means that dragging a modifier eraser over faces or objects will not do anything.

Let's take a look at how these different modifiers affect some edges in your practice file:

1. Click **Scene #2** and bring up the **Eraser**.
2. Click and drag a standard **Eraser** tool across the edges of the shape on the left.

   As expected, the edges have disappeared, and they took the faces that defined them out of the model as well.

3. Now, tap the *Shift* key and notice the change to the cursor. You have just toggled on **Hide**. Click and drag across the edges of the second shape.

   Notice that the edges have disappeared, but the faces connected to them are still here? This is what happens when an edge is hidden; the faces behave and are shaded as if the edges are still there, but the edges are hidden from view.

4. This time, tap the *Ctrl/Option* key and drag across the edges of the third shape.

   In this case, the edges have been hidden, but the faces that remain have been smoothed together as well.

5. Finally, tap the *Alt/Command* key to toggle on **Unsmooth/Unhide**. Notice how the hidden and softened lines from the last two shapes appear as different types of dotted lines. Click and drag your cursor over a few of the edges in the last two shapes.

This modifier makes it very easy to change softened or hidden edges back into regular edges. Without this modifier, you would have to redraw the edge or select it and use the **Entity** information to turn off the hidden or smoothed properties!

Paint Bucket

Odds are good that if you have used **Paint Bucket** at some point, you are aware of the *Alt/Command* modifier key. This modifier will change your cursor from the paint bucket into an eyedropper and allow you to pull any material from your model and use it as the material that the paint bucket will apply.

The other three modifiers for **Paint Bucket** control how the selected materials will be applied to your model. Let's apply some paint:

1. In your practice file, click **PAINT BUCKET** and choose the **Paint Bucket** tool.

2. Start by using the *Alt/Command* modifier to pick the red color in front of the first shape.

   Notice that when you tapped the modifier key, your cursor changed, but it changed back to the paint bucket as soon as you picked the color you wanted; there is no need to use the modifier key again to get back out of sample mode.

3. Now, apply it to the front of the cube.

   This is standard **Paint Bucket** behavior; one click applies the selected material to the selected surface. Now, let's add some modifiers.

4. Tap the *Ctrl/Option* key and then click on the blue surface of the second shape.

   Notice that not only did the face you clicked on turn red but so did the rest of the connected blue faces.

5. This time, tap *Shift*, then click on any of the green faces of the third shape.

   Not only did all the connected faces turn red, but so did the green faces in all the other shapes! This will happen to all uses of this color, anywhere in your model. If a face is visible or not, in an object out in the open, all instances of the selected color will be replaced.

6. Finally, use the *Alt/Command* modifier to pick the yellow color on the ground.

7. Now, hold down the *Shift* key and tap the *Ctrl/Option* key, then click on the red section of the third shape.

Notice that all of the red on the shape changed, but the red on the other shapes stayed:

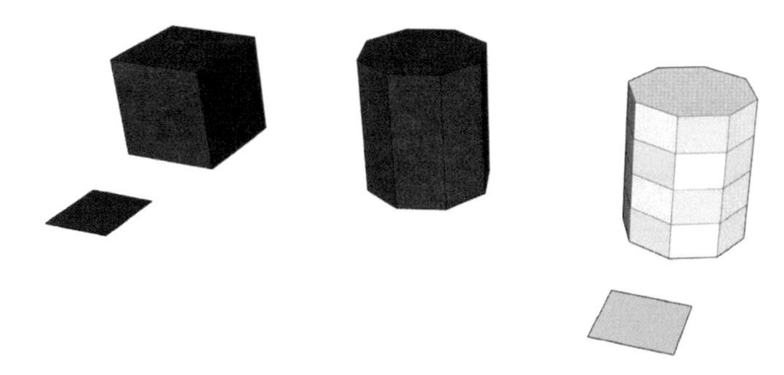

Figure 3.5 – Final colors applied using the Paint Bucket modifiers

This is because this modifier only changes colors on the current body of geometry. All other instances of the red color are safe.

> **Pre-Selecting Works Great, Too**
>
> It should be noted that using **Select** to highlight geometry that needs to be painted first, then applying the material in one click is a great way to use **Paint Bucket** as well.

Tag

The **Tag** tool is a great way to change the tags that are applied to any objects in your model. **Tag** allows you to change tags similar to painting with the **Paint Bucket** tool and, just Like **Paint Bucket**, it has a few modifiers that make it even easier to apply tags to your model.

Just like **Paint Bucket**, tapping the *Alt/Command* key will toggle on the sample mode (the eyedropper cursor) allowing you to click on an item and sample its tag. Once you do this, that item's tag becomes the active tag, ready to apply on the next click.

Tapping the *Shift* key and then applying a tag will apply the selected tag to any objects that have the same tag as the object you click on. One click and you can completely replace a tag with your selected tag!

Finally, the *Ctrl/Option* key will allow you to apply the selected tag to all instances of a component. Normally, a tag is applied to a single instance of a component (each copy of a component could have a different tag applied). With this modifier, applying it to one component applies to all copies of the component.

Move

We have spent a bit of time already talking about modifying the **Move** tool to make arrays back in *Chapter 1, Reviewing the Basics*. Copying items with the **Move** tool is just one of the modifiers you can use though.

Holding down the *Shift* key after you have selected an item to move and selecting the point you want to move from will lock your movements into a single direction. Releasing the *Shift* key will release the directional lock.

Tapping the *Ctrl/Option* key will toggle through the three move modes of the **Move** tool:

- **Move** – You can move the selected items from one point to another.
- **Copy** – Create a duplicate of the selected items and move them from one point to another.
- **Stamp** – Create a duplicate of the selected items and place a copy at each point you click.

You can tell which mode you are in by the little modifier icon on the cursor. Tapping the modifier key will move forward one in the list of modes. For example, one tap will change from **Move** to **Copy** mode. Tapping again will change from **Copy** to **Stamp** mode. Tapping the modifier key a third time will change from **Stamp** back to **Move** mode.

Tapping the *Alt/Command* key will toggle **Autofold** on and off (unless you are moving an object, in which case this will cycle through grip types, which we will cover in *Chapter 4, Taking Inferencing to the Next Level*). Autofold will control how faces connected to a face being moved will behave.

Let's get hands-on with **Stamp** in our practice model:

1. Click the **MOVE** scene.
2. Use the **Move** command to move the red pyramid forward, past the red axis.
3. Now, tap the *Ctrl/Option* key to change into **Copy** mode and move a copy of the yellow pyramid past the red axis.

   Notice that as soon as you click to place the copy of the yellow pyramid, you are automatically placed back into the standard **Move** mode.
4. This time, tap the *Ctrl/Option* key twice to get into **Stamp** mode.
5. Click on the blue pyramid to select it, then place two copies past the red axis.
6. To exit **Stamp** mode, tap the *Esc* key.

   Now, let's give **Autofold** a try.
7. Click **Move** then click on the green rectangle and move the selection around the screen.

   Notice how SketchUp prevents you from moving the selected rectangle out of the plane of the surface it was originally in? This will happen any time you try to move geometry that lies directly and completely inside of another surface.
8. Tap *Esc* to return the green rectangle to its original location.
9. Tap the *Alt/Command* key to turn on **Autofold**, then select the green rectangle and move your cursor around the screen.

Notice that the geometry of the surrounding surface is now connected to the rectangle as you move. Let's add one more keystroke to help visualize the purpose of this modifier.

10. Tap the *Up* arrow key to lock the move into the blue axis and move your cursor up and down on the screen.

    See how **Move** is allowing you to move the rectangle vertically and has automatically broken the surrounding surface at the corners? This is what **Autofold** does.

11. Move the green surface just above its original location and click to place it.

Now, you can see exactly how modifier keys can multiply out the functionality of a single command:

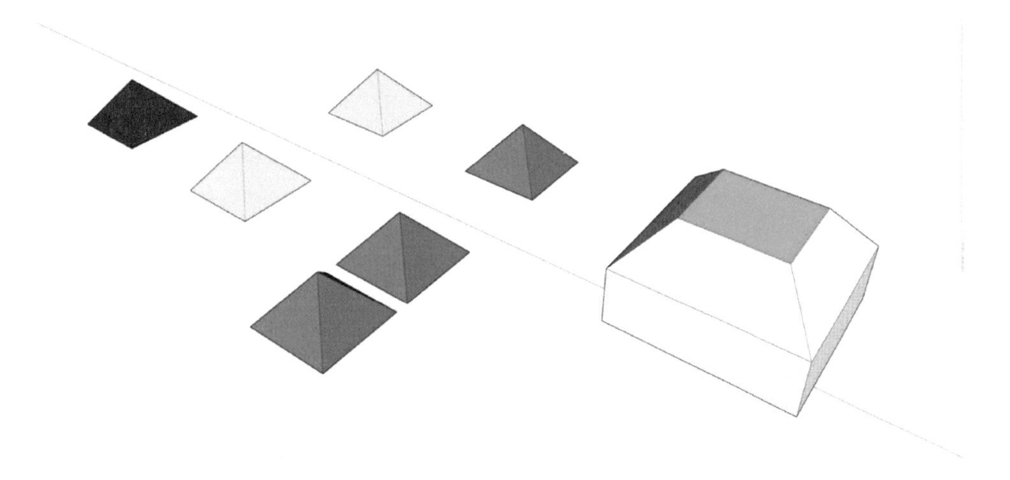

Figure 3.6 – Result of Move

That was a lot! **Move** is the perfect example of a single tool that can do so much more if you know what modifier keys to push.

Rotate

When using **Rotate**, holding down the *Shift* key while hovering the cursor over an existing face will lock you into the plane of that surface.

As we have seen in *Chapter 1*, *Reviewing the Basics*, tapping the *Ctrl/Option* key will allow you to make copies, rather than just turning the selected geometry along the selected axis.

Scale

Tapping the *Shift* key while in the **Scale** tool will change whether you are scaling uniformly. It is important to pay attention here, as some **Scale** handles will start by scaling uniformly, while others will start by deforming the selected items:

- If you click on any of the corner grips, you will begin by scaling uniformly, and tapping the *Shift* key will turn uniform scaling off.
- If you click on any of the middle or edge grips, uniform scaling will be turned off, and tapping the *Shift* key will turn uniform scaling on.

Regardless of what grip you are using to scale, **Scale** will resize to or from the opposite grip from the selected. Tapping the *Ctrl/Option* key will scale the selected items around the center point instead. As this is a toggle, you can flip back and forth during the scaling process as needed by tapping the modifier key.

Push/Pull

With **Push/Pull**, the *Ctrl/Option* key will toggle the **Create New** mode, while tapping the *Alt/Command* key will toggle the **Stretch** mode on or off. Let's look at both of these modifiers with an example:

1. In the practice model, click the **PUSH/PULL** scene, and start the **Push/Pull** tool.
2. Use **Push/Pull** to pull up the red face on the first box.

   Notice that **Push/Pull** moved the top face upward, extending the shape of the box upward.
3. Now, tap the *Ctrl/Option* key and pull up the blue face on the second box.

   Notice that **Push/Pull** pulled up the face, but this time, it left the initial face where it was and had you moving a new face upward. This is what the **Create New** mode does.
4. Now (no modifiers), use **Push/Pull** to pull up the yellow face on the third box.
5. Now, tap the *Alt/Command* key, then do the same to the green face on the third box.

Notice that while in the **Stretch** mode, **Push/Pull** will move the surface perpendicular to its original plane and stretch the surrounding geometry with it. This seems pretty similar to the way that the **Autofold** mode worked in **Move**, right? So, what is the difference? The big difference is that the **Stretch** mode will only allow you to move normally to the selected face, while **Autofold** will allow you to move in any direction. With geometry that is tilted, as in this example, **Stretch** assures the direction of the face movement.

Follow Me

If you have not preselected your path before starting **Follow Me**, tapping *Alt/Command* will automatically have your selected face follow the outline of any face you hover your mouse over. That definition is a little hard to visualize, so let's get hands-on again:

1. In your practice model, click on the **FOLLOW ME** scene and start the **Follow Me** tool.
2. Click on the vertical rectangle in the example on the left, then tap *Alt/Command*.
3. Now, move your cursor anywhere over the two conjoined circles.

   Notice that **Follow Me** automatically has the rectangle follow the shape.
4. Click to accept the new extrusion.

   While many of us will draw the outline and face connected as they were in the first example, this will also work with disconnected geometry.
5. Click **Follow Me** and click on the circle in the second example.
6. Tap *Alt/Command* and hover over the rectangle below, then click.

> **A Point about Pre-Selecting**
>
> Toggling **Use Face Perimeter** may be a time-saver in some cases, but most expert users will argue for the precision that can be attained by using a pre-selected path. Plus, this method limits you to a single face, while pre-selecting will allow you to select any number of selected edges in or out of the plane.

Offset

Tapping the *Alt/Command* key at any point while offsetting will toggle whether or not SketchUp trims overlapping geometry created by the offset:

Figure 3.7 – (Left) a standard offset, (right) the same done with Trim Overlap turned on

The overlap will only be shown after the offset has been started. Tapping the modifier key before starting the offset will have no effect.

Tape Measure/Protractor

Whether you are using **Tape Measure** to see a dimension or **Protractor** to see an angle, tapping the *Ctrl/Option* key will toggle guide creation on or off. If guide creation is off, clicking between points will give you the distance/angle. With guide creation toggled on, the second click of either command will generate a guide line.

Axes

Tapping the *Alt/Command* key while defining the axes with the **Axes** command will allow you to cycle through which axes you are currently defining. Each tap of the modifier key will change the color of the axes currently connected to your cursor.

Section Plane

Holding down the *Shift* key while placing **Section Plane** will lock to the plane of the face your cursor is currently hovering over.

With that, we have looked at all of the modifiers keys used by **Select**, the drawing commands, and the tools. Some of you may have noticed that a few tools were not in this section. There are a few tools that do not have modifiers, so I did not include them here.

Now, for those of you who identify as the **Too Long, Didn't Read** (**TLDR**) types, the next section is a set of tables that can be used for a quick lookup of modifier keys.

Reviewing the commands and tools

Looking back, I don't think that I have ever seen all modifier keys documented in a single easy-to-reference table. Since that seemed like a useful tool for people hoping to level up their SketchUp skills, I created it!

The first table is a list of all modifier keys for drawing commands specific to Windows OS:

Draw Commands (Windows)			
Command	**Shift**	**Control**	**Alt**
Line	Lock plane	-	-
Freehand	-	Decrease Edges	Increase Edges
Arc	Lock plane	With + or – to change segments	-
2 Point Arc	Lock plane	With + or – to change segments	-
3 Point Arc	Lock plane	With + or – to change segments	-
Pie	Lock plane	With + or – to change segments	-
Rectangle	Lock plane	Toggle draw from center	-
Rotated Rectangle	-	-	Set protractor baseline
Circle	Lock plane	With + or – to change segments	-
Polygon	Lock plane	Toggle interior/exterior OR With + or – to change segments	-

Figure 3.8 – Comprehensive list of all drawing modifier keys for Windows

Up next is drawing command modifiers, but this time for macOS:

Draw Commands (Mac)			
Command	**Shift**	**Option**	**Command**
Line	Lock plane	-	-
Freehand	-	Decrease Edges	Increase Edges
Arc	Lock plane	With + or – to change segments	-
2 Point Arc	Lock plane	With + or – to change segments	-
3 Point Arc	Lock plane	With + or – to change segments	-
Pie	Lock plane	With + or – to change segments	
Rectangle	Lock plane	Toggle draw from center	-
Rotated Rectangle	-	-	Set protractor baseline
Circle	Lock plane	With + or – to change segments	-
Polygon	Lock plane	Toggle interior/exterior OR With + or – to change segments	-

Figure 3.9 – Comprehensive list of all drawing modifier keys for macOS

The next table is all the modifier keys for tools, specifically for Windows:

Tools (Windows)				
Command	**Shift**	**Control**	**Alt**	**Shift + Control**
Eraser	Toggle Hide	Toggle Soft/ Smooth	Toggle Unsmooth & Unhide	-
Paint Bucket	Paint All Matching	Paint All Connected	Sample Material from Model	Paint All on Same Object
Tag	Toggle Replace Matching	Toggle Tag All Instances	Sample Tag to Apply	-
Move	Lock Direction	Toggle Move, Copy, Stamp	Toggle Autofold OR Cycle Grip Types	-
Rotate	Lock Direction	Copy	-	-
Scale	Toggle Uniform	Toggle Scale about Center	-	-
Push/Pull	-	Toggle Create New	Toggle Stretch	-
Follow Me	-	-	Toggle Use Face Perimeter	-
Offset	-	-	Toggle Trim Overlap	-
Tape Measure	-	Toggle Create Guide	-	-
Protractor	-	Toggle Create Guide	-	-
Axes	-	-	Cycle Axis	-
Dimension	-	-	-	-
Text	-	-	-	-
3D Text	-	-	-	-
Section Plane	Lock to Plane	-	-	-

Figure 3.10 – Comprehensive list of all tool modifier keys for Windows

Finally, here are all the tool modifiers for macOS:

Tools (Mac)				
Command	**Shift**	**Option**	**Command**	**Shift + Option**
Eraser	Toggle Hide	Toggle Soft/ Smooth	Toggle Unsmooth & Unhide	-
Paint Bucket	Paint All Matching	Paint All Connected	Sample Material from Model	Paint All on Same Object
Tag	Toggle Replace Matching	Toggle Tag All Instances	Sample Tag to Apply	-
Move	Lock Direction	Toggle Move, Copy, Stamp	Toggle Autofold OR Cycle Grip Types	-
Rotate	Lock Direction	Copy	-	-
Scale	Toggle Uniform	Toggle Scale about Center	-	-
Push/Pull	-	Toggle Create New	Toggle Stretch	-
Follow Me	-	-	Toggle Use Face Perimeter	-
Offset	-	-	Toggle Trim Overlap	-
Tape Measure	-	Toggle Create Guide	-	-
Protractor	-	Toggle Create Guide	-	-
Axes	-	-	Cycle Axis	-
Dimension	-	-	-	-
Text	-	-	-	-
3D Text	-	-	-	-
Section Plane	Lock to Plane	-	-	-

Figure 3.11 – Comprehensive list of all tool modifier keys for macOS

With that, we have now looked at every single modifier that can be used in SketchUp.

Summary

Hopefully, this chapter opened your eyes to exactly how modifiers can multiply the functionality of the commands you already know how to use. When presented with this level of detail, even many intermediate users will be surprised to find that there was a modifier out there that they did not know about!

You have now seen how to use modifiers to change your selections, draw geometry, and change the way that many tools function. This is another big step toward leveling up your SketchUp skills.

In *Chapter 4, Taking Inferencing to the Next Level*, we will be looking at how to become a faster and more confident modeler using SketchUp's powerful inferencing system.

4
Taking Inferencing to the Next Level

The inferencing system in SketchUp is one of the things that sets it apart from many other 3D modeling tools. When modeling in SketchUp, inferencing is always running and offering you options on where to click or how to draw in 3D space. A big difference between a SketchUp beginner and someone on track to becoming a SketchUp legend is their ability to use inferencing to draw exactly what they want.

In this chapter, we will cover three main topics:

- Inputting with inferencing
- Inferencing and tools
- Using inferencing with objects

Technical requirements

In this chapter, it is assumed that you have access to SketchUp Pro and the `Taking SketchUp to the Next Level - Chapter 4.skp` file:

`https://3dwarehouse.sketchup.com/model/46b4247c-0b54-4ce2-8b8b-da28921000ca/Taking-SketchUp-Pro-to-the-Next-Level-Chapter-4`

Inputting with inferencing

Many new users can be caught off guard by **inferencing** when they first see it. It can be confusing, with the different color lines and dots showing up and the tool tips appearing and disappearing. But once you learn what each of the colors and symbols mean, using inferencing can mean accurate input from the start and less editing of geometry after it is drawn. There is a lot to inferencing! In this section, we will look at how to use inferencing to draw along the axes and find out what different colors and symbols mean with regard to drawing in SketchUp.

Axes colors

I remember being a little bit confused when I first started using SketchUp. At that point, I had been using many other 3D modeling programs, and they all had one thing in common – *X*, *Y*, and *Z* **axes**. SketchUp, on the other hand, never mentioned any axes by these names; instead, they just colored them. Whereas I was used to a *Z* axis, they had **blue**. *X* was called **red** and **green** replaced *Y*. While this felt a little off at first, in the long run, it was so much easier.

With the colored axes, I don't have to remember what "up" is in 3D space. In fact, because I have access to the **Axes** tool, the axis that aligns with "up" can change at any point! The important point with this system is not the name or direction of each axis; it is the fact that each one has its own distinct color, and inferencing will reference and display those colors.

> **Customizing the Axes' Colors**
> Throughout this section (and the rest of the book), I will refer to the colors of the axes as red, green, and blue. If you need to change any of those colors for any reason (color blindness, for example), you can make that change in the SketchUp **Preferences** window. The first tab, **Accessibility**, has controls that allow you to change the color displayed on your screen for each of the axes.

While drawing lines along the axes tends to be a basic function, let's take a quick look at an example:

1. Open the `Taking SketchUp to the Next Level - Chapter 4.skp` file and start up the **Line** command.
2. Click just to the right of the origin. As you move your cursor around the screen, you will see a preview of the line you may end up drawing. As your cursor moves so that the line is parallel to any of the axes, the line will snap to parallel and change color. Clicking at this point will assure the creation of a line that is parallel to that axis.

   Let's draw a square using only the **Line** tool and inferencing.
3. Move your cursor until it snaps to the green axis, type `2m`, and press *Enter*.
4. Now, move your cursor to the right along the red axis, type `2m`, and press *Enter*.
5. Draw the third side by moving your cursor down the green axis, typing `2m`, and pressing *Enter*.
6. Close the bottom of the cube by clicking back to the beginning of the first line. By moving along the axes, then entering an exact length, you have created a perfect square in four steps:

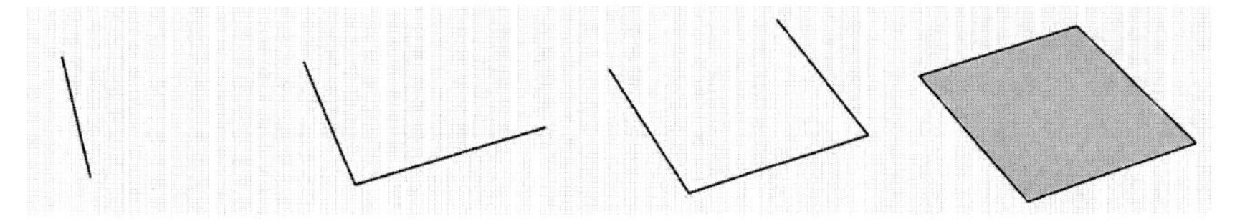

Figure 4.1 – All four steps of drawing a square using inferencing

By moving the mouse until it aligns with an axis and then typing a dimension, we are assured that the new line is exactly in line with the axis. What happens, however, when we cannot rely on the cursor to stay in line with the axis as we move the cursor through the model? This is when we must lock our inferences.

Locking axes' inferences

There are two ways to lock to the red, green, or blue axes. The first way is to tap the arrow key that aligns with the axes:

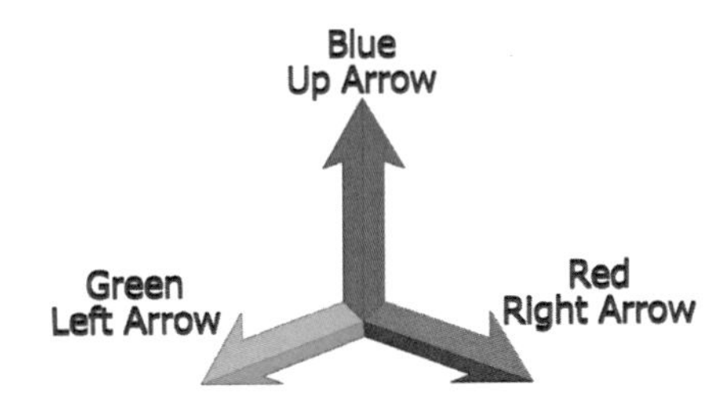

Figure 4.2 – Three axes and their inference lock keys

While the blue axis is easy to remember, I used to struggle to remember whether I should tap the left or right arrow to align to the red axis. Then, I realized that "red" and "right" both started with R, and I can now remember that the right arrow key locks in the red axis!

Let's turn this square into a cube by using the **Line** tool and locking the inference:

1. Start a line at the back corner of the square and then tap the *Up Arrow*.
2. Move your mouse upward, and then type `2m` and press *Enter*.
3. Tap the *Right Arrow* and move your cursor to the right, and then type `2m` and press *Enter*.
4. Click back down to the square on the ground to complete this side of the cube.
5. Use the **Line** tool and axes locks to complete your cube.

With that, we have a simple hand-drawn cube.

> **Seeing the Colors**
>
> If you are doing a lot of work with geometry that should stay parallel to the main axes, you may want to turn on **Color by Axis** for your edges. In the **Style** window, under the **Edit** tab, you can choose the color for the edges. In the dropdown, there is an option called **Color by Axis**. Choosing this option will color code all edges with the axis to which they are parallel.

There are probably one or more of you out there thinking that you could have done this quicker using the **Rectangle** and **Push/Pull** tools, and that is true. However, there are times when it will pay off to try using modeling techniques that you don't normally use in order to broaden your skills.

For example, I read the comments of a poster on our forum (`https://forums.sketchup.com`) who made some amazing scale train models. They were super-detailed and amazing to look at. After asking a few questions, the poster mentioned that he had never used the rectangle tool but drew everything with just the line tool. He had spent years refining his modeling process without even knowing that there was a single tool that could cut out 75% of his input time!

The point is this – whether it is in a book like this or just modeling for work or fun, do not be afraid of trying to use a new method or tool. You never know when going "outside your box" will show you a new way to model or teach you a new trick!

Leaving the axes

Knowing how to use the arrow keys to lock to an axis is great and something that you will likely use quite a bit. But being able to lock to edges that are not parallel to an axis is important as well! Let's take a look at how to lock into a line that falls off-axis, using the *Down Arrow* key to lock to the **magenta** inference:

1. Click on the **MAGENTA INFERENCING** scene in your practice file and bring up the **Line** command.

   In this example, we will add to the rectangle (which is currently 1m x 2m) to make it a 2m square. Since the whole rectangle is at an angle, you will not be able to use the arrow keys to lock to an axis:

   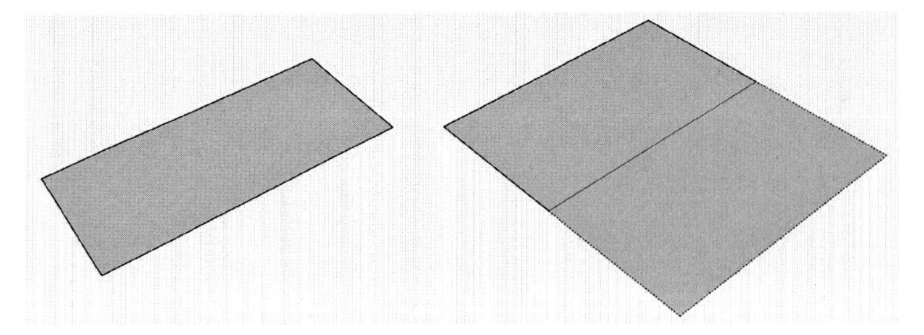

   Figure 4.3 – The rectangle on the left needs to end up looking like the square on the right

2. Click to the corner that is closest to you and move your cursor in the direction that you want to draw your first line.

   As you move your mouse in line with the existing edge of the rectangle, it will lock and turn magenta. This is what we want to lock in. To do so, we need to let SketchUp know which edge we want to inference for the magenta lock.

3. Move your cursor back along the line so that the edge you are drawing runs back up onto the short edge of the rectangle. Then, tap the *Down Arrow* key to lock to this inference.

   Note that the line you are pulling and the entire side of the rectangle have turned magenta. You are now locked to that line, as shown here:

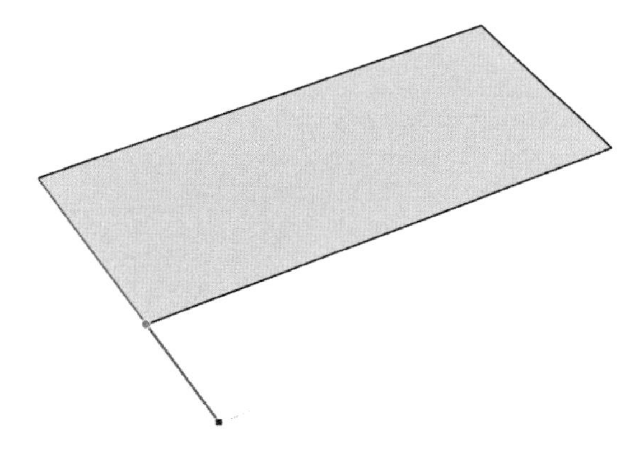

Image 4.4 – Magenta parallel reference

4. Move your cursor away from the existing rectangle, type `1m`, and press *Enter*.

   The next line you need to draw is perpendicular to the one you just completed. You can do this using the exact same key but by pressing it twice.

5. Move your cursor back over the line you just drew and tap the *Down Arrow* two times.

   The first time you tap the *Down Arrow*, you are back in a parallel inference lock. The second time you tap *Down Arrow*, you change it to a perpendicular lock. You are now locked at 90 degrees to the inferenced line.

6. Move your cursor to the right (the direction the long edge should run), type `2m`, and then press *Enter*.
7. Click back to the edge of the original rectangle to close the square.

Learning to use the magenta inference is a huge time saver if you ever model anything that is not perfectly aligned to the main axes.

The Shift Key

When drawing edges, as we have been in these examples, you can always hold down the *Shift* key to lock in the current inference. This is a nice option to have, but you lose it as soon as you release *Shift*. Using an arrow key will lock into an inference and keep you there until you finish drawing the edge or change to another inference.

Snap points

Of the things that I have always appreciated about using SketchUp is the way that the inferencing engine automatically finds and presents me with potentially useful snap points. When I first used SketchUp, I found out that the other drawing programs I had used required you to press a key to use a snap point or toggle the visibility of all snap points if I wanted to use them. Not SketchUp, though! SketchUp sees what you are doing and offers you the snap point that it thinks you might want to use.

This is an important part of modeling in SketchUp and makes it imperative that you understand the different point types you see on the screen. There are four different point types that you will see on the screen, and each one can be displayed in different colors, indicating different options:

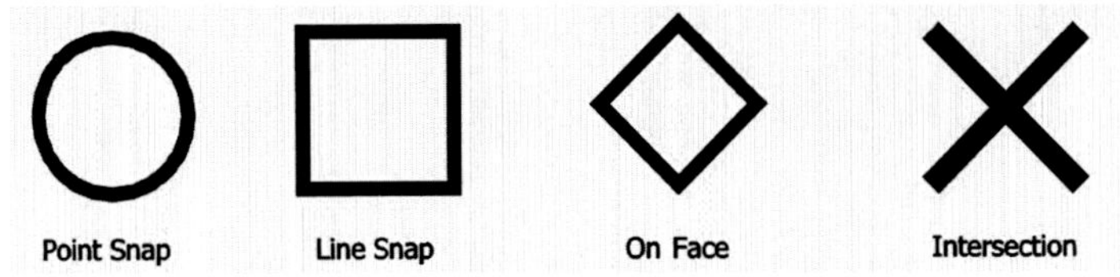

Figure 4.5 – Snap points' shapes and types

These different points may show up as you are modeling in different colors. I know it can sometimes be hard to see them, but each snap point has different potential colors and will change, based on what it is inferencing. Here is a list of all point shapes and the colors they may show up in:

Point shape	Point color	Meaning
Circle	White with a black cross	Snap to the origin
Circle	Magenta with a cross	Snap to an axis of an object
Circle	Green	Snap to the endpoint
Circle	Teal	Snap to the midpoint of an edge or arc
Circle	Blue	Snap to the center of a circle or arc
Circle	White	Snap to a guide point
Circle	Magenta	Snap to the midpoint or endpoint in an object
Circle	Gray	Object grips
Square	Red	Snap to an edge or guide
Square	Magenta	Snap to an edge in an object
Diamond	Blue	Snap on a face or a section
Diamond	Magenta	Snap on a face in an object
X	Red	Snap to an intersection
X	Magenta	Snap to an intersection in an object

Figure 4.6 – Points, colors, and snap types

Understanding that these different points will show up as you work in your model is essential for speeding up input. If you do not pay attention to what you are snapping to, it is very hard to be assured that you are actually drawing what you want, in 3D space.

Projected inferencing

An important inference that I want to look at is projected inferencing. This is its own inference that will show up as you model and is a big time saver. Projected inferencing looks at the points that are in a model and allows you to snap where the edge you are drawing to would intersect with another edge or point if the two of them met. Projected inferences show up as dotted lines along an axis, from the inferenced point to your cursor, and are a great way to align edges to existing geometry or find the center of a rectangle! Let's give that a try right now:

1. Click on the **PROJECTED INFERENCING** scene in your practice file and pull up the **Line** command.
2. Hover over the left edge of the rectangle until the midpoint snap shows up (a teal circle).
3. Without clicking, slide your cursor toward the center of the rectangle. See the dotted red line that is following your cursor from the inferenced point? That is a projected inference. If you click while that inference is visible, the point you click on is assured to be in line with the midpoint of the left edge of the rectangle:

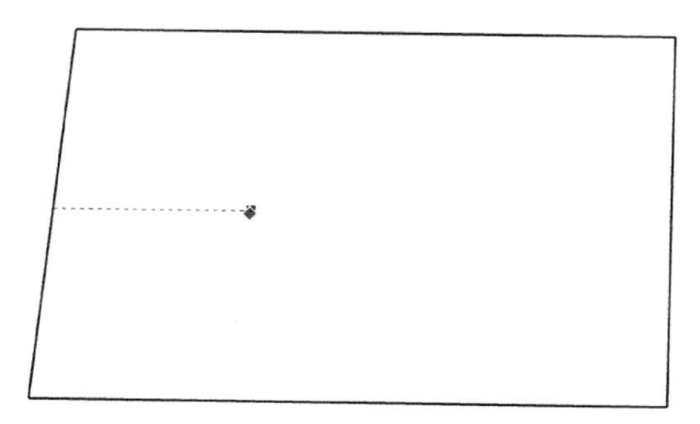

Figure 4.7 – The dotted red line shows a projected inference from the midpoint of the edge

4. Now, move your cursor up and find the midpoint of the top edge of the rectangle.
5. Once again, without clicking, move your cursor down toward the middle of the rectangle.

   This time, you have a green dotted line following the cursor, indicating that you are aligned vertically with the center of the top edge of the rectangle.

6. Continue to move your cursor down toward the center of the rectangle. When you get to the center, you should see a red dotted line from the midpoint of the left side and a green dotted line from the midpoint of the top side, meeting in the very center of the rectangle:

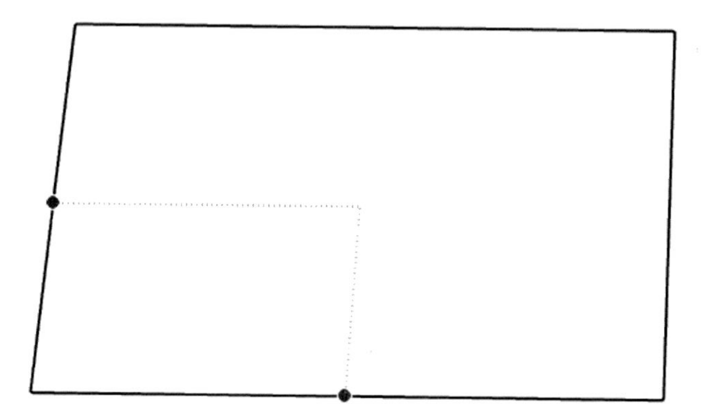

Figure 4.8 – Red and green projected inferences meet at the center of the rectangle

This inferenced point is the very center of the rectangle. This, of course, will only work if the sides of the rectangle are aligned with the axes, since the projections align to the red and green axes, but this is still a very helpful tool for aligning geometry as you draw.

> **Hover and Wait**
>
> If you are moving your cursor away from a point and are not seeing a dotted line, you may need to hover just half a second longer. SketchUp identifies points that you want to use as reference when you keep the cursor over them for a second or so.

Inferencing planes

So far, all the examples we have looked at have been about locking to edge inferences. Inferencing can be used to lock the planes as well! Let's take a look at drawing some rectangles by locking onto specific planes by finishing off a cube in your practice model:

1. Go to the **PLANE INFERENCING** scene in your practice model and start up the **Rectangle** command.
2. Tap the *Left Arrow* key to lock to the green plane. Watch your cursor as you perform plane locking. You will see that the rectangle cursor will change orientation and color as you lock to specific planes. In this case, the green plane aligns with the rectangle we want to draw to create the missing front face of the cube.

3. Click on the front corner of the square on the ground and move your mouse up and to the left.

   Note that no matter where you move your cursor, you are locked into that green plane!

4. Click on the square at the upper left to create the front side of the cube.
5. Now, tap the *Right Arrow* key to lock to the red axis and add a rectangle to the right side of the cube.
6. Finally, tap the *Up Arrow* and draw one last rectangle to form the top of the cube. That explains up, left, and right, but what about the down arrow key? Just as with inferencing for line drawing, plane inference uses the *Down Arrow* key to lock to any plane that you are hovering over, to create a magenta inference. Let's use the rotated cube on the right to see some magenta inferences.
7. Hover the rectangle cursor over one of the sides of the rotated cube and tap the *Down Arrow*. As soon as you tap the key, the rectangle cursor snaps to the plane of the face you were hovering over and turns magenta. Let's draw a rectangle on this face of the cube.
8. Click to set a start point, and then move your cursor to drag out a rectangle. Note that the rectangle stays in the plane of the face, but it does not align with the shape of the side of the cube. This is an important fact to remember about magenta inferencing. It will draw in the inferenced plane, but the actual direction of the rectangle being drawn will align, as best as it can, with the axis of the model or current context.
9. Click again to finish the rectangle, move your cursor to a different side of the cube, and tap the *Down Arrow* again.
10. Move your cursor off the cube. Note that you can move your cursor completely away from the inferenced geometry but retain the orientation of the plane. This is a key use of plane inferencing. With this process, you can duplicate things such as roof slopes or landscape gradients by inferencing one example and reproducing them elsewhere.

No Perpendicular Planes

Unlike line inferences, the *Down Arrow* will not jump to a perpendicular plane if you tap it twice. Tapping the *Down Arrow* key twice will simply release the inference.

One of the advantages of using the magenta inference is that SketchUp will remember what the last plane referenced was.

11. With your cursor in open modeling space (not over either cube), tap the *Down Arrow* again. Note that the inference has been released.
12. Now, while still in open modeling space, tap the *Down Arrow*, again. The magenta inference is back and still aligned to the side of the cube! As long as you don't tap the *Down Arrow* while hovering over another face, it will remember the last plane inferenced.

Shift Works on Planes, Too

As with line inferences, plane inferences can be held by holding down the *Shift* key. The same limitation is here, too, though. Should you release the *Shift* key at any point, you will release the inference.

That is a lot of inferencing! Note that we only really looked at two commands (**Line** and **Rectangle**), but these inferencing practices work the same for all of the drawing commands. Understanding how to use axis locking, snap points, and plane inferencing are key to speeding up your 3D modeling process and something that all great SketchUp modelers do instinctually.

Now, let's take what we have learned about inferencing for input and see how it applies to modifying geometry with the tools!

Inferencing and tools

The good news when it comes to mastering inferencing while using tools is that everything covered in the previous section (colors, arrow keys, and snap points) applies to tool usage, as well as drawing. Of course, they are used in different ways, which is what we will dive into right now.

Moving and inferencing

Let's start with a basic tool such as **Move**. Let's see how **Move** works with the different inferences and snap points when moving an edge and a face:

1. Get to the **TOOL INFERENCING** scene in your practice file and bring up the **Select** command.

   Note that this scene has the **Color by Axis** setting turned on for edges. Any edge that is parallel to an axis is shown with that axis' color.

2. Use **Select** to highlight the vertical line on the right and then bring up the **Move** command.

   We are going to use some snap points to move the selected line into the center of the circle at the back.

3. Hover your cursor over the bottom of the vertical line. Click when you see the endpoint.
4. Now, move your cursor and hover over the edge of the circle. When the center point of the circle shows up, move the line so that the bottom snaps to the center point.

   Even though you cannot see the center of a circle, SketchUp knows where it is and presents it to you as a viable snap point when the time is right, as shown here:

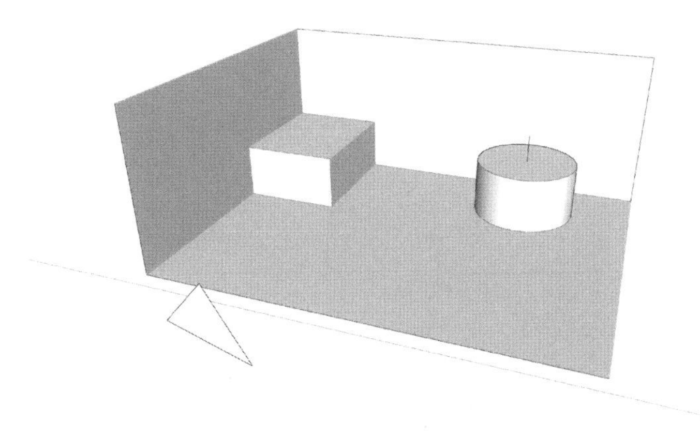

Figure 4.9 – The vertical line exactly lined up with the center of the circle

Finding the Center Point

To get the center point of an arc or circle to show up, you need to hover over the edge for a second or so. Once you do, the center point will show up with a dotted line connecting it to your cursor.

Now, let's use some projected inferences and **Rotate** to line up the triangle with the box in the corner.

5. Use **Select** to double-click on the triangle, and then bring up the **Rotate** tool.
6. Tap the *Up Arrow* key to lock rotation to the blue axis.
7. Click on the left endpoint of the triangle's base and then on the right.
8. Now, tap the *Right Arrow* key to force the triangle to align to the red axis and click again.

Since the base of the triangle and the top edge of the box in the corner are both red, we are assured that they are parallel:

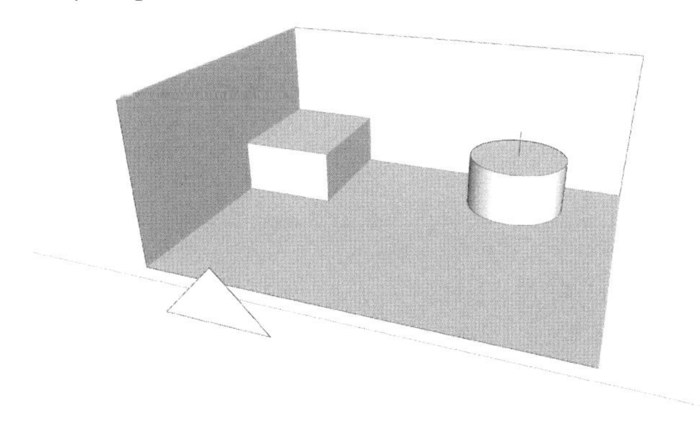

Figure 4.10 – The base of the triangle is now aligned with the box at the back

Now let's line up the triangle with that box. To practice using inferencing, we will do this in three moves.

9. With the triangle still selected, bring up the **Move** tool, choose the right end of the triangle, and tap the *Right Arrow* key.
10. Now, to line up the triangle with the box along the red axis, click anywhere on the right side of the box. You can hover over the right face of the box, on an edge that defines that face, or any of the endpoints or midpoints along those lines.

Since we moved along the red axis to the green axis of the box, we are assured that everything lines up:

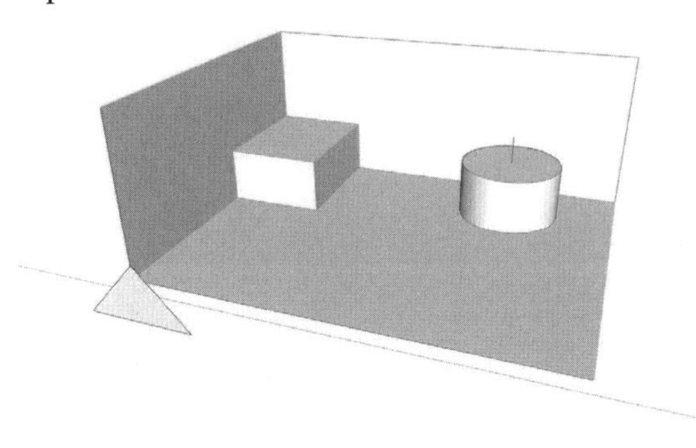

Figure 4.11 – The edges of the triangle and the box are aligned along the red axis

Now that the triangle is lined up with the face of the box, let's move it vertically so that the base of the triangle is at the same height as the top of the box. We will do this without using any snap points on the triangle.

11. With the triangle still selected and while still in the **Move** tool, click any point or edge along the bottom of the box, and then tap the *Up Arrow* key.
12. Now, move your cursor up and click on any edge, point, or face of the top of the box.

Since we have constrained our movement to the blue axis, it does not matter where we click, as long as the distance, vertically, between the first and second click is the same as the distance between the top and bottom of the box. This is sometimes referred to as a relative move, since you are using points that are at a relative distance from each other rather than picking points on the item being moved.

While we cannot go into every tool and how it uses inferencing, remember that inferencing is always happening and always available to you as you model. Almost every command allows you to lock to an axis or magenta inference at some point during its use, using either the *Arrow* keys or the *Shift* key. Knowing how inferencing applies to edges and faces is important, but if you are following the practices we set in earlier chapters, you will likely be using inferencing to modify groups and components. Let's spend a little bit of time covering how inferencing can be used when modifying objects and how it differs from moving loose geometry.

Right now, your model should look like this:

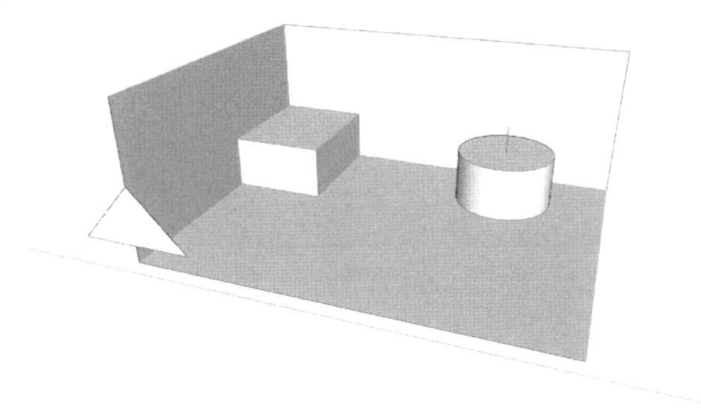

Figure 4.12 – The triangle is one step closer to being completely aligned with the box

13. Now, use the **Move** tool to select the triangle by the midpoint of the baseline and place it on the midpoint of the front of the box.

    The triangle is now perfectly aligned with the box:

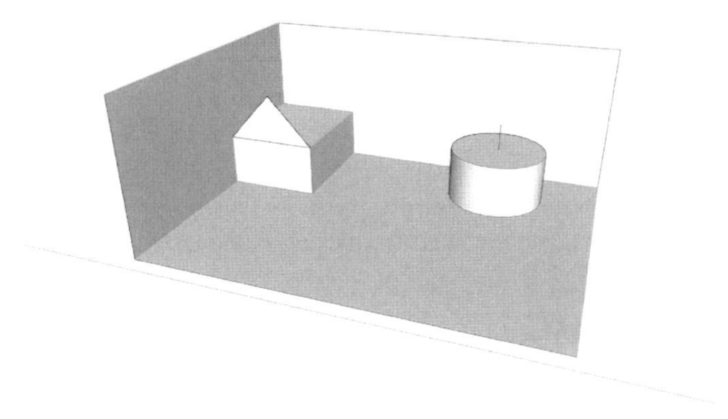

Figure 4.13 – The triangle is perfectly aligned

This is a simple example that shows the power of inferencing. If you think of geometry as existing in red, green, and blue planes, it is easy to modify it using inference locking and snap points. I know that we could have moved that triangle in one move, but I really wanted to do multiple steps to illustrate how to use axis locking to make it easier to select points while moving.

Using inferencing with objects

Whether you are modifying an object or selecting raw geometry, inferencing works the same. Picking points to use when modifying objects is, however, quite a bit different.

Since objects are considered single items rather than a collection of pieces, many tools do not work on objects. However, the tools that can be used on objects work differently, treating an object as a whole item. This gives you some capabilities that make it easier to move, rotate, or align objects that you don't have when moving just geometry. To see the options for modifying objects, let's hop back into the practice model:

1. Go to the **OBJECT INFERENCING** scene and bring up the **Move** command.
2. Hover your mouse over the cube. Note that as soon as the cursor moves over the cube, gray circles show up for all of the corners. Since the geometry of a cube completely fills the container, it may appear that the corner points are related to the geometry inside the group. Let's see how these points, referred to as **grips**, work with different geometry.
3. Move your cursor to the right and hover over the cylinder. Note that the grips are on the corners of the container and not the geometry. This is one of the big differences when it comes to moving an object. You can use any corner of a container as a handle to move and arrange groups or components. Let's move the cylinder so that it is right against the cube.
4. Pick the bottom-left corner of the cylinder, and then move that point to the lower-right corner of the cube. With that, the two containers (which happen to be the same size) are perfectly aligned. So, that works great when you can see the points, but what about the points at the back? Let's say that we want to get the cube placed in the back corner. We need the lower-left back corner to fit in the far corner of the container behind it.
5. While still in the **Move** command, with nothing selected (tap the *Esc* key to deselect any selected groups), hover over the cube.

   Note the grip floating on the front face of the cube. That is actually a snap point for the back corner of the container (the corner you cannot see from this view).
6. Hover over and click on that grip. As soon as you hover over that grip, everything in the container becomes transparent, allowing you to see the front and back sides of the geometry as well as everything behind the group!
7. Move the back corner of the cube so that it aligns with the back corner of the container behind it and click to place it. This is another advantage of moving objects over raw geometry. Since SketchUp knows what is in the object, it can selectively make it transparent, allowing you to see through to the back side. This is great for aligning things such as furniture in rooms or placing sheathing on wall framing, where things align to a corner, but what if we want to use a point other than a corner of a selected object? Let's align the middle of the cylinder with the middle of the back wall.

8. While still in the **Move** tool, with nothing selected, move your cursor over the cylinder and tap the *Alt/Command* key. Note that the modifier listed in the status bar changes from **Toggle Autofold** to **Cycle Through Grip Types** when you hover over an object. Tapping the modifier key once will toggle you through four different snap options for object grips:

    - Corners of a container (8 points)
    - Midpoints of container edges (12 points)
    - The center of the sides of a container (6 points)
    - The center of a container (1 point)

    Here is a quick diagram showing all of the possible snap options for a cube:

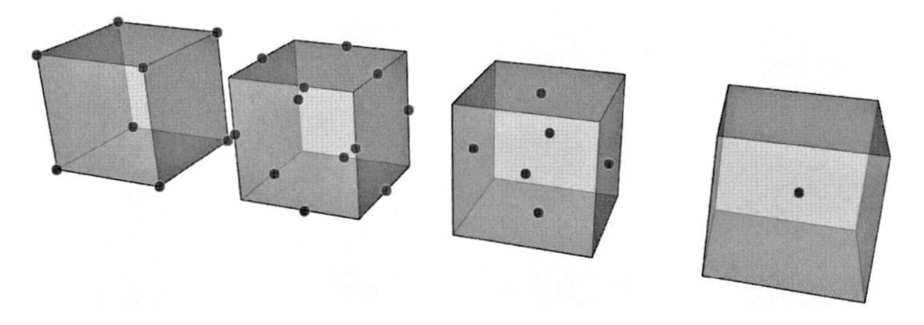

Figure 4.14 – Each of the grip sets, as they appear on a group or component

Note that toggling through this list will change the grips for all objects in the model, not just the object currently being interacted with.

9. Use the midpoint on the bottom back edge to align the cylinder with the center of the wall behind it.
10. Tap the modifier key a few more times to cycle back to corner grips, and place the sphere so that its back fits into the corner of the container behind it. With that, you should have all three objects nestled up tight into the geometry behind them:

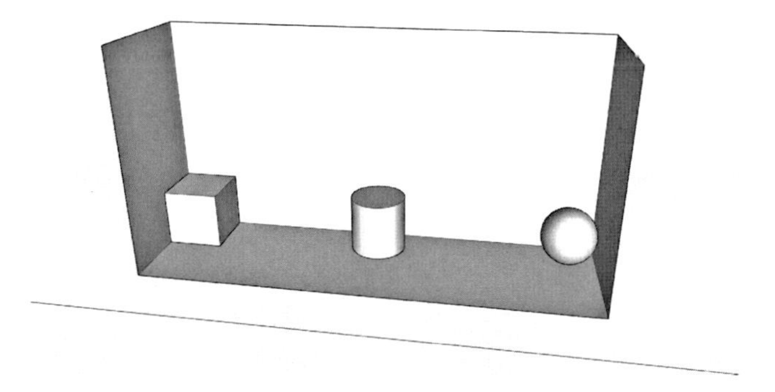

Figure 4.15 – The box, cylinder, and sphere snapped to the geometry behind

You can see how using grips on containers can be great for aligning items. Let's do one more example to see how we can use interior grips to align geometry that you cannot see:

1. Click on the **GRIP STAMPING** command, and if you are not already in it, choose the **Move** tool.
2. Hover over the sphere and cycle through grips until you get the center grip.
3. Choose the center grip and place the sphere at the end of the curved line:

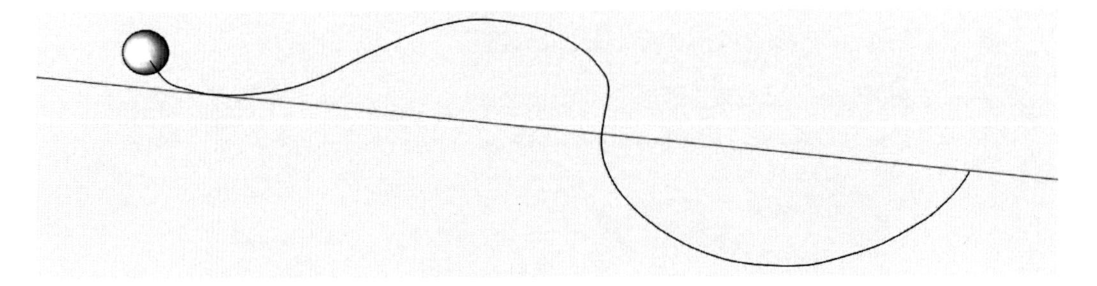

Figure 4.16 – One sphere on the path

4. Now, tap the *Control/Option* key twice to toggle to stamp mode.
5. Now, click along the line several times, placing a copy of the sphere on the line with each click:

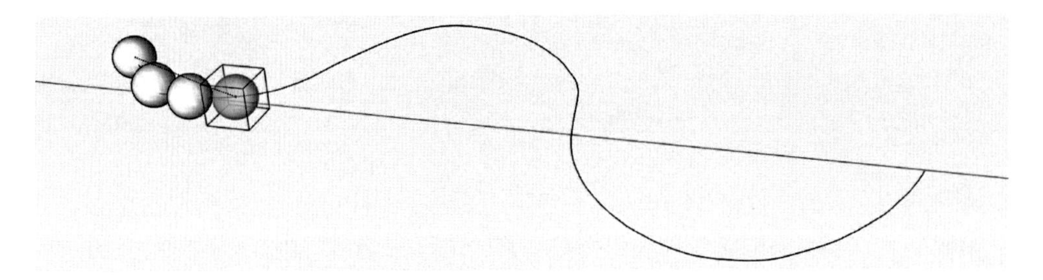

Figure 4.17 – Each click adds a new copy of the sphere to the string

6. Continue clicking along the path until you have covered the path with spheres.

Knowing how to use grips to move and arrange objects is an important piece of leveling up your SketchUp skills. You can always use the geometry inside the group as grips as well, but being able to arrange objects using the corners or sides of the bounding box helps for quick and easy alignment of groups and components.

Summary

Inferencing is something that you start using the second you hop into SketchUp. Everyone uses it, but only skilled users take advantage of everything it can do. To truly take advantage of inferencing, you need to understand not just axis and color inferences but also parallel and perpendicular inferences, which we discussed in this chapter. We also learned about snap points and how to use grips when moving objects.

In *Chapter 5, Creating Beautiful Custom Materials*, we will be looking at the process of importing imagery and creating custom materials for use with the **Paint Bucket** tool.

5
Creating Beautiful Custom Materials

Knowing how to use SketchUp to create geometry is important. But knowing how to use stock and custom materials to take your models to the next level is essential as well. In this chapter, we will dive into how good materials help to make a great model and some of the best practices for creating and working with custom materials.

In this chapter, we will cover the following:

- Importing images for custom materials
- Organizing and applying custom materials
- Deforming textures
- Applying materials to curves
- Avoiding common mistakes with materials

Technical requirements

In this chapter, it is assumed that you have access to SketchUp Pro, any image file, and the `Taking SketchUp to the Next Level - Chapter 5.skp` file:

`https://3dwarehouse.sketchup.com/model/edbc1856-f520-4140-a53a-eba645eb98f2/Taking-SketchUp-Pro-to-the-Next-Level-Chapter-5`

Importing images for custom materials

You have probably noticed that SketchUp comes with a few dozen assorted **materials** that you can use as you learn the basics of modeling. You have probably also learned that these materials are fairly limited and not very high quality. The intention of these stock materials is just to get you started. Once you have the basics of things such as using the **Paint Bucket** tool figured out, it is time for you to start creating your own materials.

Any image, assuming it is in the correct file format, can be SketchUp material. Materials are images that are saved in SketchUp and can be applied to any face using the **Paint Bucket** tool. In this section, we will talk about what sorts of images make good materials, and then how to import, save, and edit materials.

> **Materials, Textures, or Colors**
>
> These three terms seem to be somewhat interchangeable in the SketchUp **user interface (UI)**. While they generally mean the same thing, I will try to refer to the image swatches or colors that are applied to models as materials. There are, however, portions of the UI that refer to **textures** or **colors**. For our purposes, these terms are interchangeable with materials.

What makes a good material?

Before you do any importing of imagery, you should spend a minute or two thinking about what sort of image you want to import to make a material. There are a few things that you should consider when you look at an image to see whether it will make a good material in SketchUp:

- File size/resolution
- Image orientation/distortion
- Repeatability/seamless imagery

We have already mentioned that big files can bloat your model file size in *Chapter 2, Organizing Your 3D Model*, so it should go without saying that when considering an image for use as a SketchUp material, you want to be conscious of file size. As we saw, there is no reason to import a 4K image when we know that SketchUp will be displaying it at a lower resolution. It does pay to remember that the opposite can be an issue as well. Make sure that the image you are considering is not too small, either. Consider this example:

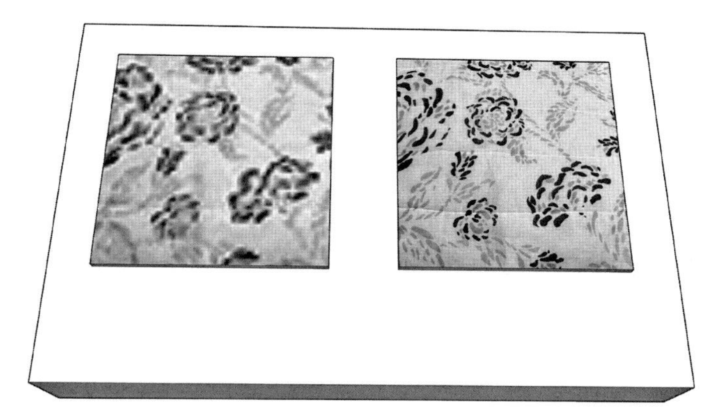

Figure 5.1 – The image on the left is a too low resolution

When thinking about an image that is going to be imported into your model, file size is important but it does also have to look good. Those of us who are super conscious of file size may be tempted to import a very low-resolution image. But if that image does not serve to make our model better, is it worth saving a few KB of disk space?

While the image on the left in *Figure 5.1* looks blurry at the size it is at, it could be used assuming it was small enough. While it would look terrible and out of focus if you tried to use it as a material for a bedspread, it may work as a small pillow. In most cases though, the image on the right would be preferable as a SketchUp material.

Another important aspect to think about when importing an image for use as a material is the orientation or distortion of the image. We will be talking about tools that can be used to manipulate imported images later in the *Deforming textures* section. So, keep in mind that we can make changes to rectify small issues with images. However, when choosing which image to import, consider using the one that will require the least correction. Here are a few images considered for importing for use as a wood grain material:

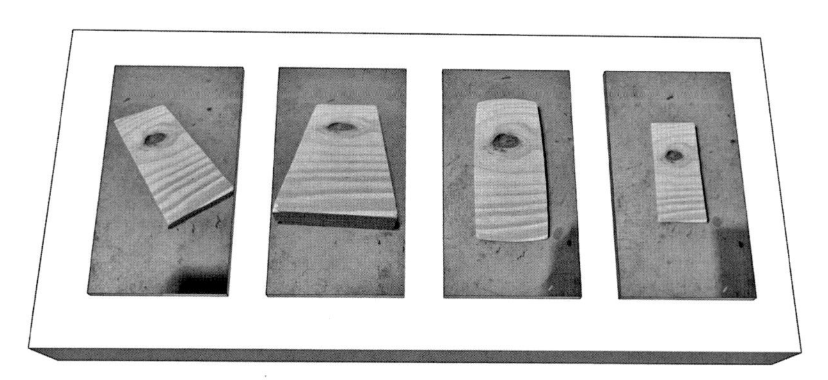

Figure 5.2 – Given the choice, the image on the far right would make the best material

The first image, while not ideal, could work in a pinch. Later, in the *Deforming textures* section, we will talk about editing materials using the **Position Texture** command, which could help this image. But, whenever possible, it is better to have a material that is not at an angle. Ideal images for use as materials squarely face the camera.

The second image in *Figure 5.2* is probably the worst option. The main subject of the image is in focus at the bottom and slightly out of focus at the far end of the block of wood. While we could probably use **Position Texture** to get this image to a rectangular shape, the top would still be blurry. Ideal images for use as materials are in even focus and evenly lit throughout.

The third image in *Figure 5.2* was taken with a fisheye lens or adjusted in an image editing software or something similar. Regardless of how the warped perspective was added to this image, this makes it a problem for use as a material. Ideal images for materials are not distorted or corrected from their actual shape.

The final image in *Figure 5.2* is the ideal image for use as a material in SketchUp. It is small, as it does not take up the whole image, but the texture of the wood can clearly be seen, it is in focus, and not distorted from its original shape (which happens to be a rectangle and is perfect for a material).

Seamless images

Seamless images are images that have been edited so that the opposite sides connect seamlessly to each other. Seamless images can be tiled from end to end and side to side, and look like one big continuous image. Nowadays, seamless images are fairly easy to find online, and there are many online tutorials that show you how to use your favorite image editor to make your own seamless images. In fact, there are even a few online tools that will allow you to upload an image and it will make a seamless image for you!

When it comes to using seamless images, there are certain cases where you do want to consider using them. The perfect case for considering a seamless image is a floor or wall covering. Situations in which you have large surfaces that the material needs to cover are ideal for seamless materials, as shown here:

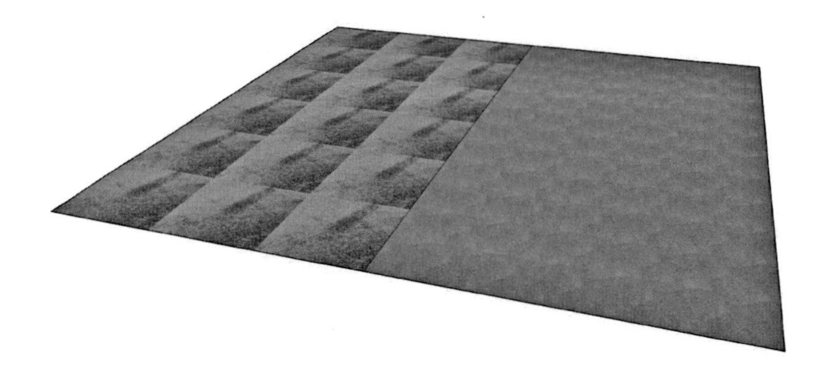

Figure 5.3 – Seamless texture on the right

Not every image should be seamless, though. The process of making a texture seamless can reduce the quality and detail of the image. When you don't need to have an image that seamlessly tiles, it is often better to stick with a regular image. Tiling a seamless material makes sense on carpet, but not on a wood board, as seen in the following:

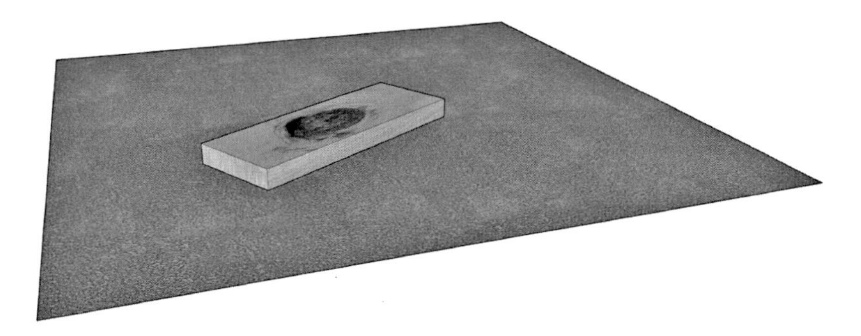

Figure 5.4 – Seamless material only on carpet

> **Using Your Own Pictures as Materials**
> Seamless or not, many people enjoy finding images online or downloading textures from websites for use as materials. Remember, you can always take your own photos and import them for use as materials.

Now that we have established what makes a good image for use as a SketchUp material, let's dive into the process of importing these files.

Importing files as textures

Regardless of where you get image files from – downloaded, purchased, or taken yourself – at some point, you will need to import the files into SketchUp before you can use them as materials. This is done using the **import** command.

Importing an image file to use as a material will require you to bring the file into SketchUp and place it on a face. This face can be a part of the model you are working on, or it can be a "throwaway" rectangle. As soon as an image is loaded into the model, it is saved as a material, so the geometry it was placed upon on import can be removed and the material will still exist. Let's walk through this, step by step.

First, you will need your own image file to import. This can be an image of anything for this example and can be downloaded or an image that you have taken yourself on your phone. The important part is that it is in a format that SketchUp can import. SketchUp can import any of these file formats:

- Bitmap (`.bmp`)
- JPEG (`.jpg` or `.jpeg`)

- Portable Network Graphics (`.png`)
- Photoshop (`.psd`)
- Tagged Image File (`.tif` or `.tiff`)
- Targa file (`.tga`)

For this example, the file needs to be somewhere that you can find on your computer.

Importing images in Windows

Let's first look at the process to import an image for use as a texture for those of you running SketchUp on a Windows computer:

1. Open `Taking SketchUp to the Next Level - Chapter 5.skp`.
2. Choose **Import** from the **File** menu.
3. Navigate to your image file.
4. To limit the selectable files to only images, select **All Supported Image Types** from the dropdown to the right of the **File name** field:

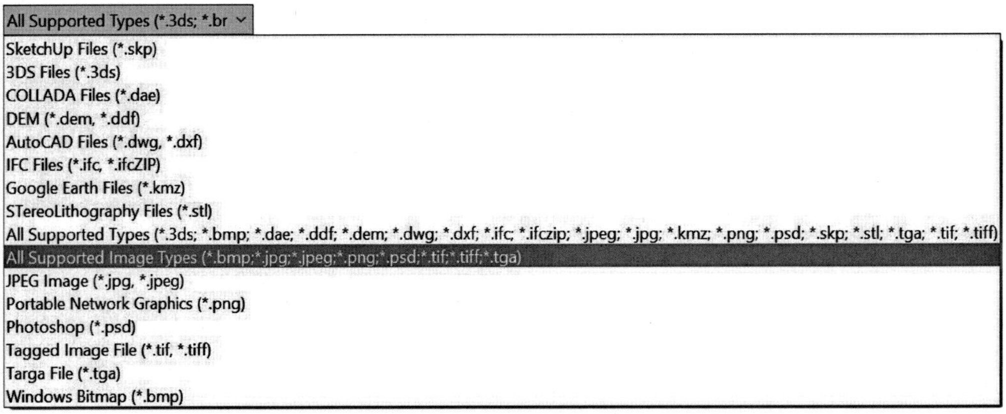

All Supported Types (*.3ds; *.br
SketchUp Files (*.skp)
3DS Files (*.3ds)
COLLADA Files (*.dae)
DEM (*.dem, *.ddf)
AutoCAD Files (*.dwg, *.dxf)
IFC Files (*.ifc, *.ifcZIP)
Google Earth Files (*.kmz)
STereoLithography Files (*.stl)
All Supported Types (*.3ds; *.bmp; *.dae; *.ddf; *.dem; *.dwg; *.dxf; *.ifc; *.ifczip; *.jpeg; *.jpg; *.kmz; *.png; *.psd; *.skp; *.stl; *.tga; *.tif; *.tiff)
All Supported Image Types (*.bmp;*.jpg;*.jpeg;*.png;*.psd;*.tif;*.tiff;*.tga)
JPEG Image (*.jpg, *.jpeg)
Portable Network Graphics (*.png)
Photoshop (*.psd)
Tagged Image File (*.tif, *.tiff)
Targa File (*.tga)
Windows Bitmap (*.bmp)

Figure 5.5 – All Supported Image Types

5. Click on the file you want to import.
6. To ensure that this file will be imported as a material, select **Texture** from the **Use Image As** list (notice that this list appears after you have selected a valid image file):

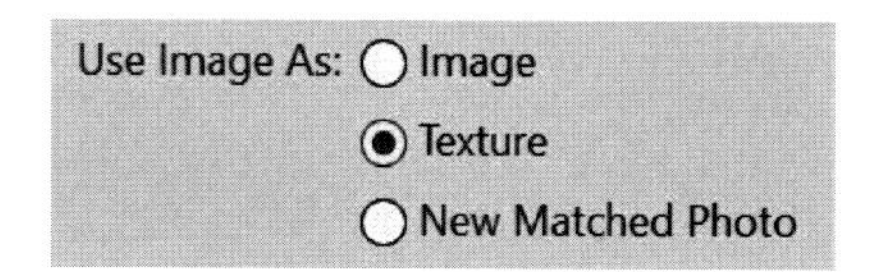

Figure 5.6 – Use as Texture

7. Click the **Import** button.

   At this point, the image is ready to be placed in your model. The initial placement will require a face to apply to (the square on the ground). This initial placement will also set the scale of the image as it will be used as a material.

   The material is attached to your cursor by the lower-left corner of the rectangular image. Start placing by clicking on a face in the model. For this example, we will use the square in the first scene.

8. Click on the lower-left corner of the square (the corner nearest to the origin).

   SketchUp now wants you to choose the top-right corner of the image. Defining this rectangle will set the scale at which the image will be used as a material. For our example, we will stretch it to the square.

9. Move your cursor so that it snaps to the right edge of the rectangle and click again.

   At this point, you have created a brand-new material with your image. If you look at the **Colors in Model** list in the **Materials** window, you should see your image:

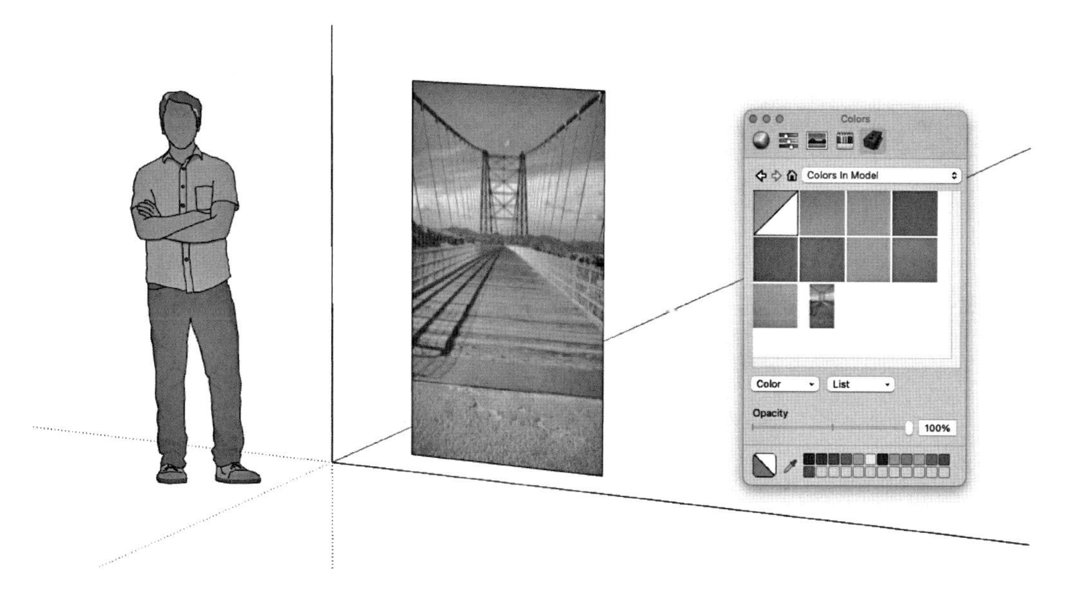

Figure 5.7 – The imported image in the model, and in the Colors window

Let's drop this new material onto a few more faces really quickly:

1. Start the **Paint Bucket** command and use the Sample Material modifier (*Alt*) to sample your new material, then click on the **APPLYING** scene.
2. Click on the shapes in the next scene to apply your new material.

Importing images in macOS

Now let's see what the process to import an image for use as a texture looks like for those of you running SketchUp on a Mac:

1. Open `Taking SketchUp to the Next Level - Chapter 5.skp.`
2. Choose **Import** from the **File** menu.
3. Navigate to your image file.
4. To limit the selectable files to only images, click on the **Format** drop-down menu, and select **All Supported Image Types**:

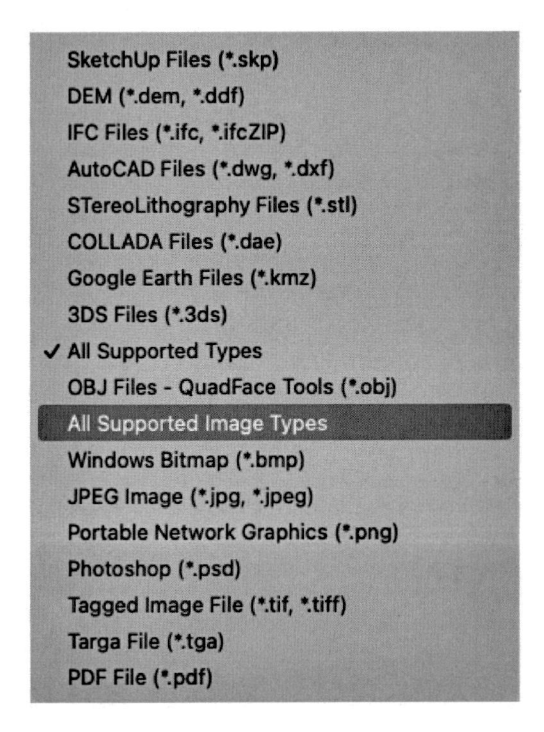

Figure 5.8 – All Supported Image Types

5. Click on the file you want to import.

6. To ensure that this file will be imported as a material, select **Use As Texture** from the dropdown format (notice that this dropdown only becomes available after you have selected a valid image file):

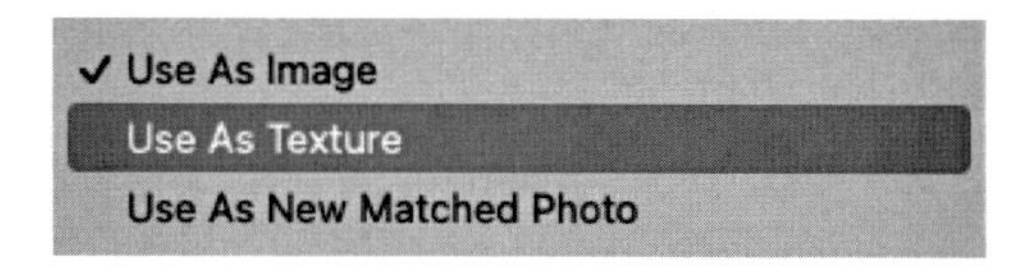

Figure 5.9 – Use As Texture

7. Repeat *Steps 7* to *9* for the image import process for Windows users.
8. Start the **Paint Bucket** command and use the Sample Material modifier (*Command*) to sample your new material, then click on the **APPLYING** scene.
9. Click on the shapes in the next scene to apply your new material.

Modifying Texture Placement

We spent all of *Chapter 3*, *Modifying Native Commands*, talking about how to use modifier keys. Notice that you have access to modifiers when placing a new texture, as well. Double-clicking on a surface will place the image at full size. After placing the first corner, tapping *Ctrl/Option* will size the image about the center point, while holding down *Shift* will allow you to distort the image with non-uniform placement.

Importing an image for use as a material is not just a good skill to have, but an essential skill for anyone hoping to move beyond basic modeling. You simply cannot make properly detailed models without knowing how to bring images into SketchUp for use as custom materials. Something else you will need to know is what to do with those materials once they are created. Let's cover that in the next section!

Organizing and applying custom materials

At this point, you have seen how to import images and how they show up as SketchUp materials. You have even sampled those materials and applied them to a face or two. What you may not have noticed, though, is that the image you imported into the practice file only existed in that practice file. If you look at the **Colors In Model** list in the **Colors** window right now, you will see your imported image. If you start a new file though, you will not see that imported image anywhere in the **Colors** window.

.smk Files

When an image is imported into SketchUp and a new material is created, that material exists as a `.smk` file. This is a proprietary file that includes everything that SketchUp needs to use the image as a material.

Custom materials exist in the model where the image was imported. To have those new materials show up in other models, we must tell SketchUp we want to use them outside of just this model. There are three ways to bring custom materials into other models:

- Apply the new material to a component, save the component as its own `.skp` file, then import it into a new model. The new material will then show up as a **Color In Model** item in the new model.
- Save the model with the new material as a template. When starting a new model and using this template, the new material will show up as a **Color In Model** item.
- Add this new material to a new or existing color list in SketchUp's material library.

The first option is not really a way to load materials as much as it is a side effect of loading components into an existing model. The second is a viable option but will only work for new models and only for a single template. Of these three options, the last one is ideal for a material that we will want to use regularly in the future or we would want to use in existing, as well as new, models. Let's look at the process of making a custom material list, loading it with custom materials, and saving it as a part of the material library.

Creating custom material lists

Let's create a list of materials that can be used in any model we open or create in SketchUp:

1. In `Taking SketchUp to the Next Level - Chapter 5.skp`, click on the scene called **IMPORTED**.
2. Bring up **Paint Bucket** and bring up the **In Model** materials in the **Color** window.

This scene shows you eight tiles with custom materials applied to them. These materials are saved to this model as materials in the **In Model** list. They do not, however, exist anywhere else in the library. Let's create a custom list called `Level Up` and put all these materials in there.

Lists or Collections

On macOS, the term *list* is used as a name for a group of materials shown in the **Colors** window. On Windows, the term *collection* is what the groups of materials in the **Materials** window are called. For our intents, they are the same thing.

Since the UI for adding lists is different depending on your operating system, we will go through this twice, once for Windows, and then again for macOS.

Creating a custom material list on Windows

Let's create a custom material list in SketchUp for Windows:

1. In the **Materials** dialog in the default tray, click the **Details** icon (the icon just below the eyedropper icon) and choose **Open or create a collection...** as shown here:

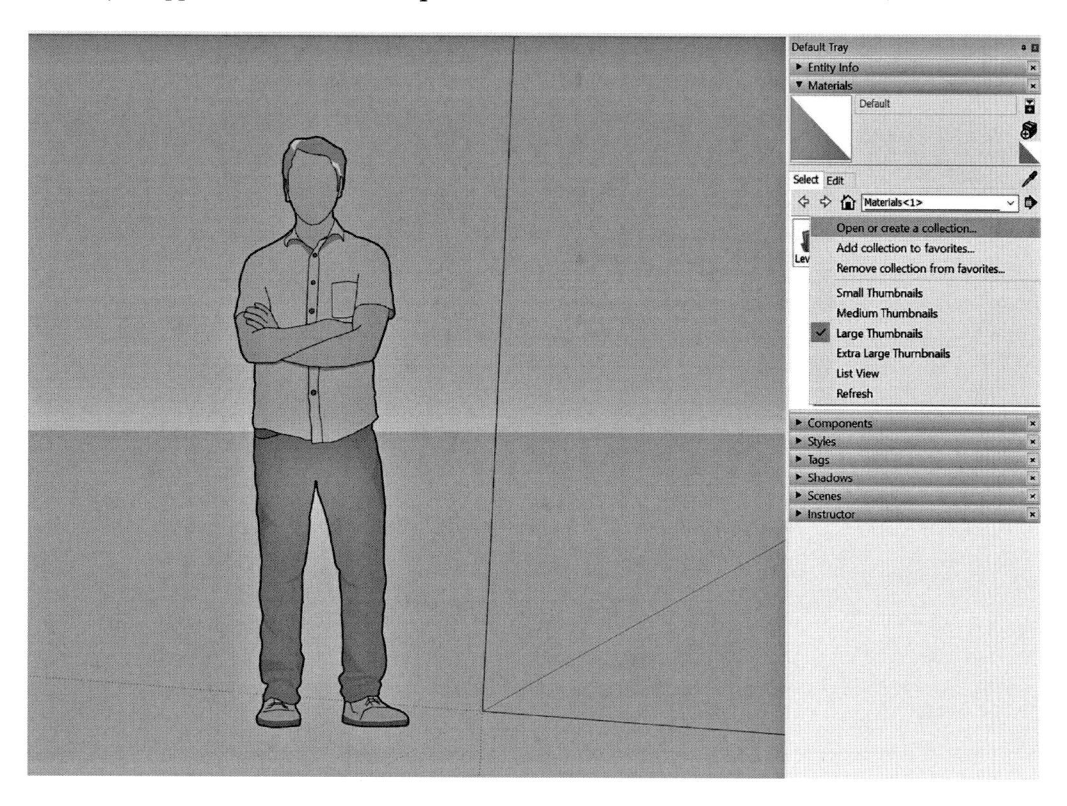

Figure 5.10 – Windows Materials window details

This will launch a File Explorer window showing the contents of the `Materials` folder as defined in your **Preferences**.

2. To create a new collection, right-click in the window, select **New**, then select **Folder**. Name the new folder `Level Up`, then select it, and click **Select Folder**.

   With that, you have created an empty collection of materials. On Windows, material files are simply saved into folders. SketchUp knows to look for folders at this location, and any folders will be shown as collections. On Windows, this means you can very easily create or move collections around outside of SketchUp, since they are just folders. Now that we have a new collection created, let's load it up with custom materials.

3. Click the **Display the secondary selection pane** icon at the very top-right corner of the **Materials** dialog.

   This will open a second pane containing a copy of the material collections in the **Materials** dialog.

4. In the second pane (the lower one) click the **Home** icon to bring up the **In Model** collection of materials.

5. Click and drag the first material (*Blue Curtain*) from the bottom pane to the top.

> **All Materials Have Names**
>
> Even though they may not be on the screen when you pick them, every single color, stock material, or custom material has a name. If you are seeing your materials as tiles, hovering your cursor over a tile for half of a second will show you the name in a tooltip.

You have now added a single material to your custom collection:

Figure 5.11 – The Materials dialog with two panes

6. Click and drag the other eight custom materials (they should be the top six materials and the last two) into the **Level Up** collection.

   With that, you have created a collection of eight custom materials that will show up any time you run this installation of SketchUp:

Figure 5.12 – The Level Up material list with eight custom materials

You can see that adding a material collection on Windows is not a difficult process, especially once you have seen how it is done.

Creating a custom material list on macOS

Now, let's see how we would go about the same process on macOS:

1. In the **Colors** window, click the **List** drop-down menu and choose **New...**.
2. Type `Level Up` into the text field, then click **OK**.

   Believe it or not, that is all there is to creating a custom material list on macOS. There is, however, the process of getting materials into the list. Since macOS does not have the ability to open two material lists at once like on Windows, this can take a little bit more time. We will look at a couple of different ways to get some materials into our list.
3. Click the **Home** icon to bring up the **In Model** materials.
4. Double-click the *Pillow* material (the one that looks like a white star on a blue background).

   Double-clicking on a material will bring up the editing screen for that material. While this screen is up, you can change material lists while keeping the material visible in the editor:

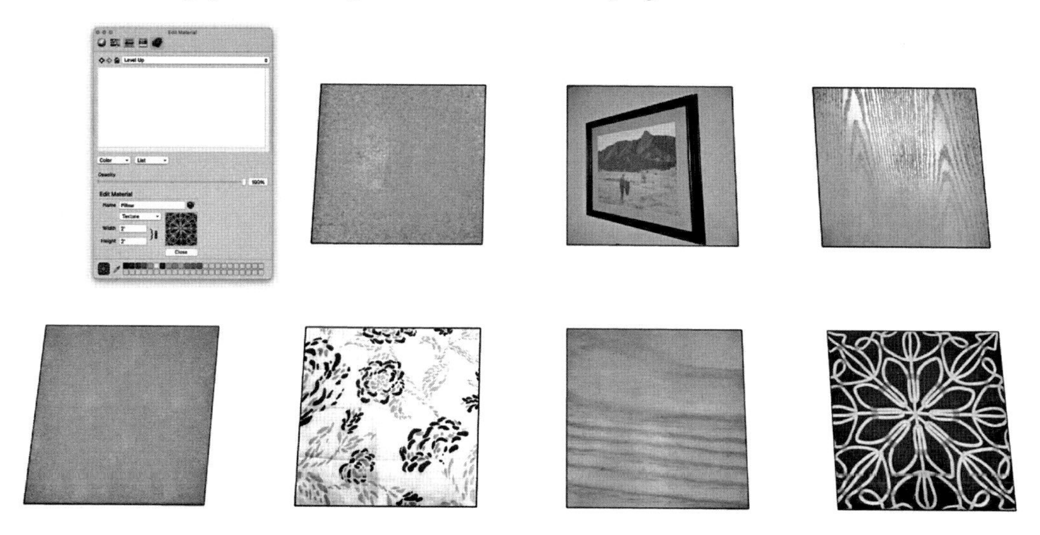

Figure 5.13 – The Color window is split between the editor and material lists

5. Change to our new **Level Up** material list.
6. Now, click on the material tile in the editor and drag it up and drop it into the top of the window (the empty list of materials in the **Level Up** list):

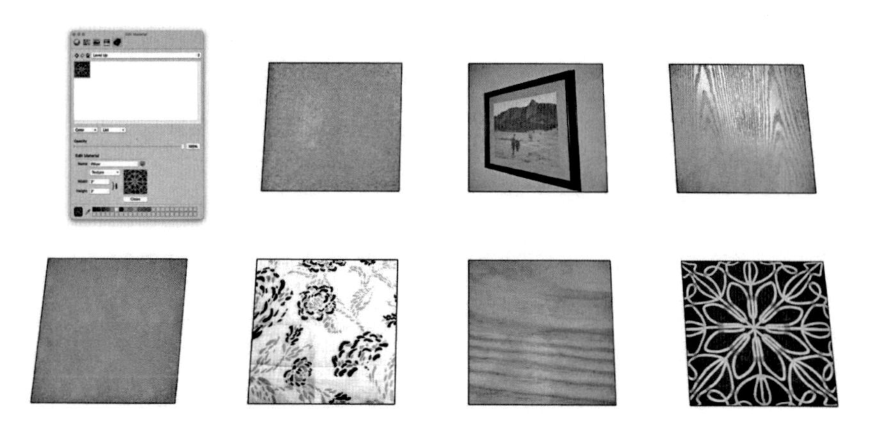

Figure 5.14 – The Level Up material list now has a single material

That material is now a part of our custom material list. To add more materials from the **In Model** list, we would need to switch back to that list. Double-click the next material to bring it up in an editor, then switch back to our custom list to add it. Let's look at a way to add these materials with fewer clicks!

7. While still in **Paint Bucket**, tap the *Command* modifier key to bring up the eyedropper cursor.
8. Click on one of the tiles in the model (not the *Pillow* material).

   Notice that selecting a material from the modeling screen automatically loads it into the editor. You now have that material loaded into the editor while still in the **Level Up** material list.
9. Click and drag the material into the **Level Up** material list.
10. Repeat the process for all the tiles.

At this point, you have created a list of materials that will be available any time you open a model on this installation of SketchUp Pro:

Figure 5.15 – Your custom material list with eight custom materials

At this point, we have seen how to create custom material lists on both Windows and macOS. In both cases, the lists live in the SketchUp Pro installation in which they are created. Let's check out how to move one or more material lists into another installation. This could mean moving a material list from one computer to another, or from one version of SketchUp to another.

Moving custom materials

Since moving these files from one location to another will require you to get to files on your computer, it will be different based on the computer that you are using. Unlike creating material lists, moving custom-made material lists is the same, regardless of the operating system. The fact is it is just as simple as copying a folder from one location to another. The trick (if there is one) is in getting to the folder.

The easiest way to do this is through **Preferences**. The **Preferences** window can be accessed through the **Windows** menu on Windows, and the **SketchUp** menu on macOS. Let's run through the steps to copy a material list from one copy of SketchUp to another:

1. In the installation of SketchUp that you are copying from, open the **Preferences** window and click the **Files** tab.
2. To the right of **Materials** in the list is a file folder icon. Click this icon to launch **File Explorer/Finder**.
3. Select and copy any custom material lists you want to copy.
4. Open the installation of SketchUp you want to copy to and perform *Steps 1* and *2* to open the **Materials** folder.
5. Paste the folder from the first installation into the **Materials** folder.
6. Close and restart SketchUp for it to find the new **Materials** folder.

If you are moving these folders from one computer to another, you may have to include a step where you paste the copied files onto a flash drive or via a network folder. The important part to note is that copying the folder full of material files can be done across any recent version of SketchUp and is even the same across operating systems!

What about Stock Materials?

You may have noticed that the only custom material lists were in the **Materials** folder, but you have access to hundreds of other materials through the **Paint Bucket** tool. Where do those come from? Stock materials are installed as part of the SketchUp installation and exist as program files. It is possible to modify or remove these stock materials. But, as they are considered program files, they will reinstall any time your installation is reinstalled or repaired. For this reason, it is best to leave them alone and focus your material customization on the files in the **Materials** folder.

With that, you not only know how to create custom materials, but also how to organize them and move them from one instance of SketchUp to another. Up until now, we have assumed that materials were as good as they were when imported. What happens when we need to change the orientation, angle, or scale of a material? Fortunately, that is exactly what the next section is all about!

Deforming textures

There are a lot of great places to get images for use as materials out there. There are free and paid sites where you can download high-resolution, seamless-tiling images of just about anything. There is also a chance that you are an expert photographer who can perfectly square up a texture in the real world, take a picture, and have it applied as a perfect material once it is imported into SketchUp. There are also a lot of good-looking images that will need a little bit of editing once they get imported. This section is for those.

Modifying textures

To modify a texture, it must be in the model space. It cannot simply be saved somewhere in the model but must actually be visible on the screen. To edit the material, select and right-click the face that it is on. If the face you need to edit is in a container, you may have to double-click to enter the container before you can get to the face with the material. Let's hop into our practice model and modify a few materials using the **Texture Editor** and **fixed pins**:

1. Click the **FIXED PINS** scene.
2. Select the top surface, right-click, then click **Texture**.
3. From the pop-up menu, select **Position**.

   This will bring up the texture editor with the custom material tiled at its initially imported size and orientation. Notice the four *pins* at each corner of the center tile:

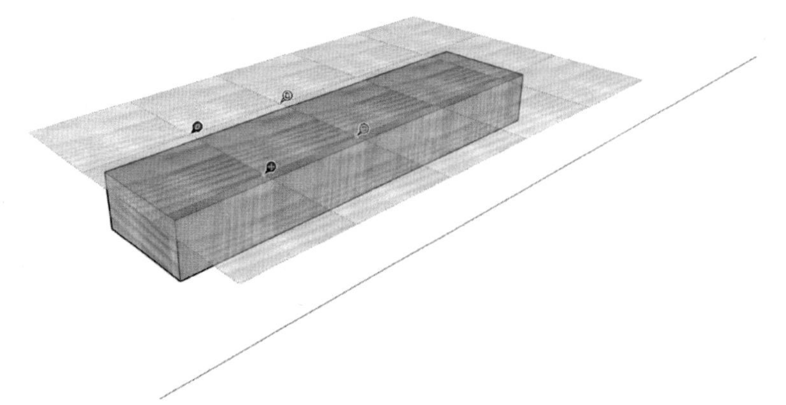

Figure 5.16 – A different colored pin on each corner of the material

These pins will allow you to distort the image. Each color has a different function:

- **Red pin** – This handle will allow you to move the material and the tiled copies along the plane of the face. Note that clicking and dragging any place on the face of the image will allow you to move the material.
- **Blue pin** – This handle will allow you to shear the material. Clicking and dragging this handle will bring up a compass showing the angle of the shear. Note that the shearing is uniform, in that the opposite side and all copies of the material shear the same.
- **Yellow pin** – The yellow pin is the distortion pin. Clicking and dragging the yellow pin will allow you to stretch the entire tiled grid.
- **Green pin** – The green pin will allow you to rotate and scale the center material. Clicking and dragging the pin will bring up a compass that will allow you to specify exact rotation by moving around the pin. Pulling the pin away from its origin will allow you to scale the material.

For this example, we need to scale up the material to cover the block. Since the grain of the wood is already running in the direction of the block, there is no need to rotate.

4. Click and drag the green pin down the board (along the red axis) until the center tile is longer than the board:

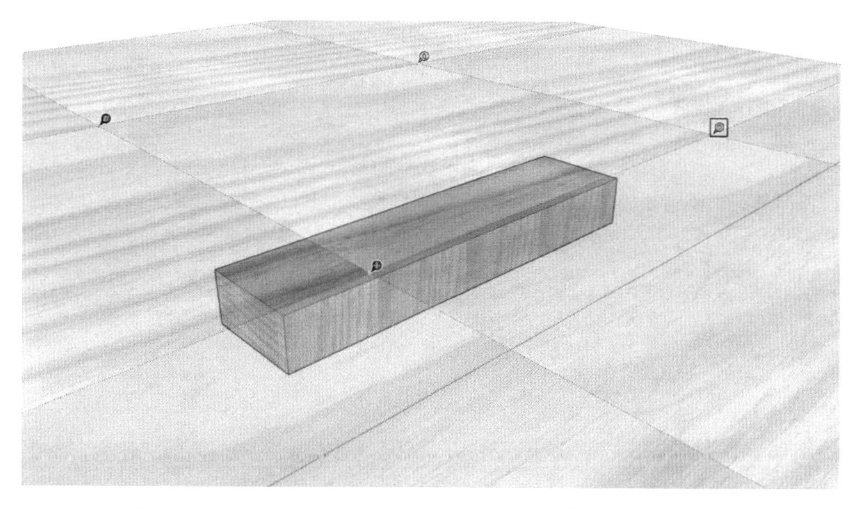

Figure 5.17 – Using the green pin to scale up the wood material

Now that the material is big enough, we just need to move it so that we do not see the seam at the edges of the image.

5. Click on the material (or the red pin, if you prefer) and drag it back up the red axis until the center tile covers the top face of the block. If you did not make the material large enough in the last step, use the green pin to scale up again:

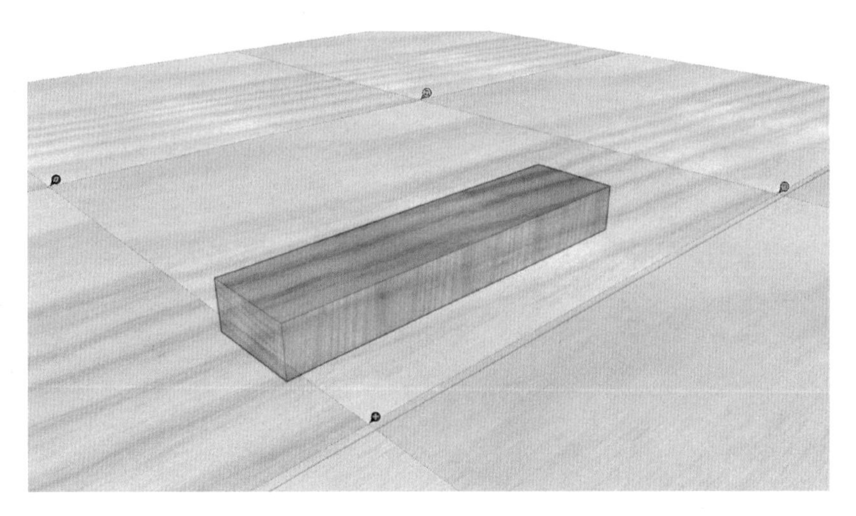

Figure 5.18 – Dragging the material so that it covers the top face

6. Once the material covers the top face, click anywhere outside of the material tiles to place the material.
7. Select the side of the board, right-click, then choose **Texture**, and then **Position**.
8. This time, start by grabbing the green handle and turning it around the compass as shown here:

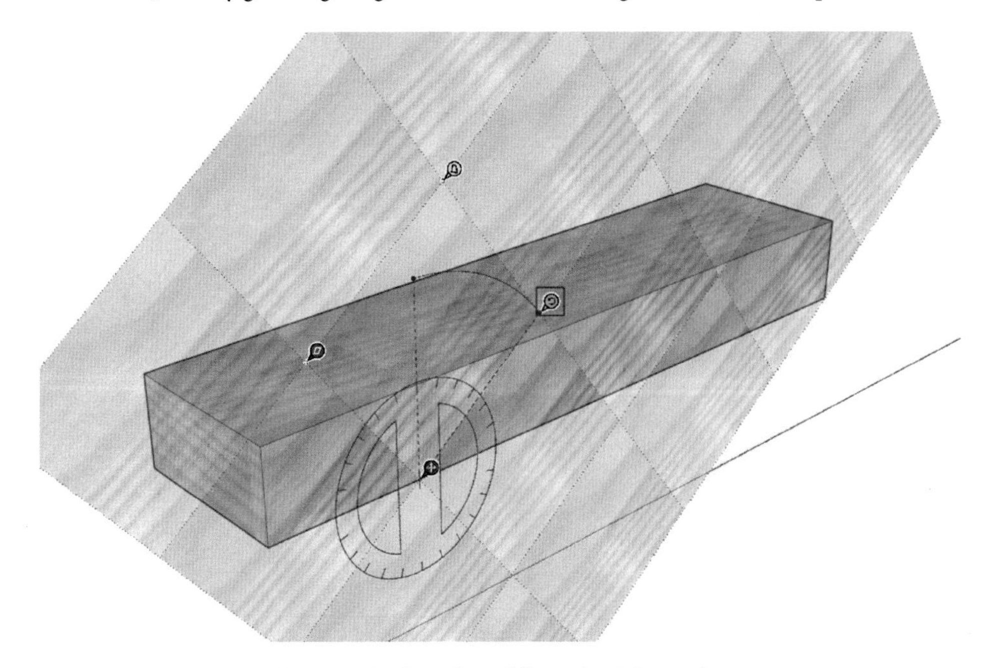

Figure 5.19 – Rotating while maintaining scale

9. Rotate the material 90 degrees (so that the grain runs in the same direction as the board).
10. Use the green handle to scale up the material, then move it as needed.
11. Click outside of the material grid to close the texture editor.

    The last material that needs to be edited is the end. Normally, if I were to texture a board like this, I would have a separate material just for the end. In this case, we can use the material we have, and we just need to be creative. In this case, you can use the green pin to rotate and scale the material until the grain is fairly large, and rotate it so that it does not run perfectly parallel to the edges of the face, similar to the following:

Figure 5.20 – The same material rotated and scaled for use as end grain

12. Use the texture editor and the green pin to scale and rotate the material until you have a material like the one in *Figure 5.20*.

    The last step is to copy the edited materials from the three sides to their opposite faces.

13. With the **Paint Bucket** tool, use the *Alt/Command* modifier to sample the top material, then orbit under the board and apply it to the bottom. If needed, right-click and move the materials to prevent the material edge from showing up.
14. Do the same for the side and end faces.

With that, you have successfully used the texture editor to apply a single material to six sides of a model. Some of you are probably asking, "*What about the yellow and blue pins? When are we going to try them out?*" My answer would be to go ahead and play with them. To be totally honest, I have used the blue pin once on an odd material and have never used the yellow pin on a model. If I ever have a material that needs to be edited, I use **white pins**.

For this example, we are going to take an image and use the white pins to stretch it, so that it fits into a totally new shape:

1. In your practice model, click the **WHITE PINS** scene.

   In this example, we have a wall and a picture hanging on it. Notice that the picture has an image of a framed artwork that was taken at an angle. What we want to do is use the white pins to stretch the image inside the frame so that it aligns with the geometry in the model.

2. Select the face with the image on it, right-click, then choose **Texture**, and then choose **Position**.

3. Right-click again and click on **Fixed Pins**.

   This will toggle fixed pins (the colored pins) off. Notice that the colored pins are replaced by white pins:

Figure 5.21 – White pins at each corner of the material

   We will use these pins to stretch our image to match the geometry. What we want to do is move the four pins so that they are on the four corners of the photo in the image.

4. Select the lower-right pin by clicking on it once (click and release).

   This will attach the pin to your cursor. Now, you will need to place it in the lower-right corner of the photo.

5. Move the pin and place it so that the tip of the pin is at the lower-right corner of the photo.

6. Repeat *Steps 4* and *5* with the lower-left and upper pins.

   At this point, you should have a pin in each corner of the photo in the image:

Figure 5.22 – Pins aligned with the four corners of the photo

Now comes the fun part. Next, we will use the pins to stretch the image so that the corners of the photo align with the corners of the geometry in the model.

7. Click and drag the lower-right pin to the lower-right corner of the rectangle.

   Notice how the geometry stretches when you move the pin? This can be a little bit disorienting, but when we get all four pins aligned, it will look perfect.

8. Click and drag the remaining pins to their respective corners.

   With that, you have stretched the material so that the orientation of the photo is back to a rectangle:

Figure 5.23 – Rectangular image aligned to rectangular geometry

9. Click outside of the material tile to close the texture editor.

The white pins are amazing tools to make images that are not perfectly aligned to your model fit where you need them. Obviously, this works great for something such as artwork, but can also be used to align facades on the front of buildings, or even stretch landscape imagery.

At this point, you have seen how to import, organize, and modify materials. Up next, let's take a look at a couple of different ways to apply materials when your geometry is more than a single face.

Applying materials to curves

In SketchUp, there is a one-to-one relationship between a face and its material. Any given face will have one material applied to it, and how the material falls on that face is relative to just that face. This is different from processes such as UV mapping, where materials are applied to a model as a whole.

This is not a big deal when you are applying colors or when custom materials are going on flat surfaces but can become a bit of an issue when the faces you are applying materials to are more complex, such as rounded shapes or wavy meshes. Fortunately, there are a few tools and techniques in SketchUp that can help you to apply materials to these sorts of geometries.

Projecting materials

One method to apply materials to curved or wavy surfaces is to use **projected textures**. When you apply a material to a face, that face applies normally to the face. This means that the material is aligned so that it is in the same orientation as the face and then lies flat on that face.

A projected material, on the other hand, is applied flat to one surface and is then projected onto other faces. This can cause a bit of distortion to the material but will allow for materials to flow across curving or wavy meshes. Let's practice this in an example:

1. In your practice file, click the **PROJECTED** scene and bring up the **Paint Bucket** tool.
2. Sample the blue material on the left and apply it to the wavy mesh in front of it.

   That is a mess. When you apply a material to a mesh like this, you are not applying it flat to a single face, but you are applying it flat on each face that makes up the mesh. Let's look at exactly what is happening here.
3. Click the **View** menu and toggle **Hidden Geometry** on.

   See all those dotted lines? Each of the faces created between those dotted lines has the material applied to it. You can see that SketchUp tries to connect the material when faces are edge to edge, but it cannot connect the material across all the faces:

Figure 5.24 – A material applied directly to a wavy mesh

This is where projected textures come into play. Let's try applying the material on the right using a projected texture. First, we need to turn that **Hidden Geometry** off.

4. Click the **View** menu and toggle **Hidden Geometry** off.

   This step is taken not just because it will be easier to see, but also to show the wavy mesh as a single surface. When applying materials with **Paint Bucket**, it will apply to the face that is clicked on. This means if **Hidden Geometry** is turned on, **Paint Bucket** will apply materials to faces one at a time, rather than the whole surface.

5. Right-click on the material on the right, click **Texture**, then select **Projected** from the popup.
6. Sample this material, then apply it to the wavy surface in front of it.

   Notice how the material flows evenly all across the wavy surface? This is what a projected material will do. It will slightly deform the material on each face so that all faces on the surface appear to be connected:

Figure 5.25 – Project texture applied to wavy surface

This is the method to use when you are placing a material on a curved or wavy surface like this. There are some cases, though, where you may want the material to wrap around a surface, rather than project onto it.

Wrapping with a material

In this example, we will wrap a rectangular material around a cylinder. Just applying it will result in something like the blue material in the previous example, and projecting will end up stretching the materials at the sides:

Figure 5.26 – Can 1 – Material applied directly, can 2 – Projected material applied, and can 3 – What we want

To get a material to wrap, we will sample the material and apply one face at a time in sequence, so that the result is the appearance of the material wrapping around the geometry:

1. In your practice file, click the **WRAPPING** scene and start up the **Paint Bucket** tool.
2. Sample the *Cola* material to the right of the can.

   At this point, applying it directly to the can will result in the material being all messed up like can 1 in *Figure 5.26*. We need to apply it directly to the faces that make up the surface. To do this, we need to see the hidden geometry.
3. Toggle **Hidden Edges** on in the **View** menu.
4. Click on one of the faces in the middle of the can to apply the material:

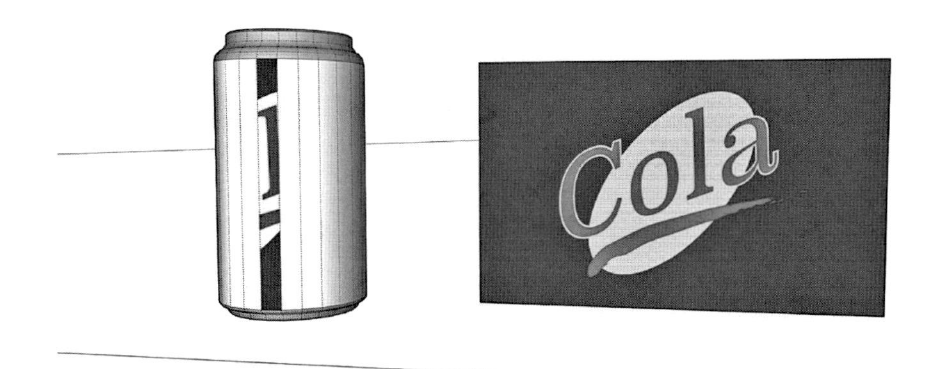

Figure 5.27 – Material applied to a single face of the can surface

5. Click on the face directly to the right of that face. Continue to click on faces to the right until you are only applying blue (half of the *Cola* logo is finished):

Figure 5.28 – Half of the can wrapped in the material

6. Now, go back to the center and start applying to the faces to the left of the center.
7. You will have to orbit as you go, but you should be able to click faces until the image wraps completely around the can:

Figure 5.29 – The can with the sampled material wrapped all the way around

In this case, the image used was long enough to completely wrap around the can. If the image were shorter, you would have had to re-sample the material as you apply or restart the image partway through.

With that, we have learned how to use native tools to apply materials to curved or wavy surfaces. If this is something that you do a lot, you may benefit from a UV mapping extension or another painting tool (we will talk about extensions in *Chapter 11, What Are Extensions?*).

Up next, let's take a look at some of the most common mistakes when it comes to creating and using custom materials.

Avoiding common mistakes with materials

We have learned a whole lot about creating and using materials in our model. I figured it was worth mentioning a handful of issues that people run into when using materials in SketchUp. Some of these are improper processes or assumptions, but most power users have run into these mistakes on the road to leveling up their SketchUp skills.

Relying on materials to add all the details

I see this all the time. Modelers want to cut corners or save time, so they apply a photo texture to the front of a building instead of modeling details. Now, this can work in some cases, but a lot of times, this will end up giving you a flat-looking model that cannot be used out of a single context.

Take this model, for example:

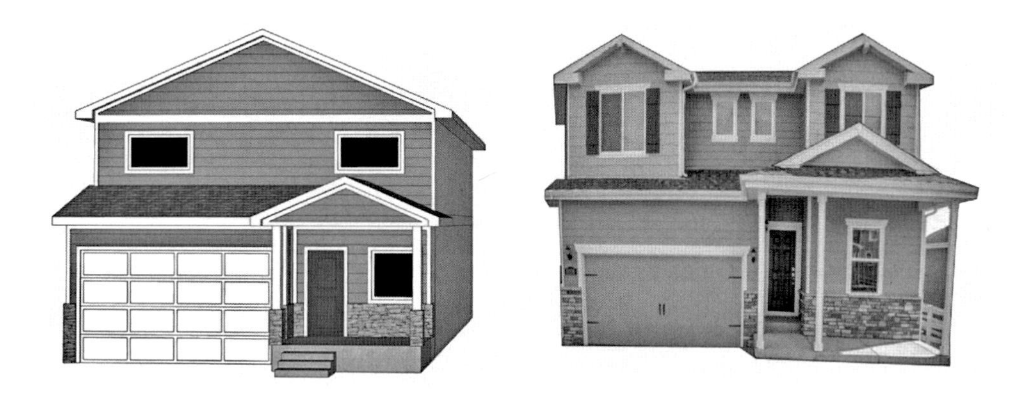

Figure 5.30 – The left house has modeled details and the right house is mostly materials

In this example, the house on the right was a 2D cutout of an image that was extruded. It looks OK in the current view. In fact, some might even prefer the way it looks to the flat colors used in the model on the left. However, the appearance starts to fall apart when the camera moves, or shadows are applied:

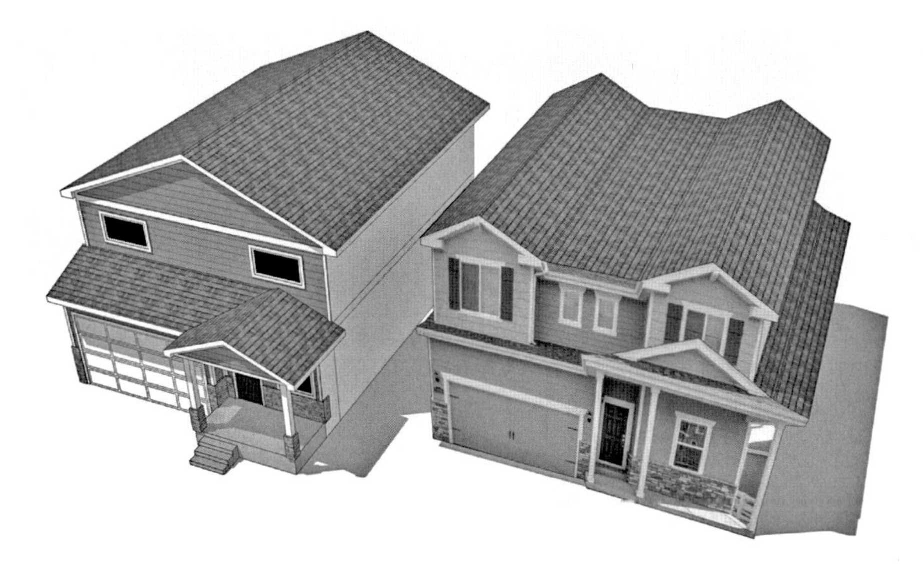

Figure 5.31 – The material façade looks less impressive from this angle

While the model on the right took a fraction of the time to model, if it is used for anything other than background geometry, it just does not look very good.

Import as a texture

OK, I know this one sounds a little ridiculous, but I have made this mistake and I have seen newer users asking about this. When you are importing, make sure you have **Use As Texture** selected. I would venture to say that all SketchUp users have, at one point or another, imported an image as a material, then been confused when it shows up and cannot be loaded into **Paint Bucket**.

Import proper-sized files

One of the tricks of mastering SketchUp is understanding when and where to add detail. Adding too much detail can slow down a model, and adding too little can make your model look bland or boring. The same goes for imported imagery. We have seen that we can use the texture editor to move, scale, rotate, or stretch out images, but it is important to be conscious of the intended use of a material before it is imported.

Look at this house model with a gingerbread house material added to it:

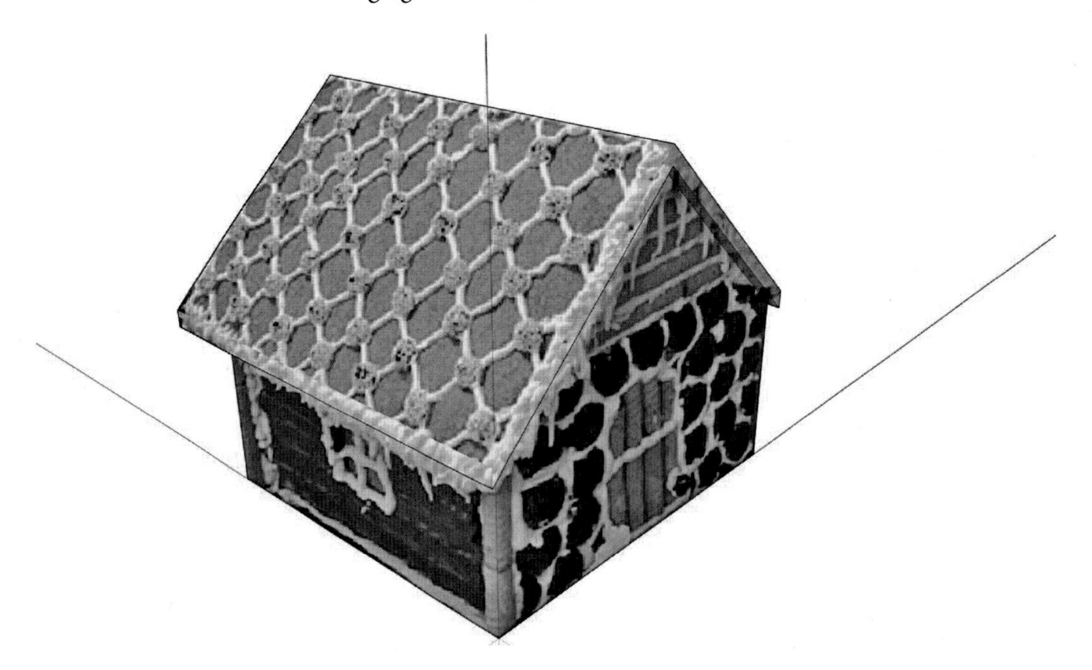

Figure 5.32 – A simple model with a simple material applied

The level of detail on this model seems OK, but getting even slightly closer shows how low quality the image used is:

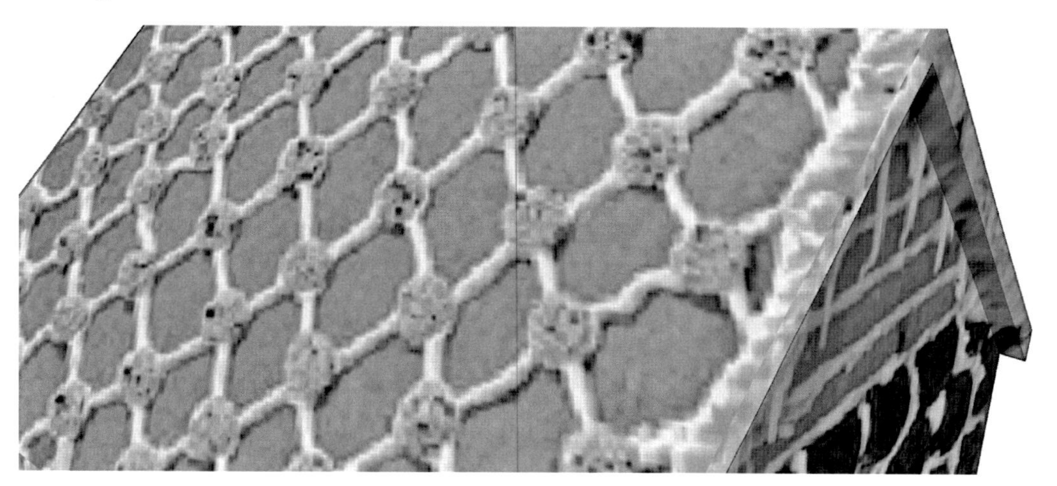

Figure 5.33 – Zooming in shows the lack of quality in the image

Does that make this material unusable? Of course not. Low-quality images can be used in items that do not need to stand up to up-close scrutiny. This model would work just fine as a model of an actual gingerbread house sitting on a kitchen counter. If Hansel and Gretel were visiting, and the model had to be life-size, it would probably be a good idea to get some higher-quality images.

Apply to the outside

One of the keys to leveling up your SketchUp skills is learning to model for a specific purpose. All models are made for a reason, and the reason you are creating a model should help you to make decisions on how your model is built. A lot of models end up in layout and as construction drawings. Some models are exported directly from SketchUp for use in game engines. Many models head out to rendering engines. If your model is being exported for use in another software or rendering, it is important to make sure that you are not hiding **reversed faces** with your materials.

When you create your geometry in SketchUp, you see front and back faces (white and gray, respectively). Beginner users will model without concern about what a front face is and what a back face is, then cover the whole thing up with materials and never think about it again.

In this example, the reversed faces (seen on the right) are hidden by the material applied on the left:

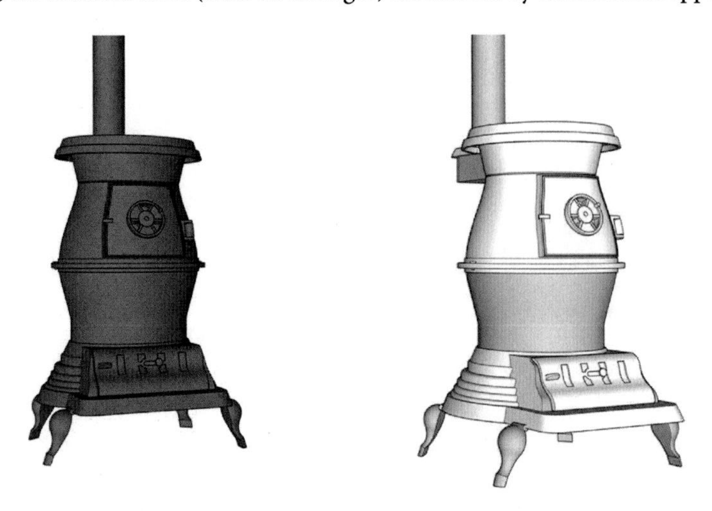

Figure 5.34 – The materials are covering up the reversed faces in this model

This can cause some real issues when your model leaves SketchUp. Reversed faces may not get rendered by your rendering engine or your model may not be accepted as a solid if you hope to have it 3D-printed.

Orient materials

It can be very tempting to slap a material onto a face and call it good, but leveling up means raising the bar for what makes a good model. In this case, it may mean modifying materials on each face as you apply them. Consider the wood materials added to this model of some wall framing:

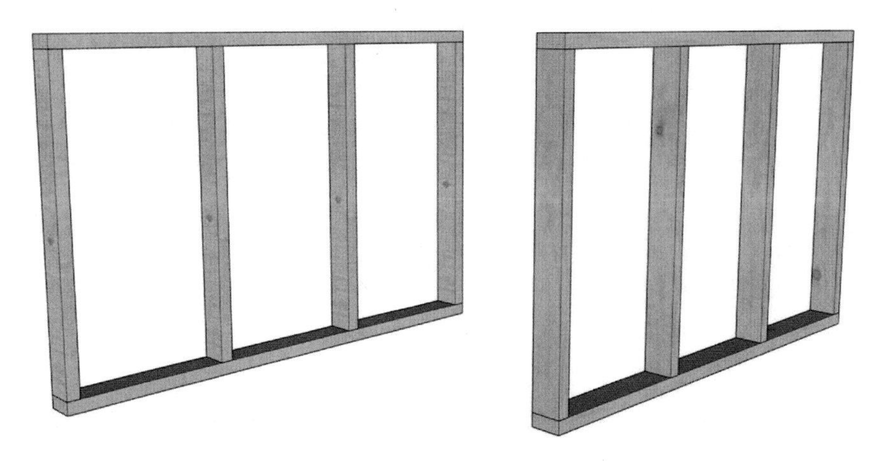

Figure 5.35 – The left wall has the same texture on every piece, and the right has textures that have been edited

In this example, you can see that the wood grain runs in the wrong direction on the studs in the example on the left. Additionally, notice that the knot in the wood is the same on every piece. In the example on the right, the wood grain runs in the direction of the studs and has also been shifted on a few pieces to give a little visual variety.

Take a picture

This was mentioned in the *Importing images for custom materials* section but bears repeating. If you need a specific material in your model and cannot find a good image, take a picture yourself. Nowadays, most phones have cameras that are more than capable of taking pictures that are great for use as materials. Yes, it may be a few extra steps, but if you start taking pictures of good-looking materials and save them in custom lists, over time, your collection will grow, and eventually, you will be filling your models with beautiful custom materials without having to leave your computer.

Don't use materials at all

This seems a little weird as a tip at the end of a chapter dedicated to materials, but not every face of every model requires a material. In some cases, applying a color will work just fine. For example, take this model of a mini steam engine:

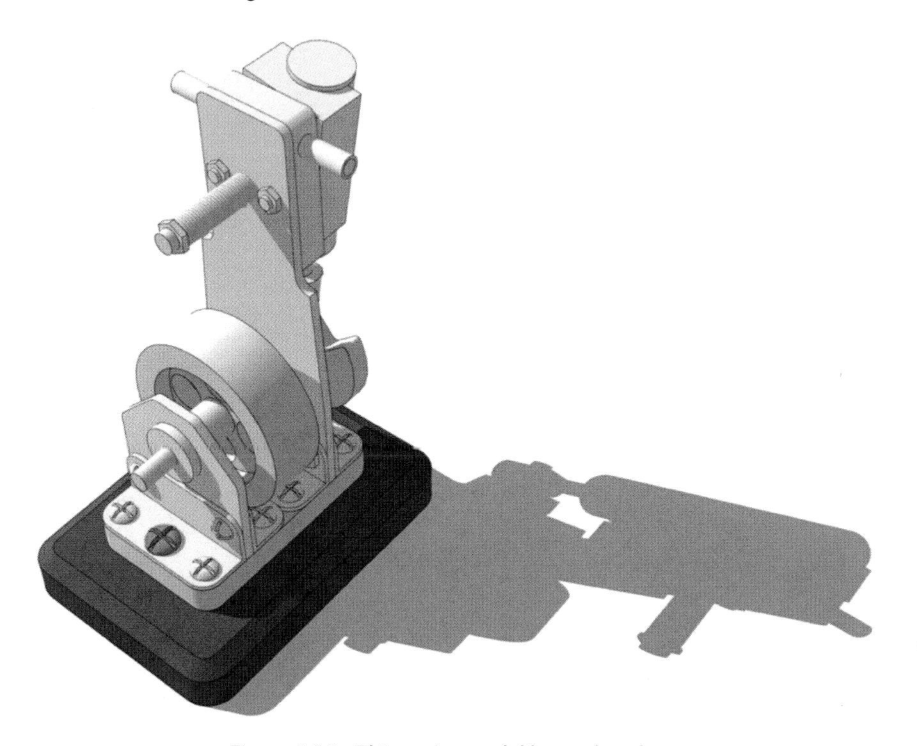

Figure 5.36 – This entire model has only colors

Again, modeling for purpose means that this model looks just fine using colors alone since its purpose is to demonstrate how a steam engine works. Adding images of reflective metals may look good at one angle, but as the viewer moves around the model, the images will look odd and out of place. In this case, colors are the better choice.

If you have already stumbled on one or more of these mistakes, know that you are in good company. Well, you are in my company at least. This list comes from specific examples that I have made over the years in an effort to get the most out of materials. Remember, it's OK to make mistakes, as long as you learn from them!

Summary

Materials are a great way to make a good model great, and a great model amazing. You have now seen how to import your own images for use as custom images and how to organize them into custom material lists. We also covered how you can edit your materials using the texture editor and covered how to put materials onto wavy or curved surfaces. With these skills, your models should be looking better than ever before!

In *Chapter 6, Knowing What You Need Out of SketchUp*, we will be looking at your specific needs, and how you can develop your own custom workflows tailored to those needs.

Part 2: Customizing SketchUp and Making It Your Own

Practicing and getting better at using a tool is one thing; developing a plan so that you can truly grow in your capabilities as a 3D modeler is something different, entirely. Here, we will explain the process of moving from a beginner to an intermediate user and developing your own modeling process.

This part contains the following chapters:

- *Chapter 6, Knowing What You Need Out of SketchUp*
- *Chapter 7, Creating Custom Shortcuts*
- *Chapter 8, Customizing Your User Interface*
- *Chapter 9, Taking Advantage of Templates*
- *Chapter 10, Hardware to Make You a More Efficient Modeler*

6
Knowing What You Need Out of SketchUp

How can you learn what you don't know until you know what you don't know? This is the conundrum of the self-taught. It is impossible to know what you have left to learn until you know the things that you have not yet taught yourself!

In this chapter, we will run through and take inventory of the skills you have already developed and identify which ones need to be strengthened. By doing this, we will try to identify the new skill you want to develop and give you an idea of the things that you can do to customize SketchUp for your specific workflows.

In this chapter, we will be covering the following topics:

- Identifying what you are modeling now
- Discovering what you will be modeling next
- Establishing your leveling up goal

Technical requirements

In this chapter, you will need access to SketchUp Pro and the models you have created over the last year or so.

Identifying what you are modeling now

The first step in making a plan to get better at something is to figure out how good you are at it right now. Depending on your working environment, this may be easy to figure out or it could be a bit of a struggle. While there are quite a few SketchUp users who have had formal training and work in an office with other designers, many SketchUp users are self-taught and work on their own.

For this reason, it's not a simple "Are you a beginner, intermediate, or advanced user?" question. Likewise, since SketchUp can be used across so many different industries, it is hard to identify what skills or competencies must be in place to establish a user as being a "good" SketchUp designer.

So, what do we do here? First off, let's agree to drop simple labeling. I know that there is something in us that wants to categorize and rank our abilities, but the fact is that it does not matter. Who cares if you are a "level 32 furniture modeler" or an "advanced architectural modeler" in the grand scheme of things? What is important is knowing what you are capable of now and then using that information to identify your own personal opportunities for growth and making a plan to improve. If that sounds like a good idea, then read on!

Let's break down your current modeling process by looking at four questions:

1. Why are you using SketchUp?
2. How are you using SketchUp?
3. What tools are you using in SketchUp?
4. What is your SketchUp workflow?

Once we have answered these four questions, you will have a better understanding of where you are currently with your SketchUp skills.

Outlining Where You Are at

While you are more than welcome to read through this chapter and treat it as "food for thought," I do recommend getting more physical with it. Grab a piece of paper and write the answers down. Create a document on your computer and type in the answers. Being able to refer to them in future chapters will be important, and taking the time to stop, reflect, and actually write them out will be key to creating a real plan to get better at SketchUp. This will end up creating an outline of sorts, so make sure to give yourself extra space to add information, and don't be afraid to edit or start over.

Why are you using SketchUp?

I don't mean this question as "Why SketchUp instead of something else?" but rather "Are you using SketchUp to model buildings at work, or are you using SketchUp to make models to print on your 3D printer?" What you should look at here is everything you are currently using SketchUp for – business or pleasure, big or small.

Write `Current` on a piece of paper. This will be your list of how you are using SketchUp right now. List out all the ways you are using SketchUp at present. This should not be a list of specific models you have created but the type of work you have done in SketchUp.

This list should be very high level. That's why I put "why" into the questions. What we are looking for here is the purpose of the models you are creating. Examples of possible answers could be something like these:

- Site plans for work
- Bedroom layout for my partner
- Scale train cars for 3D printing
- Models of flowers to learn subdivision modeling

On a piece of paper, list the reasons why you model in SketchUp. Put down the purpose you have found for using SketchUp. List the kind of modeling you are called upon to do for work. Put the modeling you have done while watching YouTube videos. Include the stuff you play around with at home or during your lunch hour. Write one reason per line, giving yourself plenty of space between points (five or six lines) to add detail in the next steps.

Do not be embarrassed or worried about the stuff you have tried and failed at either! Throw it on the list! If you wanted to try to get into modeling cars, just for fun, but gave up after a few attempts, write that down. At this point, you need an inventory of the reasons you have used SketchUp.

How are you using SketchUp?

Up next, you are going to get more specific. For each of the "whys" listed, jot down as many examples of "how" you used SketchUp as you can think of. This is where we get a little more specific. You want to look at real examples of models you have created. If it helps, open SketchUp and look at some of the models you have worked on. What you want to end up with here is one or two actual examples of modeling you have done for each of the points from the previous *Why are you using SketchUp* section (with space between them for even more detail). At this point, your outline may look something like this:

- Site plans for work:
  - Fairground model
  - 1322 Blake Street
- Bedroom layout for my partner:
  - A model I made before we bought a new bedroom set
- Scale train cars for 3D printing:
  - A Flying Scotsman model
  - A red caboose model

- Models of flowers to learn subdivision modeling:
    - An attempt at modeling a rose
    - A daisy model

There is a very good chance that you may have more than two examples. If you have more examples of models, look for the ones that are different from others. What we want to do with this list is break down the things that you know how to do in SketchUp, currently. If you have two dozen models of sunrooms that you model for work, what is the one that was the gold standard for how it should be done? Then, what is the one that was done in a way that strayed furthest from your standard workflow?

What we need here is a list of models that shows the breadth and depth of your current modeling capability. Consider this to be the portfolio of your current SketchUp skills.

Remember in the *Why are you using SketchUp* section where I said to include those efforts that did not work out? The same goes for example models. Any time we try something, even if we do not achieve the result we were hoping for, we still succeed if we learn something. To that end, include the model that you were really hoping would turn out better than it did. Include the one that almost made it or the model that makes you proud, even though you don't really want to show it to anybody else.

What tools are you using in SketchUp?

Time to dive into the details! Now, go back under each job and jot down the commands you primarily used to create each model. I know – there is a good chance that you used dozens of commands on each mode, but do your best. At the very least, try to identify the "big ones." List out the main tools you used to create or edit when you were working on each model. You can probably skip commands such as **Save** or **Select**, as you cannot get by in SketchUp without using these, but try to think of the drawing commands you used the most often and the tools you used more than others. If you used any extensions in any models, go ahead and add those to the list as well.

The idea here is to be honest about what tools you are using in the hope that you will see a pattern in your work. Maybe as you create your list you will see a specific tool coming up too often or see a tool missing from the list. At this point, your outline should look something like this:

- Site plans for work:
    - Fairground model:
        - **Rectangle**, **Eraser**, **Sandbox Tools**, **Import**, **Push/Pull**, and **Follow Me**
    - 1322 Blake Street:
        - **Rectangle**, **Circle**, **Eraser (Smooth)**, **Sandbox Tools**, **Push/Pull**, **Follow Me**, and **Move (Array)**

- Bedroom layout for my partner:
  - A model I made before we bought a new bedroom set:
    - **Rectangle**, **Push/Pull**, **3D Warehouse**, **Paint Bucket**, and **Dimensions**
- Scale train cars for 3D printing:
  - A Flying Scotsman model:
    - **Rectangle**, **Arc**, **Weld**, **Push/Pull**, and **Move** (**Stamp**)
  - A red caboose model:
    - **Line**, **Arc**, **Weld**, **Push/Pull**, and **Move** (**Stamp**)
- Models of flowers to learn subdivision modeling:
  - An attempt at modeling a rose:
    - **Polygon**, **Push/Pull**, **SubD**, and **Move**
  - A daisy model:
    - **Artisan**, **Move**, **Rotate**, and **Paint Bucket**

Now, before we push on, look at your list. Can you find a favorite command? Is there something that you fall back on with most or all of your models? This may not be a bad thing. In the example list, **Push/Pull** shows up in almost every model. It is safe to say that **Push/Pull** is a command that most SketchUp users would use to create 3D geometry. It's something worth noting, though.

> **My Own Opportunity For Growth**
>
> I have to admit that I have a habit of using the **Line** tool along with inferencing to draw rectangles, rather than using the **Rectangle** tool. It is a bit of a bad habit that I have developed over the years. I know that using **Rectangle** will save me time, and to improve as a modeler, using it is something that I need to be aware of so that I can add that tool to my workflow. If I were making a list, I would call attention to the fact that I lean on the **Line** tool a little too much.

Thinking back to the previous chapters, are there any commands that you read about that you do not see in your list? Think about the commands you have used and where you could save time modeling by implementing a tool that you already have available to you. It may seem silly to think that changing a command may only save you one or two mouse clicks to create the same geometry, but those mouse clicks add up!

How Much Does a Click Cost?

Let's do a quick breakdown of my example about using **Line** instead of **Rectangle** – if I draw a square using the **Line** command, it will take me four mouse clicks at a minimum, while using **Rectangle** will allow me to create the same geometry in two clicks. If I have a model in which I need to draw 20 rectangles, that is the difference between clicking 80 times and 40 times.

I know that this seems like a simplistic example, but think about some of the other commands that are out there. What is the difference between using **Cut** and **Paste** to create an array of lights on a street versus using the **Move** command? How much work is it to cut and weld geometry together manually rather than using the **Solid Tools**? In some cases, using the proper tool for the job can mean hours of time saved over its lifetime.

What is your SketchUp workflow?

Okay, this is the big one. This is where you are going to dive in and really think about how you use SketchUp to create 3D models and take a look at your **workflows**.

When we talk about workflows, we are talking about a high-level overview of what you do in SketchUp to create your model. This is not a step-by-step list for a specific model but a more general overview of the process that you use. When thinking about workflows, you want to identify the most important steps in your process.

You do not need to do this for every example on your list, but maybe pick two or three that have distinct workflows. If all the models that you create for work are modeled more or less the same way, call that one workflow. Maybe your hobby modeling has a couple of different potential workflows; maybe see whether you can consolidate them into one master workflow.

An example of a high-level workflow might be something like this:

1. Import a site plan and scale.
2. Trace a building footprint, pull up, and group.
3. Create a landscape mesh.
4. Model concrete.
5. Add vegetation.
6. Create scenes for output.

This seems pretty simple, but if this is your workflow, each step in the preceding list should trigger you to think through all of the actual steps you use. When you model the concrete, are you using **Line** and **Arc**, or are you modeling a series of rectangles and circles? Does each step end with you tagging the new geometry? Are you searching 3D Warehouse for trees and bushes for each job, or are you pulling from a custom component library?

This is your opportunity to get introspective and see where you have ineffective modeling techniques. It is possible that you do not know any way to model other than what you are doing right now. That is fine, too! You should still spend some time looking at what you are doing and identifying the pain points in your process. Look for areas of your workflow that:

- Are monotonous or repetitive:
  - Anything that is done more than once can probably be sped up or consolidated into fewer commands.
- Require you to go back later to modify:
  - If you are modeling something knowing that you will be coming back later to fix it, what can you do to make it right the first time?
- Take more time than the final result is worth:
  - Remember to model at the level of detail the model requires. Do not spend hours modeling details that will not be seen in the final output.
- Could be done easier elsewhere:
  - Are you spending time in SketchUp adding details that are better off being added in LayOut? Are you spending time modifying textures that would be easier to clean up in a photo-editing program? Do your best to use all the tools available to you to get the job done.

The goal of this section was not to make you feel like you are not doing a good job or to prove to yourself that you can do better. The goal here was to stop modeling, step back, and look at what you are doing currently so that you can think about how you might be able to make some changes and level up your SketchUp skills.

By this point, you should have a good idea of what you are doing now. By looking at the models you have created, you know what sort of work you are currently capable of. By looking at the commands you are currently using, you now know where you might be able to take advantage of other tools in SketchUp. By walking through your workflows, you know where a change to your modeling process may lead to you becoming a more efficient modeler.

Now that you know where you are, we can start thinking about where you can go next.

Discovering what you will be modeling next

In this section, you will create another outline. This one will be all about what you want to do with SketchUp. I am making an assumption here – that, for sure, you have some sort of plans to go beyond what you are modeling now (if not, you probably wouldn't have picked up this book).

Let's take a look at the sort of work you want or plan to do in SketchUp and break it down into these questions:

1. Do you want to model something new?
2. What new skills or tools do you need to know?
3. What would a new workflow look like?

I know there is a little bit of fortune telling here, as you may not know exactly what you will be modeling in the future, or if you do, you may not know how to get that modeling done. Do your best, and remember that this is an exercise that can be done and redone as often as you like. If you do not know what type of model you will need to create for work tomorrow, do this exercise now and then do it again tomorrow!

Do you want to model something new?

One of the rules of brainstorming is that there are no bad ideas. Jot down your ideas on the types of models you would like to be creating but are not. This can be modeling you want to do for fun, or something you have needed to learn to do for work but have not had the time. If there is a project at home that needs a 3D model to help you visualize or if you saw something online that you want to try to model in SketchUp, write them down!

While you want the items on this list to be attainable, you do not need to make a list that you are certain to complete. These should be models that will push the boundaries of what you are capable of and could be and that are beyond what you know to do right now. Do not list a couple of models that are just a little bit different from what you do now if you can help it. The whole purpose here is to level up, and you cannot do that if you do the same thing over and over!

Grab a fresh piece of paper and write `New` at the top. This will be your list of new skills and abilities.

Now, you should end up with a list that looks something like this:

- Model more architectural details (doors and windows gutters) for work models.
- Model and save custom Face Me components for entourage.
- Model some spaceships from *Star Wars*.
- Model building additions with material lists.

Not a bad list. All these things are likely things that I have thought about doing in the past but put off because it was beyond what I was capable of, or I did not have the time to learn. Each of these projects has something involved in their completion that is beyond what I can just sit down and do today. So, each one will require me to go beyond my current skill set.

The next step is to break these new models down and figure out what is needed to make them happen.

What new skills or tools do you need to know?

Making a list of the things that you do not know is always a hard task. How do you write down the things that you have yet to learn? For this step, you will need to do some research.

It is easy to think that something that is new to you is new to everyone and no one has ever done anything like this before. While that may be true in some cases, there is a very good chance that someone out there among the millions of SketchUp users has done something similar to what you want to learn.

It is also possible that you put a type of modeling on the list specifically because you know what skills it will take and they are ones that you want to develop. In this case, it will probably be easier to make a list of the skill or tools you need to learn but likely still worth reaching out for an idea on how to get started.

The question remains – how do I find out how to make these new types of models?

Unfortunately, as much as I want to, I cannot offer a comprehensive list of everything that can be modeled in SketchUp and what tools each one requires. I can, however, offer a few suggestions about where to get an idea:

- **Coworkers** – I know that not everyone works in an office with other SketchUp users, but if you do, you should take advantage of the knowledge there. Even if you and your coworkers learned SketchUp at the same time, there is a good chance that they have picked up a trick or found a way to use a tool that you are unaware of. If you do work in an office with other SketchUp users, try to absorb any information you can from them. In my experience, most designers who use SketchUp are happy to share their experience and knowledge.
- **Books** – I know that you were hoping that this would be the only SketchUp book that you would ever need, but there are always more books! If you are hoping to learn a new modeling technique and there is a well-reviewed book out there, it is probably worth looking into. You can keep it on the shelf next to this one!
- **YouTube** – There are a lot of videos on using SketchUp. I have even made a few myself. Odds are good that if you want to model something, someone somewhere has made a video that could help. A word of warning here – YouTube is like the Wild West. Anyone of any skill level can post anything. Just because someone recorded themselves modeling something and posted it does not mean it is the right way to model. As with any time you seek information on the internet, look at other videos from the person who posted the video, read the comments, and pay attention to how old the video is. While good videos can teach you an amazing amount in a short amount of time, poor videos can set you down a path of bad habits.
- **Forums** – One of the things that sets SketchUp apart from any other modeling software out there is the community of users that helps to support it. I am constantly amazed at the amount of time strangers will spend helping each other to learn new skills and overcome obstacles. It is honestly the best and most supportive software community I have ever experienced. There are two main forums that I would recommend using:

- **Official SketchUp forum** – This is the forum that is hosted by Trimble. It is moderated by members alongside Trimble staff and is vibrant and active. Most questions posted on the forum are responded to minutes after posting. You will need to create a free account in order to post on the SketchUp Forum. You can go to this forum following this link: `https://forums.sketchup.com`.
- **SketchUcation** – This is a third-party forum that is completely run and moderated by SketchUp users. In addition to the forum, SketchUcation has free and paid extensions, books, and access to modeling services. SketchUcation offers free or paid memberships to access their content. You can go to this forum by following this link: `https://sketchucation.com`.

Use these resources to learn what you can about the models that you want to learn to create. Add any tools that you believe will be needed to complete these models to your outline, as follows:

- Model more architectural details (doors, windows gutters) for work models:
  - **Import**, **Rectangle**, **Push/Pull**, **Follow Me**, **Paint Bucket**, and **Components**
- Model and save custom **Face Me** components for entourage:
  - **Line**, **Rectangle**, **Polygon**, **Component**, **Always Face Camera**, and Adding to a **Component Library**
- Model some spaceships from *Star Wars*:
  - All the drawing commands, **Follow Me**, **Move** (Stamp), and **Component Window**
- Model building additions with material lists:
  - **Rectangle**, **Move**, **Push/Pull**, **Rotate**, component organization, and some material list extension

It is possible that, at this point, your list of new skills may be a list of things to learn more about or questions that need to be answered. This is perfectly fine! What we are looking for at this point is where you can possibly expand your abilities.

What would a new workflow look like?

Time to put what you learned in your research to work! Do your best to write out what a new workflow would be for the new modeling you plan to do. Remember that we are still brainstorming here, so don't worry about getting this perfectly right on the first try.

You might be able to map out how something should be modeled based on a video you watched, or maybe you are tweaking an existing modeling workflow and feel pretty good that you know what changes need to be made off the top of your head. If you know that you want to model something but are unsure of what the process will be, start writing down the steps that you think you need to follow. If you hit a point where you are unsure what to do next, hop back online and ask a question on a forum or search for a video.

Learn by Doing

Failing everything else, you can always open up SketchUp and try the workflow you are developing. I know that sometimes people can be hesitant to just "jump in" and want instructions first, but exploratory modeling can be fun and a great way to learn.

Regardless of how you do it, you should have a few workflows documented on how to create these new models. Here is a potential workflow for creating custom Face Me entourage:

1. Import an image of a person.
2. Scale the image to the proper size.
3. Trace an outline of a figure.
4. Add details and then color with **Paint Bucket**.
5. Make a component.
6. Make the component a Face Me component.
7. Store the component in a custom collection of Face Me entourage.

I want to stress again that these do not need to be tested and refined workflows, but they need to be a place to start. They just need to exist at this point. Time and practice will help to refine them. For now, they will serve as a direction to move as you continue to level up your SketchUp skill set. Speaking of which, let's start working on a plan to develop these new skills!

Establishing your goal

At this point, you have taken an inventory of the skills you currently employ and have made a list of the new skills you need to develop in order to level up. All we need to do now is to figure out the difference between the two and make a plan to start developing some new skills! We will do this in three steps:

1. Identify new skills.
2. Make a plan to learn new skills.
3. Implement your plan.

As is the custom in this book thus far, let's start the list with the first step.

Identify new skills

Take your outline of existing models and lay it down right next to the outline you made for new models. Looks through the list and identify the tools or processes that are in the new models list but not in the existing models list. Grab a third piece of paper. Label this sheet `To-dos` and write out the skills, tools, or extensions that you will need to learn to level up your SketchUp skills.

Make this list absolute statements. Use "I will" statements! This is your chance to tell yourself exactly what it is that you are going to do, no matter what. Your list should look something like this:

- I will learn to import images as full-size references.
- I will get better at using **Follow Me**.
- I will learn how Face Me components work.
- I will master custom component collections.
- I will find out how to use movie screenshots as a reference.
- I will find out whether I can use components to add detail to a spaceship.
- I will find a material list extension.
- I will get better at naming and organizing components.

I know this feels like a self-help book with positive affirmations, but stating what you will do helps to solidify this as a goal that you will accomplish, rather than an idea that you may or may not pursue.

Note that the *To-dos* list does include a few investigative tasks. In many cases, I know that I can do something with SketchUp, but do not know how (yet). In other cases, it is OK to set a goal of learning about these skills. In the end, you may find that you cannot do something the way you thought but may end up finding another solution!

After all this, we now know what skills we need to develop in order to level up. The next step is to figure out how to develop those skills

Make a plan to learn new skills

Remember your *New* list from the *Discovering what you will be modeling next* section? Pull that out and set it next to your *To-dos* list. Do all of the items on your *To-dos* list exist in one or more of your workflows? If not, is there a place that you could fit them in?

What we need to do now is develop workflows that you can start developing, which includes the skills you have identified in your *To-dos* list. If there are skills that you need to develop but they are not in any of your new workflows, you may need to add them.

Go through your workflow and mark the steps that connect to the items on your *To-dos* list. Highlight them, circle them, or underline them (I have shown these steps in *italics*). Identify these steps, as these are the skills that will help you to level up. These are the steps that you are going to need to focus on. If we take the example of a custom entourage, it may end up looking like this:

1. Import an image of a person.
2. *Scale the image to the proper size.*
3. Trace an outline of a figure.
4. Add details and color with **Paint Bucket**.
5. Make a component.
6. *Make the component a Face Me component.*
7. *Store the component in a custom collection of Face Me entourage.*

The idea at this point is to have an intention and a plan to start doing something that you do not currently do, or do not currently know how to do, on your list. Now that you know what you need to start doing to level up your game, you need to find a way to start doing it!

Implement your plan

Believe it or not, this is the hardest part of everything you have done in this chapter. In order to develop new skills, you have to push yourself outside of your current process. You need to come up with a plan to do something new on a regular basis in order to learn and develop new workflows.

Developing new skills really comes down to only two steps:

1. Learn to do something new.
2. Do it repeatedly.

That sounds simple, right? Unfortunately, anyone who has tried to learn to play an instrument by themselves, figure out how to paint a landscape, or teach themselves a foreign language knows that it is far from simple. Teaching yourself is a difficult process and requires dedication, focus, and time.

Time is often the toughest one. You are six chapters into this book, so I believe that you have the dedication you need to get better at SketchUp. You have made it most of the way through this chapter without any pictures or actual modeling, so that indicates a good level of focus. How do you find the time to level up your SketchUp skills?

In my own experience, time is rarely something that we can find. It is not lying around, ready to be used up for something new. Waiting to find the time you need to practice a new skill is a way to never learn that new skill. In order to learn your new workflows, you will need to be intentional and make the time.

Something that I have recommended to my students for years is to be very intentional about how you go about trying to practice something new. If you are modeling for fun, this may be as simple as putting your current project on hold and spending your fun modeling time trying something new. The trick, in this case, is to keep at it until you have developed the new workflows you have been thinking of into real processes that you can repeat and create 3D models with.

Finding time as a professional modeler

Since many of the people I have taught have been practicing professionals, they have had their plates full of paying work. I can tell you right now that a paying job is the last place you want to try to learn or implement new skills. Sure, something small like trying out a new tool that you rarely use could work, but a whole new workflow should not be used on a job that you are doing in order to pay the bills. The reason? You need to work on something where you can fail.

Developing a new workflow or learning to use a new tool is a process of experimentation. If you knew how to do this new thing, you would already have implemented it into your process! Since you do not know what you are doing, you need to attempt this new process in a place where you can fail, try again, and change how you go at it. Paying work is simply not the place for this to happen. Paying work should be set apart. Do your paying work the way that you have been. In fact, stick to doing it the quickest way you know so that you can finish ahead of schedule and have time left over to work on your new workflows!

So, if you are spending all of your work time modeling the way you used to, how do you ever develop new skills? This is the trick that will set master modelers apart from the rest. If you want to become a great 3D modeler, you will need to practice morning, noon, and night. No, I don't mean practice SketchUp all day long! I mean look and find time during the morning, noon, and night to get more SketchUp time under your belt! Here are a few things to try.

Morning

I saw a post from a graphic designer I follow on Instagram that suggested you should get up 30 minutes earlier every morning to practice a new skill. If you did this, for the next year, just on weekdays, that would add up to 130 hours of practice time. Imagine how good you could become at modeling organic shapes, for example, if you dedicated 130 hours to learning that!

Try this – for the next two weeks, get up 30 minutes earlier than you normally would. Grab yourself a cup of coffee, tea, or cocoa (whatever you like to drink in the morning) and work on modeling something that you would not model for work – something fun, such as a spaceship, a boat, or your dream house. That's it, though! You only get to work on it right when you wake up. When you finish your modeling time, save the model and close it, and do not work on it again until the next morning. If you picked a good model and are learning and getting better at SketchUp, you may find that it is easier to get up and get into that model. In fact, you may look forward to your morning modeling sessions!

My Morning Routine

The truth is that most of this book was written at 6:00 in the morning. I found that the first hour of the morning was a great time to focus and learn to write (this is the first book that I have written like this) and get something done that I could not fit elsewhere into my schedule. I thought that I did not have extra time in my schedule, but it turned out to be the most productive time of my whole day!

Noon

Most of us take some sort of break from work in the middle of the day. Many of us eat something during this break. I know that there are a lot of people who go to lunch with coworkers or friends, but not the 3D modeling elite! They pack themselves a sandwich or leftovers and work on modeling light fixtures from product catalogs!

In all seriousness, giving yourself time at midday to stop modeling for work and model something that you enjoy is not only a great way to develop some new SketchUp skills but also a way to keep yourself from burning out on work modeling. Keeping a fun, rewarding model in the hopper to work on lets you step away from the day-to-day routine and is a great way to keep SketchUp enjoyable.

Lunch Modeling at Trimble

The truth is that when I started working at Trimble as a SketchUp sales engineer, most of the work I did involve me working with other people's models or workflows. I did not really get to use SketchUp the way I wanted in my daily work. This was when I started fun modeling at lunch. I would pick something that I did not know how to model and commit 1 hour in the middle of the day to see how much I could model. It was a great way to push myself outside of my current skill set and keep SketchUp as something I could enjoy, rather than just a business tool.

Night

It can be very difficult to dedicate nighttime to extra 3D modeling. If you use SketchUp during the day, it can be hard to come home and do more modeling. If that works for you, that is great and will be time well spent. If you are feeling burnt out and do not want to jump back in, take this time to do some extra learning instead. Rather than watching *The Simpsons*, watch a couple of YouTube videos on modeling in SketchUp. Rather than reading a novel or magazine, crack open a book on SketchUp or LayOut. While getting your hands dirty in a 3D model will always be the best way to learn, there are opportunities when you can't actually be in SketchUp!

Weekends

Okay, I know this may be controversial but hear me out. The last thing I want you to do is spend so much time in SketchUp that you are sick of it, so stepping away from the computer on the weekend is probably a healthy practice. However, if you want to be an amazing SketchUp modeler and cannot find the time to develop new skills during the week, you may want to set aside a little bit of time at the weekend. I am not saying that you should be spending 48 hours in a 3D modeling marathon, but a few hours set aside on a weekend can be a good way to find some uninterrupted modeling time.

Get it done

Alright, now you have the workflows and new skills that you need to practice. You have time set aside to make it happen. The only thing left is to do it! No matter what you are trying, there always comes that point where you just have to jump in and do it. Your plan will help you to develop the new skills or workflows that you have identified. Depending on your existing ability and the complexity of the new workflow, you will reach a point where you are ready to implement these new skills into your standard workflow.

At what point do you move from practicing to production? That is completely up to you. At some point, you will feel comfortable with what you are doing. You may not have the new workflow 100% down, have a few steps that are still a little clunky, or feel like you have a bit more polishing to do on a modeling process, but you can implement it when you think the time is right.

Before we wrap up this chapter, it is important that you see this process as a part of a self-improvement cycle. Try not to think of this as a one-time process that ends with you adding a new tool to your modeling toolbox but instead a consistent stream of skill development that will help you to become a better modeler each day. Improving your SketchUp abilities is not something you do once and then call it quits; instead, it is a mindset that has you always ready to add new skills, find new workflows, and regularly find a way to level up.

Summary

Hopefully, this chapter was helpful for you. It was a bit different from the previous chapters without any images and lacking hands-on practice. I do believe, though, that this may be one of the most important chapters as far as importance in your SketchUp growth. You now have an inventory of the SketchUp skills and tools you currently employ. You have an idea of the workflows that you want to develop. Most importantly, you have developed a plan to develop and implement those workflows.

In *Chapter 7, Creating Custom Shortcuts*, we will take a look at stock shortcuts and decide how to best customize your shortcuts to your current and future SketchUp workflows.

7
Creating Custom Shortcuts

A question that most SketchUp modelers with a desire to level up their skill will ask is, "How do I model faster?" We have covered a few concepts that can speed up modeling, but a definite solution is by mastering the use of shortcuts. Shortcuts can save you time and energy if you learn how to set them up and get committed to using them.

In this chapter, we will be doing the following:

- Discovering shortcuts
- Learning how to create custom shortcuts
- Identifying which commands need shortcuts
- Figuring out what keys to use for shortcuts

Technical requirements

In this chapter, you will just need access to SketchUp Pro.

Discovering shortcuts

People like toolbars. They like to set up their programs, especially 3D modeling programs, with all of their toolbars out on display.

It makes it easier to click on a button if the toolbar is visible at all times. It also looks nice! It is usually colorful and reminds you of all the power you have available to you at the tip of your finger.

But it is also a waste of space and an inefficient way to call up commands.

If you want to be a master modeler, you need to be able to jump from one command to the next, taking only a fraction of a second to change commands. Relying on finding a button for a command on a toolbar takes too long for a SketchUp ninja, and adding a bunch of toolbars to the screen shrinks your usable modeling area. Look at this SketchUp workspace:

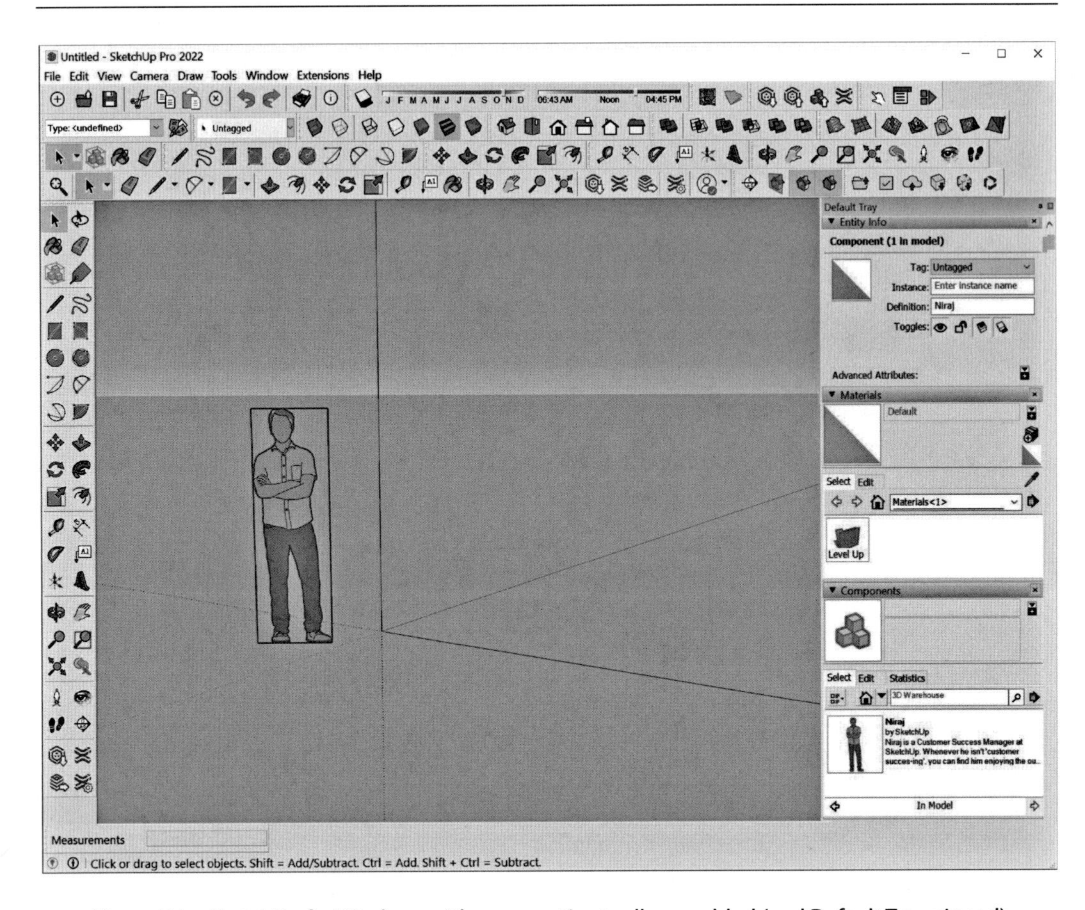

Figure 7.1 – SketchUp for Windows with every native toolbar enabled (and Default Tray pinned)

Not only does it eat up my usable space in SketchUp – it also creates a quilt of icons that are distracting (especially if you like pretty colors as I do), and with so many visible at one time, it is hard to find the one for the command you need. Now, compare that to this workspace:

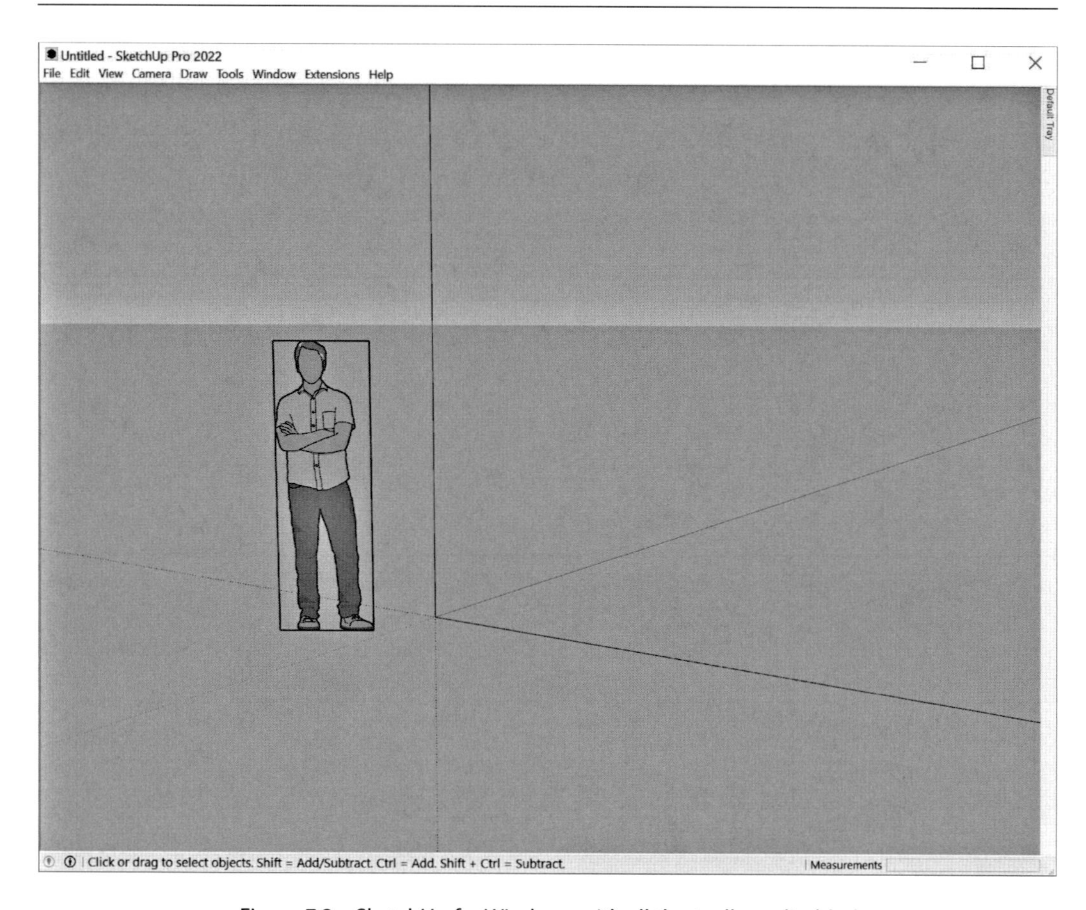

Figure 7.2 – SketchUp for Windows with all the toolbars disabled

Look how the whole screen is available for modeling rather than full of icons! Some of you reading this are shifting uncomfortably in your seat right now. "Without those buttons, how will I use SketchUp? Do I need to go into the menus for everything?" You have, of course, already figured out that the answer to that question is, "No, you will learn to use **shortcuts**!"

When we say shortcuts, we are referring to keyboard shortcuts. These are amazing time-saving macros that allow us to call up a command from anywhere in SketchUp by pressing a single key. Rather than hunting through a sea of toolbars or pulling down menu after menu to find your command, you can simply tap a key to get into a drawing command or trigger a tool.

"Wait, a single key? There are only 40-something keys on my keyboard! I use more commands than that!" Yes, there are more commands in SketchUp than there are keys on the keyboard, and adding in extensions, you are looking at even more. Fortunately, you can use the same keys you use to modify SketchUp commands to multiply the number of shortcut keys you use.

This means that a single key (let's use the *Z* key as an example) can trigger more than one command. Pressing the *Z* key can trigger one command, while holding down *Shift* and tapping *Z* can trigger a totally different command. Plus, you can stack modifier keys, meaning that *Control* + *Shift* + *Z* can trigger a third command.

Let's do some quick math on the number of shortcut keys we have available to us. On my keyboard, I have 26 letter keys that I can use for shortcuts. I am on a Mac, so I have three modifier keys available to me (*Shift*, *Control*, and *Option*; the *Command* key cannot be used in shortcuts). If I was to look at all the combinations I can make with the *Z* key, I would get the following:

- *Z*
- *Shift* + *Z*
- *Control* + *Z*
- *Option* + *Z*
- *Control* + *Shift* + *Z*
- *Option* + *Shift* + *Z*
- *Control* + *Option* + *Z*
- *Control* + *Option* + *Shift* + *Z*

That is eight possible shortcuts assigned to a single key. That is a total of 208 shortcuts, just using the alphabet keys! Additionally, if that was not enough, there are another seven keys on the keyboard that are available as well (*/ ; ' [ ] \ `*) and could be used. With modifiers, that is another 56 shortcuts, giving us a grand total of 264 potential shortcuts.

Now, do I recommend assigning every single command to a shortcut? No, I do not. In fact, I would recommend against using triple modifier keys in shortcuts. There is no point in requiring hand yoga to activate a command, as these are supposed to be time-saving steps after all. The main point here is that there is more than enough room on your keyboard for all the important commands you use in SketchUp.

We will talk more about using those 264 shortcuts later when we discuss creating custom shortcuts. For now, let's take a look at the shortcuts that are mapped by default in SketchUp.

Discovering the defaults

When you first install SketchUp, on Windows or Mac, you will have a set of default shortcuts. There is a set of system commands that use the same default shortcuts used by other software on your operating system. These commands are hardcoded and cannot be changed (we will learn that many other commands can be customized when we talk about customizing shortcuts in the *Learning how to create custom shortcuts* section):

Hardcoded shortcuts		
Command	**Default Windows shortcut**	**Default Mac shortcut**
Help	F1	-
Context Help	Shift + F1	-
New	Control + N	Command + N
Open...	Control + O	Command + O
Save	Control + S	Command + S
Print...	Control + P	Command + P
Copy	Control + C	Command + C
Cut	Control + X	Command + X
Paste	Control + P	Command + P
Undo	Control + Z	Command + Z
Redo	Control + Y	Command + Shift + Z

Figure 7.3 – Hardcoded system shortcuts

The nice thing about these system shortcuts is that there is a good chance that you already know them or use them. While they are not specific to SketchUp or drawing commands or tools, using these keys instead of hunting for an icon or menu item will save you time and energy as you model.

Up next are the SketchUp-specific shortcuts. These shortcut keys relate specifically to SketchUp commands and can be modified or reassigned (which will be discussed in the *Learning how to create custom shortcuts* section) as needed. Many of these shortcuts are mapped to keys that make them easy to learn (*C* for **Circle**, *M* for **Move**, *Z* for **Zoom**), while others are assigned due to the ease of access to a specific key (*Spacebar* for **Select**). While not every command gets a shortcut key (in fact, there are only 29 commands on this list), many of the most commonly used commands receive default shortcuts:

SketchUp shortcuts		
Command	**Default Windows shortcut**	**Default Mac shortcut**
Select	Spacebar	Spacebar
Lasso Select	Shift + Spacebar	Shift + Spacebar
Select All	Control + A	Command + A
Deselect All	Control + T	Command + Shift + A
Invert Selection	Control + Shift + I	Command + Shift + I
Orbit	O	O
Zoom	Z	Z
Zoom Window	Control + Shift + W	-

Zoom Extents	Shift + Z	Shift + Z
Pan	H	H
Back Edges	K	K
Line	L	L
Eraser	E	E
2 Point Arc	A	A
Circle	C	C
Rectangle	R	R
Make Component...	G	G
Move	M	M
Push/Pull	P	P
Offset	O	O
Rotate	Q	Q
Scale	S	S
Paint Bucket	P	P
Tape Measure	T	T
Next Scene	Page Down	Page Down (Fn + Up Arrow)
Previous Scene	Page Up	Page Up (Fn + Down Arrow)
Image Igloo	I	-
Search	Shift + S	Shift + S

Figure 7.4 – The default SketchUp shortcut keys

If you have, up to this point, been using toolbar buttons or grabbing commands from the menus, then learning these standard shortcuts will be time-saving, for sure. If you already know about and use these default shortcuts and are looking for a way to speed up your modeling process even more, then read on and learn about setting custom shortcuts!

Learning how to create custom shortcuts

In this section, we will take a look at the process of connecting a command to a key (or keys). The good news is this is the easiest part of the process! Shortcuts have their own tab in the **Preferences** window. For this example, let's create a shortcut together for **Follow Me**:

1. Open the **Preferences** window (the **Windows** menu on Windows or the **SketchUp** menu on Mac).
2. Click the **Shortcuts** tab. This will display a list of every command that can have a shortcut applied to it:

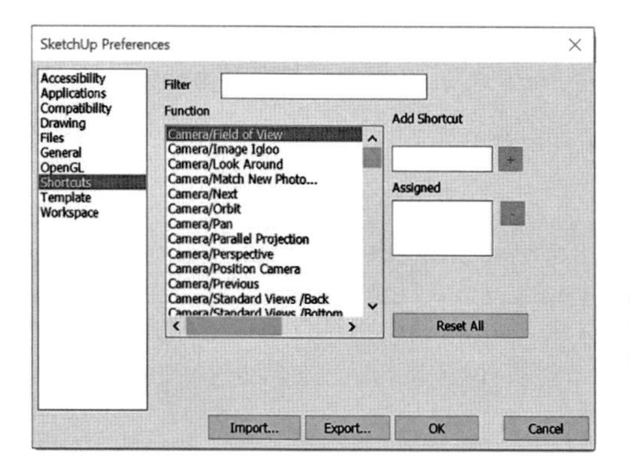

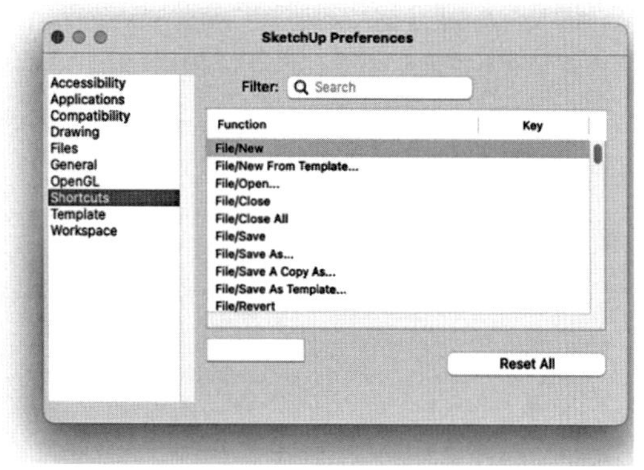

Figure 7.5 – The Windows shortcuts (left) and the macOS shortcuts (right)

As you explore this list, you may find that, while you can apply shortcuts to most commands, there are some functions in SketchUp that do not show up here. These may be fields in the tab bar or context-sensitive commands, but there are a few functions that cannot have shortcuts assigned to them.

> **Context-Sensitive Commands**
>
> Before SketchUp Pro 2020, the list of commands shown in the **Shortcuts** tab was context-sensitive, meaning that you would only see commands that could be run in the current model in the current state. Since that version, all commands are available in the list.

Seeing that this is a big list of commands, you may want to narrow it down using the **Search** command.

3. Type `follow` into the search bar. Since there is only one command in SketchUp that contains the word `follow`, only one command shows up in the list. Let's assign a shortcut to this command.
4. Select **Tools/Follow Me** from the list.
5. Click on the shortcut key field:

   I. On Windows, this is the field directly to the right of the list.

   II. On Mac, this is the field below the list.

6. Hold down the *Shift* key and press *F*.

   Note that, in the list, the *Shift* + *F* shortcut showed up to the right of **Follow Me**. At this point, you have successfully assigned a shortcut for that command! Any time you are in SketchUp and type *Shift* + *F*, it will be the same as if you had selected **Follow Me** from the **Tools** menu.

The list of shortcuts includes all commands and shows which ones have shortcuts assigned to them. If you have never edited your shortcuts, then you will be able to see only the default shortcuts.

No Changing Hardcoded Shortcuts

What you will not see in the list are the hardcoded or "system" shortcuts, as mentioned in the *Discovering the defaults* section. These commands, often reserved at an operating system level, cannot be edited, and thus are not available through the **Shortcuts** tab. There are, however, commands that can have both hardcoded and custom shortcuts applied. If you don't like using *Control* + *S* to save and you want to use the *D* key instead, you can create that custom shortcut, which will function alongside the hardcoded shortcut.

The **Shortcuts** tab in the **Preferences** window will show you every shortcut you have set and allow you to modify them as needed. If you upgrade from your current version of SketchUp to a new version, your shortcuts will automatically show up in your new version as long as you are upgrading on the same computer. How much work it is to transfer your shortcuts depends on what operating system you are using. If you are using Windows, the process is pretty easy:

1. On the Windows computer that you want to copy shortcuts from, click the **Export** button in the **Shortcuts** tab of the **Preferences** window.
2. Enter the name and location of the file you want to export, then click **Export**. The default file is named `Preferences.dat`, which is saved to your desktop. If you are going to transfer this file via a USB drive or network folder, you can navigate there to save yourself the bother of moving the file later.
3. On the Windows computer that you are moving shortcuts to, click the **Import** button in the **Shortcuts** tab of the **Preferences** window.
4. Navigate to the file saved in *Step 2*, select it, and click **Import**.

With that, both computers should have the same shortcuts. If you want to move your shortcuts from one Mac computer to another, you will need to manually transfer the file. The `SharedPreferences.json` file can be copied from one Mac computer to another to move all your shortcuts (and a few other presets).

On Mac, find your presets at this path: `Users/Username/Library/Application Support/SketchUp####/SketchUp/SharedPreferences.json`.

Folder Names

Note that the `Username` folder is the name of the user account that you are using; most likely, this is the account that you log into your computer with. The `SketchUp####` folder is the version that you are copying to or from (SketchUp version 2022 is in the folder named `SketchUp2022`). SketchUp folders will always contain the name **SketchUp** followed by the year of release.

At this point, you should have a good idea of what shortcuts are, what the default shortcuts are, and how to create your own shortcuts. The question now is, what commands should you set shortcuts for? Read on because that is exactly what we will be covering in the next section!

Identifying which commands need shortcuts

Obviously, you can set shortcuts for whatever commands you want. In fact, you could probably create a shortcut for every single command you use. This might now, however, be the ideal way to spend your time in SketchUp and probably would be difficult to remember. Shortcuts that you cannot remember will not get used and not end up helping you to be better at SketchUp.

What you need to do at this point is identify the commands that you use most often and make sure that they have shortcuts assigned to them. After that, we need to look at the commands that you plan to learn and hope to use more often and get some shortcuts applied to them.

Assuming that you are reading this book chronologically and not skipping anything, you should have a pretty good idea of what those commands are from *Chapter 6, Knowing What You Need Out of SketchUp*. If you skipped that chapter or were just so excited about shortcuts that you jumped right to this chapter, you should probably go back and read it (my editor, Rashi, told me it was really well written).

In *Chapter 6, Knowing What You Need Out of SketchUp*, we went through and listed all of the commands you are currently using and labeled the list *Current*. In that list, you wrote down the commands that were most used in your models. If I were to pull the commands from the example list that I made, I would end up with the following list:

- **Rectangle**
- **Eraser**
- **Sandbox Tools**
- **Import**
- **Push/Pull**
- **Eraser (Smooth)**
- **Follow Me**
- **Circle**
- **Move (Array)**
- **3D Warehouse**
- **Paint Bucket**
- **Dimensions**

- **Arc**
- **Weld**
- **Line**
- **Polygon**
- **SubD**
- **Artisan**
- **Rotate**

The next step is to identify the commands that do not already have shortcuts applied to them. If I remove all of the commands that have default shortcuts, my list gets a lot shorter:

- **Sandbox Tools**
- **Import**
- **Follow Me**
- **Eraser (Smooth)**
- **Move (Array)**
- **3D Warehouse**
- **Dimensions**
- **Weld**
- **SubD**
- **Artisan**

Modifying Default Shortcuts

You may have a desire to modify existing shortcuts, and that is perfectly fine. We will be talking about optimizing shortcuts based on a key's location on a keyboard in the next section, *Figuring out what keys to use for shortcuts*

Looking at this list, there are three things listed – groups of commands, modified commands, and standard commands. Not everything listed here would benefit from a shortcut. Let's look at why these groups are different from one another, as it relates to assigning shortcuts.

Groups of commands

In my list, **Sandbox Tools**, **3D Warehouse**, **SubD**, and **Artisan** are not individual commands but suites of tools or commands. Shortcuts are applied to individual commands, so I cannot apply them to a group of related commands. What I want to do in this case is dive into the groups and see whether there is a specific command in the group that it would make sense to apply a shortcut to.

In **Sandbox Tools**, maybe I use the **From Scratch** command and **Smoove** the most often, so they would be candidates for shortcuts. **Artisan** and **SubD** are extensions, each of which has multiple commands. Maybe the command I use the most is **Toggle Subdivision** from **SubD**, so that may be the command that should get a shortcut. Finally, there are a handful of commands that relate to **3D Warehouse** in SketchUp. If I look back at my workflow from *Chapter 6*, *Knowing What You Need Out of SketchUp*, I can see that I was importing models from **3D Warehouse**. A great step to add to certain workflows, but defining a shortcut for this command will not likely make me a more efficient modeler.

Commands That Do Not Need Shortcuts

The main point of setting shortcuts for commands is that it reduces the amount of time you have to spend moving your mouse to a menu or toolbar to activate a command. This works great for commands that you can switch between while in the midst of geometry creation or manipulation. Certain commands break you out of the modeling flow, no matter what. Opening **3D Warehouse**, for example, does not benefit from a shortcut because, as soon as it is open, you need to stop modeling and use your mouse to navigate the warehouse.

Focusing on individual commands that are or can be a part of a modeling workflow will help reduce our list of potential shortcuts.

Modified commands

Way back in *Chapter 1*, *Reviewing the Basics*, we looked at how many different ways there are to use the **Move** tool. In *Chapter 3*, *Modifying Native Commands*, we saw that modifiers can be used on almost all SketchUp commands. Modifiers add a lot of additional functionality to commands and are essential in most workflows. Modified states of commands, however, cannot have separate shortcuts applied to them.

This means that **Move** can have a shortcut, but modifying **Move** to copy will not. A shortcut is responsible for activating a command initially, while modifying is something that happens while you are in the command. For this reason, two of the commands that are on our list (**Eraser** (Smooth)) and **Move** (Array)) are redundant, as **Move** and **Eraser** already have shortcuts applied to them.

Commands ready for shortcuts

When we refine our list, we end up with the following commands:

- **From Scratch**
- **Smoove**
- **Import**

- **Follow Me**
- **Dimensions**
- **Weld**
- **Toggle Subdivision**

This is a solid-looking list! I know it is not huge, but it is a start. Creating shortcuts is, for many, an ongoing process. Even now, after more than a decade of using SketchUp, I will occasionally add a shortcut to a command that I start using more or to a tool in an extension. This list is a list of tools that will save time and energy given the right shortcut keys.

Speaking of shortcut keys, choosing the right key as a shortcut for a command is an important step and something we will be exploring in the next section.

Figuring out what keys to use for shortcuts

We have a list of commands that we know we want shortcuts for, and we know how to apply shortcuts to commands. At this point, it seems like we should just hop in and set the shortcuts! Yes, that is an option, but I would recommend taking just a minute or two to think about what keys you actually want to use for these shortcuts.

In setting your shortcuts, there are basically two schools of thought – mnemonics or location. Let's look at how each works and what your keyboard might look like for each.

Assigning shortcuts based on mnemonics

Simply put, mnemonics is a system that helps you to remember. In the case of keyboard shortcuts for SketchUp, this means choosing keys that will help you to remember which key command they are connected to. This is the way that most of the default shortcuts are assigned.

In the default shortcuts, the *M* key is the shortcut for the **Move** command because it is easy to remember that the word *move* starts with M. If you look at the list of default shortcuts, you will see that most of them are letters in the actual command, with a few exceptions.

The exceptions, though, are mnemonic in their own way. The **Pan** command has *H* as the shortcut because the cursor for **Pan** is a hand, and *hand* starts with the letter *H*. The **Rotate** command could not use *R* because that was already assigned to **Rectangle**. The shortcut key for **Rotate** is *Q* because the letter Q looks like the compass that is used on screen to rotate geometry.

Mnemonics are a great way to learn shortcuts and were used when the default shortcuts were assigned. For most users, having a device that helps them to remember a few dozen shortcuts is a good thing. There are, however, downsides to this method.

Not enough keys

The first issue is that we can only assign one command to a key, so certain letters can get crowded. Let's consider the *S* key. Tapping the *S* key, by default, will bring up the **Scale** command because that is one of the most common commands that starts with the letter *S*. But what about the **Section** command? If we use that command a lot, we can assign that to *Shift* + *S*. Perfect! Now, what about the **Shadows** command? That can be *Option* + *S*! Then, we can use *Command* + *Shift* + *S* for **Send to LayOut**, then *Command* + *Option* + *S* for **Subtract**, and *Shift* + *Option* + *Command* + *S* for **Soften Edges**. Okay, now what about **Save As**?...

Yes, this is a bit of an extreme example, but remember that there are many more commands than there are keys on the keyboard, and we have seen how the modifier keys can help. But eventually, it ruins the mnemonic device, for I have to remember which *S* command I activate if I hold down *Option* and *Command* or *Option* and *Shift*.

Shortcuts everywhere

The second issue you can run into when using mnemonic keys is the location of the key on the keyboard. The main goal of using shortcuts is that it is quicker to tap a key on the keyboard than it is to use your mouse to find a command. If you end up having to hunt for the key associated with the command you need, it can reduce the benefit of the shortcut.

Now, odds are good that, even if you have to look across your keyboard for the key and you need to move your hand slightly to get to the key you need, it is still going to be quicker than clicking on a menu to start a command. But it can be a minor inconvenience to some, especially those who are not master typists.

To see what I mean, let's take a look at a keyboard with the default keys highlighted:

Figure 7.6 – The default shortcut keys on a keyboard

Additionally, some keys are easier to use with modifier keys than others. It is pretty easy to hold down the *Shift* key and tap *S* with your left hand, keeping your right hand on your mouse. Conversely, it may be a stretch, depending on the size of your hand and your keyboard, to hold down *Shift* and get to the *G* key.

Mnemonics are great and make shortcuts even more friendly and easy to remember, but let's look at another way to assign keys to your commands.

Assigning shortcuts based on location

The biggest benefit of using shortcuts is that you limit the amount of time you stop your modeling flow to change commands. No matter what key you use or how many modifier keys you add, a keyboard shortcut is almost always faster than finding a command in a menu and usually quicker than moving your mouse off the model to click on an icon.

One of the only default shortcut keys is **Select** being assigned to the *spacebar*. There is no mnemonic association between spaces and **Select**. The *spacebar* is simply the biggest key on the keyboard, and **Select** is the most commonly used command by far in SketchUp. Plus, the *spacebar* is centered on the keyboard, making it equally easy to get to for right-handed and left-handed modelers.

Some users have taken this concept even further by moving their shortcuts to one side of their keyboard, meaning they have to spend little to no time moving their keyboard hand to find a shortcut. If you are thinking about assigning shortcuts based on location, you need to first look at where you keep your hands when using SketchUp.

Looking at where your hands are

Most users have their dominant hand on their mouse and their non-dominant hand on the keyboard. Generally, the non-dominant hand resets on the associated side of the keyboard. A right-hand dominant user would most likely look like this:

Figure 7.7 – A dominant right hand on the mouse and a left hand on the left side of the keyboard

Conversely, a left-handed user would likely use SketchUp like this:

Figure 7.8 – A dominant left hand on the mouse and a right hand on the right side of the keyboard

In some cases, users may prefer to slide their non-dominant hand to the opposite side of the keyboard. This might happen, for example, if you are right-handed and using a keyboard with a 10-key pad on the left side. To have quick access to the ten-key pad for inputting measurements, you may choose to slide your left hand to the right side of the keyboard:

Figure 7.9 – A dominant right hand on the mouse and a left hand on the right side of the keyboard

One more setup to consider is those who use a touchpad. In this case, your dominant hand is likely to be on the touchpad while your non-dominant hand will be on the keyboard:

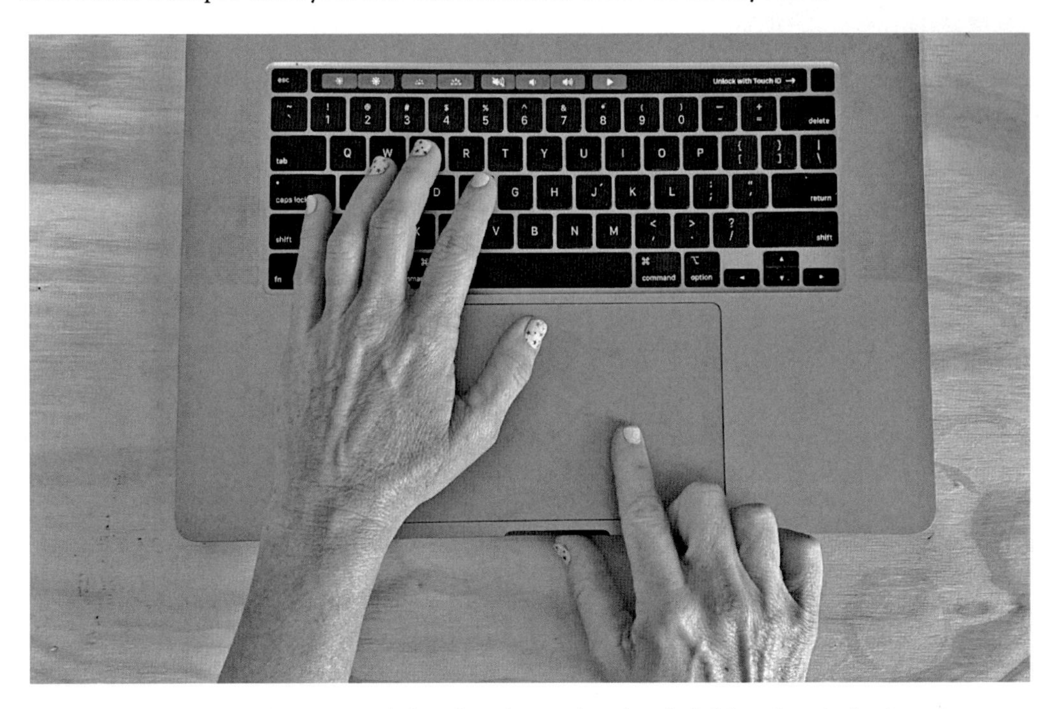

Figure 7.10 – A dominant right hand on the touchpad and a left hand on the keyboard

> **3D Mice, Tablets, and More**
>
> Yes, there are additional input devices besides the keyboard and mouse, and we will talk about how to add them to your setup in *Chapter 10, Hardware to Make You a More Efficient Modeler*. These additional setups can affect how you set up shortcuts, but we will touch on that later.

None of these setups are any better than another, but it is important to know where you prefer to keep your hands while using SketchUp before you start assigning shortcuts.

Picking keys for shortcuts

Once you know where your hands rest while using SketchUp, you can consider where you want to place your shortcut keys. For this example, let's assume you are right-handed, with your left hand resting on the left side of your keyboard.

With your non-mouse hand on the left side of the keyboard, you may want to consider keys that fall on the left side for your most common shortcuts. Ideal keys for this setup may be these keys:

Figure 7.11 – Ideal keys for the left side of the keyboard

This gives us 11 keys for shortcuts. As we add modifier keys and combinations, this ends up being 88 shortcuts. If you have larger hands, you can even add the *T*, *G*, and *V* keys into the mix; with modifiers, that could be an extra 32 shortcuts!

The downside of clustering shortcut keys

If you are the type that looks for every optimization that you can possibly take advantage of, locating your shortcuts together like this may sound great, but know that, just as with mnemonics, there is a downside to this method as well.

The first issue is that it may not be super-easy to remember your shortcuts. Having to remember that *Shift* + *D* is **Move** may be more difficult than remembering *M*. Using non-mnemonic shortcuts may even require you to write down your shortcuts so that you can refer to them until you have them firmly implanted in your memory.

The second downside will only be an issue for some people. Having a non-standard shortcut setup like this will make it more difficult for others to use your installation of SketchUp, or for you to teach others to use it. If you are a heads-down production modeler, this will likely not be an issue for you.

Which is the best?

The best solution is the one that works for you! Like all things, you may have to try and fail before you find the perfect solution. For many people, like myself, a combination of these methods is best.

I personally still have all of the defaults in place. I have added a few mnemonic shortcuts added for extensions that I use, and a few shortcuts assigned to empty keys on the left side of the keyboard because I am right-handed.

To find your perfect setup, assign a few shortcuts and try them out. If you are struggling, go in and change them. Remember that you can assign and reassign shortcuts as many times as you need. The important part is that, in the end, you have a set of tools that help you to become a better SketchUp user.

Summary

Shortcuts are amazing time-saving tools that you can customize in any way that you need. The important thing to remember is not to attempt to get every single command assigned to a key (the commands that you don't use often will still be safe and sound in their menu or toolbar, ready for occasional use), and be considerate about what keys you use for your shortcuts. Getting a good set of shortcuts can knock a lot of time off of your modeling time and help you to level up your SketchUp skills!

In *Chapter 8, Customizing Your User Interface*, we will look at how you have SketchUp set up and see whether there are any efficiencies to be gained by rearranging what tools you have on screen.

8
Customizing Your User Interface

User Interface (**UI**) is a software development term that means *all the buttons and menus and stuff on the screen that allow you to interact with a computer program*. As with most software, you can customize some of the UI in **SketchUp** to suit your modeling workflows.

In this chapter, we will look at what can (and cannot) be customized and try to figure out what changes to the UI will be most beneficial in helping you to level up your SketchUp skills. This means diving deep and learning exactly what UI elements are available to you in SketchUp as well as what elements can and cannot be customized. Specifically, we will be covering the following:

- Reviewing screen layout
- Deciding how many toolbars are needed
- Setting up trays and dialogs
- Organizing a custom UI
- Creating a next-level workspace

Let's start by looking at the different elements and seeing how they can be customized. Once we have a firm understanding of what we can use, we'll figure out how to set it all up to make a custom workspace specifically designed for your workflows.

Technical requirements

In this chapter, you will just need access to SketchUp Pro.

Reviewing screen layout

Before diving into the fun stuff (the actual UI elements: toolbars, trays, icons, and more), we need to take a look at what you are working with and figure out where to ideally place your UI. One of the nice things about SketchUp is that it can run with fairly minimal screen real estate. Many users run SketchUp all day on a single laptop screen.

At the same time, other users prefer to spread SketchUp across multiple displays. Some users, like myself, end up using SketchUp in different configurations, depending on the day. The first step in coming up with the perfect workspace starts with taking inventory of your hardware.

Operating system

The first thing to consider when thinking about your workspace is how SketchUp works and can be customized based on the operating system you are running. Of course, you have two options (Windows or macOS), and you should, at this point, know which you are using.

As we continue through this chapter, some sections will be split into two sections, the first showing Windows and the second showing macOS. If you are dedicated to one and only one operating system, feel free to skip the portions that do not apply to you. If you are a rare user with access to both operating systems (or if you really like reading), then please feel free to read everything!

> **So...Which Is Better?**
>
> It is a rare occasion where Windows and macOS are listed side by side and the question of preference is not raised. Does SketchUp run smoother on one operating system? What operating system does the development team use? Which operating system is best? Most users start using SketchUp on the operating system that they are already using. Occasionally, when it is time for a new computer, a user may switch operating systems to take advantage of SketchUp on Windows or macOS. This chapter will run through all of the UI options of both operating systems, so you can decide.

Monitor(s)

When fine-tuning your UI, you have to be conscious of the space that you are working on. If the icons and windows in SketchUp are the tools you work with, then your monitor is your workbench. One monitor is essential (obviously). Whether this is a laptop screen, a 19" LCD, or some huge curved 8K mini-movie screen, you need to have at least one monitor to run SketchUp.

Many users take advantage of multiple monitors to get even more out of their UI. While SketchUp was not necessarily designed for multiple-monitor use, there are advantages to adding an extra screen to your setup.

When running SketchUp, the main modeling window must be on one screen. Whether you have two, three, or six monitors connected to your computer, the modeling window can get as big as one display will allow. Additional monitors, however, can be used to store additional UI or for displaying reference materials.

Using additional displays with Windows

If you are running SketchUp on Windows, you can take advantage of the docking toolbars and trays within the main modeling window (more on that in the *Deciding how many toolbars are needed* and the *Setting up trays and dialogs* sections of this chapter); additional monitors can be used to store floating trays and toolbars. You cannot snap them to the side of a secondary monitor, but you can move them out of your primary display, giving you more room to work on your model.

Using additional displays with macOS

Unlike Windows, SketchUp running on a Mac will not allow you to dock toolbars or dialogs. Whether you are running a single-display setup or have multiple additional monitors, your UI will be floating over the top of your modeling window. Moving it to a secondary monitor will just get those floating tools off your modeling screen, making more of it visible.

Why Do I Need a UI?

I know, we just spent the previous chapter talking about how shortcuts should be used in place of toolbars because it is quicker to tap a key than to hunt down an icon on a toolbar. The fact is, even if you get all of your commonly used commands tied to shortcut keys, there will still be some that you don't use often enough to justify a shortcut. Plus, trays and dialogs need to be used to display information about selected entities or let you set properties of items as you model. Shortcuts are great, but do not eliminate the need for all UI.

Using additional displays for reference

Putting aside the potential need to see other programs while in SketchUp (it's okay to have your messenger and music app up while you model, right?), you may want to use additional displays to show you some information about what you are modeling. SketchUp will allow you to import reference images or import images as watermarks so you can see them, but there are plenty of times when you may want a dimensioned drawing of a plot of land or a PDF of a part you are modeling available to look at.

When I start modeling something from reference information, I will try to import data first and reference that. There are, however, plenty of times that I am modeling from a series of photos or a plan with lots of little, tiny dimensions. In these cases, I will open the images or PDF files on my second monitor so I can get the information I need when I need it. Yes, in some cases it is worth it to import an image into SketchUp and lay it on the ground next to my modeling space, but then I have to zoom out or pan over to the image each time I need to reference it. With the image sitting on my second monitor, all I have to do is turn my head a few degrees to the right and I can see everything that I need.

The fact is, if you have a decent-sized second display, it is pretty easy to get everything (toolbars, trays/dialogs, reference imagery, and even your music player) up on your secondary display, thus reserving your entire primary display just for your modeling screen!

Let's take a look at two different ways that I have my displays configured. I do most of my work on a MacBook. My main setup uses just the display of the MacBook for everything. My primary way of using SketchUp looks like this:

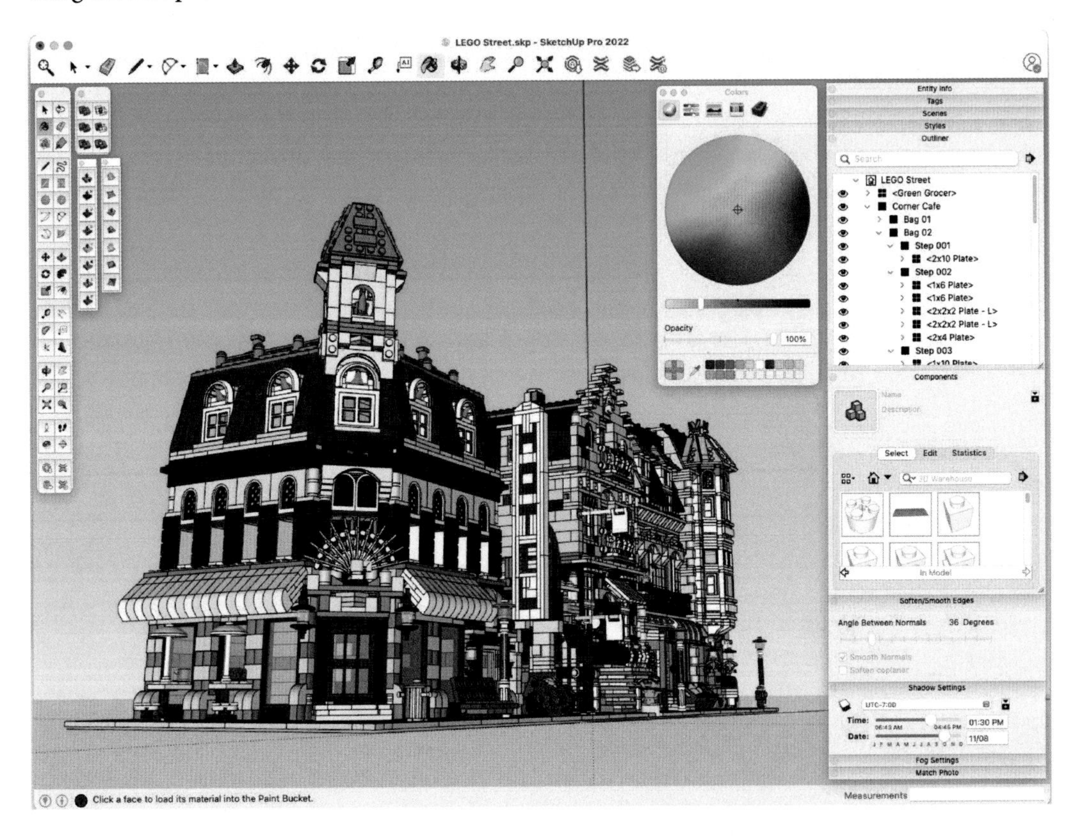

Figure 8.1 – Everything needed to run SketchUp on one display

This setup works and would be totally usable if I were traveling or did not have access to a second monitor, but it is not ideal. Running a single display like this, I can only have a few dialogs open at any given time and I lose way too much screen space to the **Colors** window and the floating toolbars. Plus, if I wanted to open a reference image, I would have to switch back and forth between the image and SketchUp as I model.

Alternatively, I could set up two displays, my MacBook display and a secondary monitor. I would have something like this on my MacBook display:

Figure 8.2 – Primary display with just the modeling screen

The primary monitor would be reserved for modeling. The secondary (or any additional monitor) would be used to display additional content and the UI. Here is the secondary monitor display:

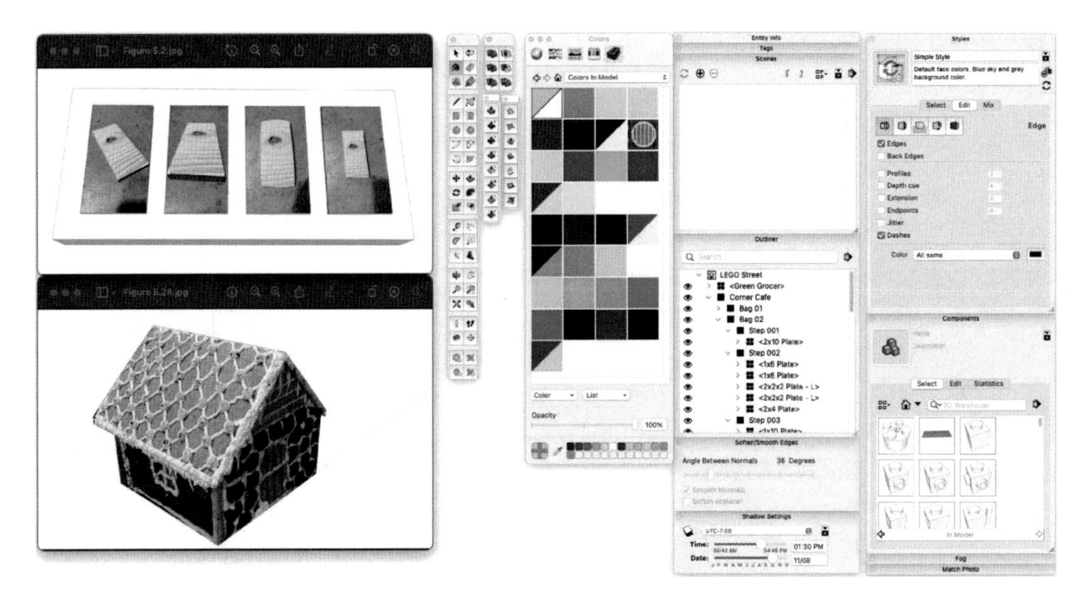

Figure 8.3 – Secondary monitor with reference imagery, dialogs, and toolbars

With these two displays, I can have all of my dialogs open and expanded, plus have plenty of space for other applications. I can display reference images or have up my messaging app and music player without sacrificing any of my modeling screen or having to switch between apps.

This setup will appeal to some more than others. There are those who want to see everything open at once, while others prefer to only display what they need when they need it. Neither of these scenarios is the right one, and it really comes down to personal preference. What you need to take away from this section is that you have options to set up SketchUp to work across multiple displays or on a single display.

Now that we have discussed how we can use displays to present our UI, let's dive into what UI we actually need to have visible.

Deciding how many toolbars are needed

I love **toolbars**. I love **icons**. I think that having a single **button** to launch a command is great UI. Plus, a well-designed icon stands on its own and tells you what it does, regardless of what language you speak. Does this mean that I turn on every single toolbar when I run SketchUp? It does not. After reading *Chapter 7, Creating Custom Shortcuts*, do I keep all toolbars turned off and rely solely on keyboard shortcuts? No, this is not ideal either. I believe that the ideal use of toolbars lies somewhere between these two setups.

In order to level up your SketchUp skills, you do have to rely on shortcuts as much as possible. Tapping a key to bring up a command will always be faster than clicking a button on a toolbar. There are commands, however, that may not be everyday commands and so do not get a shortcut. You may have commands that you only use in every fourth model, so learning shortcuts for them will be a challenge because you don't get a chance to practice using them. For these sorts of commands, it is a good idea to lean on toolbars.

Before we decide which toolbars are the ones you need, let's take a look at how toolbars work. Because SketchUp relies on the operating system to display and control the toolbars, this is different for Windows and macOS.

Using Windows toolbars

To edit your toolbars in SketchUp for Windows, click on the **View** menu, then **Toolbars...**. This will bring up the **Toolbars** window. From here, you simply have to check the box next to a toolbar to enable that toolbar.

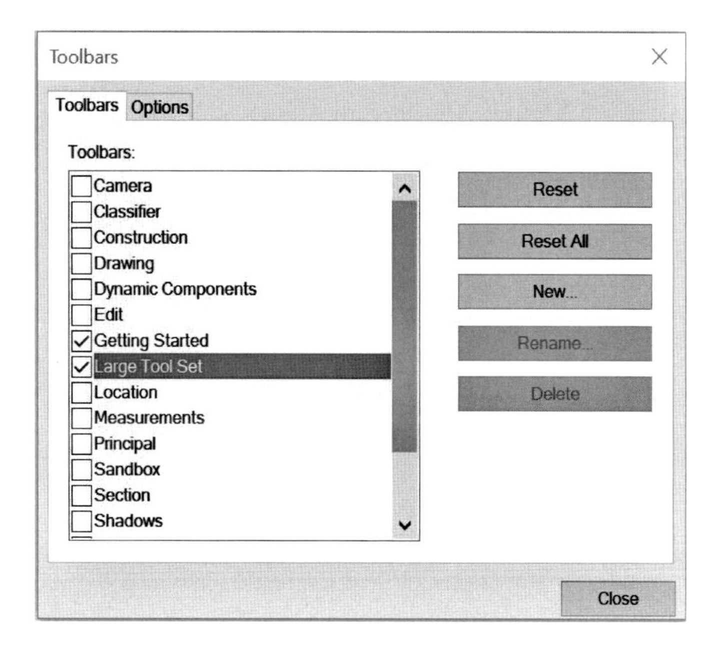

Figure 8.4 – Toolbars window in SketchUp for Windows

By default, the **Toolbars** window shows all of the toolbars that can be enabled in SketchUp. If you have extensions installed, and those extensions have toolbars, they can be turned on or off here as well (more on that in the *Organizing a custom UI* section). When you first turn on a toolbar, it will be displayed at the last location it was used (if you have never used a toolbar before, it will show up at its default location). The previous or default locations are not always very useful, but you do have the ability to move toolbars anywhere you want.

In SketchUp for Windows, you can click and drag any toolbar anywhere you want. Dragging a toolbar toward the top, bottom, left, or right edge of your modeling screen will automatically dock the toolbar to that edge. Dragging a toolbar and dropping it anywhere away from an edge will create a floating toolbar that can be moved anywhere on any of your displays. It is important to note that you can only move toolbars if the **Toolbars** window is closed.

A Single Exception

OK, I lied, just a little! Every toolbar can be docked to any side of the drawing screen, except one. The **Large Tool Set** toolbar will only dock to the left or right side. Due to its unique layout (two buttons wide), it will not automatically reorient to a horizontal toolbar.

Customizing toolbars on Windows

The default toolbars are logical groupings of commands based on their function or the menu in which they reside. That works for most people, but what if you don't want to show all of the commands from a certain menu? What if you want access to buttons for the **Tape Measure** and **3D Text** tools, but don't want to see the rest of the commands on the **Construction** toolbar?

Fortunately, SketchUp for Windows makes customizing your toolbars extremely easy. Let's create a custom toolbar. Let's say that I want to create a toolbar that has my most commonly used commands. These commands are as follows:

- **Line**
- **Eraser**
- **Rectangle**
- **Push/Pull**
- **Follow Me**
- **Make Group**
- **Zoom Extents**

These commands can all be found on other toolbars, but by placing them on one custom toolbar, I can save space on my display. Plus, I can change the order of the buttons on my custom toolbar to mimic the order of operations when I model. Let's create this toolbar:

1. Click on the **View | Toolbars...** options.
2. Click on the **New...** button.
3. Type `My Toolbar`, then click **OK**.

This will create a brand-new empty toolbar. Now we need to add some buttons. This is done by dragging buttons off other toolbars onto our custom toolbar. To do that, we need to have a toolbar to drag buttons from. Many of the commands we will be placing in our custom toolbar are on **Large Tool Set**, while **Make Group** is found on the **Principal** toolset.

4. Turn on **Large Tool Set** and the **Principal** toolbar.

   All that is left to do is drag the buttons we need into our custom toolbar. This is as simple as clicking and dragging the buttons you need. This function only works while the **Toolbars** window is open.

5. Drag the **Line** icon (the red pencil) from **Large Tool Set** to our custom toolbar and drop it.
6. Do the same for **Eraser**, **Rectangle**, **Push/Pull**, **Follow Me**, and **Zoom Extents**.
7. Drag **Make Group** from the **Principal** toolbar.

   With that, we have a custom toolbar! Now, let's take advantage of toolbar customizability and rearrange the buttons in the following order:

   I. **Make Group**
   II. **Line**
   III. **Eraser**
   IV. **Rectangle**
   V. **Push/Pull**
   VI. **Follow Me**
   VII. **Zoom Extents**

8. With the **Toolbars** window still open, click and drag the buttons on the custom toolbar until it looks like this:

Figure 8.5 – Your custom toolbar

You can see the advantage of creating custom toolbars, toolbars that only have the tools you need. With the right custom toolbars, you could turn off the stock toolbars and have the commands that you need on the screen and nothing more.

What if we want our custom toolbars and we want access to the stock toolbars? If you look at the stock toolbars (**Large Tool Set** and **Principal**), you can see that they are missing the buttons that we put onto the custom toolbar. Let's get those buttons back before we wrap up this process.

9. In the **Toolbars** window, select **Large Tool Set** from the list, then click the **Reset** button and click **Yes**.

   All of the buttons are back on the **Large Tool Set** toolbar! This works great for resetting a single toolbar at a time, but there is a way to set all toolbars back to the default.

10. Click on the **Reset All** button, then click **Yes** to put all buttons back on all stock toolbars.

    Notice that it will put back the stock buttons but not mess with your custom toolbar.

    Finally, since we only turned on the **Principal** toolbar to get access to the **Make Group** button, let's go ahead and turn that off, then exit the **Toolbars** window.

11. Un-check the **Principal** toolbar.
12. Click on the **Close** button to close the **Toolbars** window.

> **Shortcut to the Toolbars Window**
>
> Next time, rather than pulling down the **View** menu to get to the **Toolbars** window, right-click on any toolbar. This will display a flyout menu listing all the toolbars. At the very bottom is a **Customize...** option, which will open the **Toolbars** window.

With that, we have turned on some extra toolbars and created our own custom toolbar in SketchUp for Windows. Now, for macOS users, let's create something similar.

Using macOS toolbars

Unlike SketchUp for Windows, SketchUp for macOS has very few stock toolbars. In fact, it really only has two: the toolbar at the top of your drawing screen and **Large Tool Set**. The rest of the toolbars that you can turn on are a result of extensions (toolbars such as **Solid Tools** or **Sandbox** tools are native extensions).

To turn on **Large Tool Set** in SketchUp for macOS, click on the **View** menu, then hover over **Tool Palettes**. This will bring up a flyout menu listing all of the toolbars you have available to you. The toolbars that are currently active have a checkmark next to them. Turn a toolbar on or off by selecting it from the list.

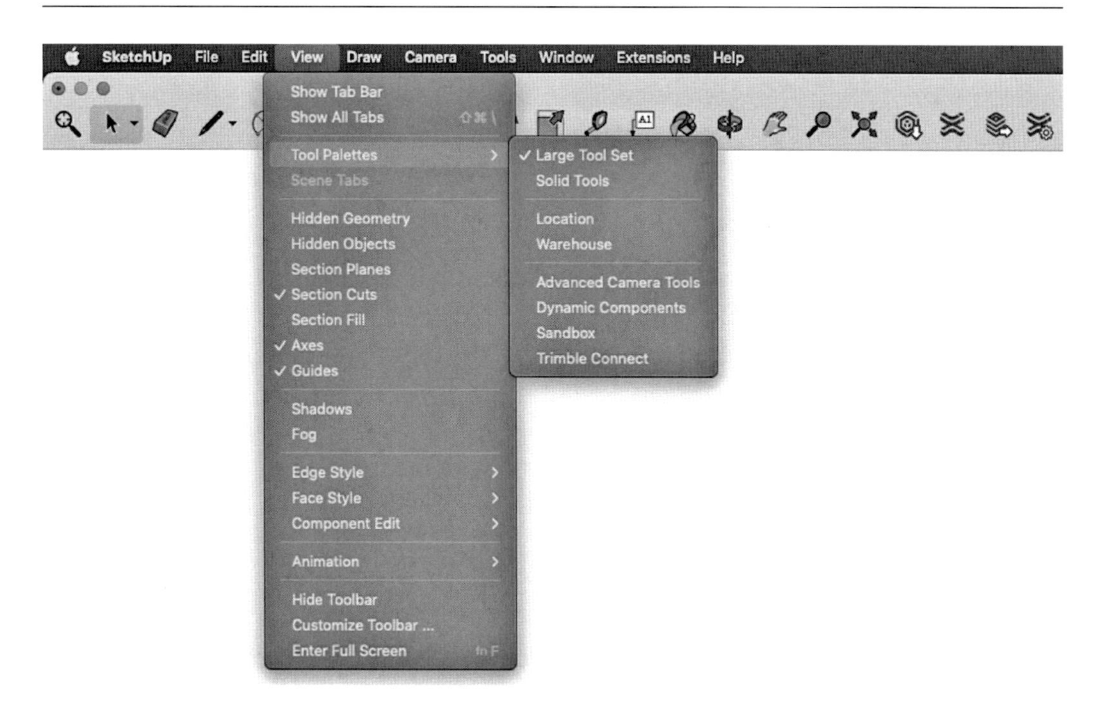

Figure 8.6 – Tool Palettes flyout in SketchUp for macOS

This list will show all of the toolbars that can be enabled in SketchUp. If you have extensions installed, and those extensions have toolbars, they can be turned on or off here as well (more on that in the *Organizing a custom UI* section). When you first turn on a toolbar, it will be displayed at the last location it was used (if you have never used a toolbar before, it will show up at its default location). The previous or default locations are not always very useful, but you do have the ability to move toolbars anywhere you want.

In SketchUp for macOS, all toolbars are floating toolbars. Any toolbar can be placed anywhere on any display.

The Main Toolbar Is Always There

The toolbar at the top of the main drawing screen is always on. There are no controls to turn it on or off. It also cannot be moved and will always reside across the top of your screen. It can, however, be customized to display the tools that you want.

Customizing the top toolbar on macOS

Unlike SketchUp for Windows, there is only one customizable toolbar. Fortunately, you can add any buttons to it in any order you like, and it can be as big as your display will allow. Additionally, you can add any buttons from any toolbar directly to it without having to move buttons from one toolbar to another.

Let's customize the top toolbar. Since the top toolbar has many of the most commonly used commands already, for this example, we will add additional buttons for commands we may want to add to our workflow. Let's say that I want to add a group of buttons to the custom toolbar. These commands are as follows:

- **Dimensions**
- **From Scratch**
- **3D Text**
- **Section Plane**
- **Smoove**
- **Outer Shell**
- **Position Camera**

What we will do is add a divider (an empty space) between the default buttons and our own custom tools in the top toolbar. Let's get in there and edit the toolbar:

1. Right-click on the top toolbar and click on **Customize Toolbar...**.

   This will display the toolbar customization window. This window will display icons for all commands that can be placed on the toolbar, including those from extensions (which we will discuss more in the *Organizing a custom UI* section). To add a button to the toolbar, you just drag it from the window and drop it onto the toolbar at the top of your screen (or in the preview of the toolbar at the bottom of the window).

2. Drag the **Dimensions** icon from the window to the toolbar and drop it in the empty space to the right of the existing icons.

3. Do the same for **From Scratch**, **3D Text**, **Section Plane**, **Smoove**, **Outer Shell**, and **Position Camera**.

   Now that we have the buttons we want, let's get them a little better organized. Let's start by separating them from the default buttons. Let's place an empty space between the stock buttons and our custom commands. Notice that one of the options in the button list is something called **Space**. This is just an empty space that you can use to separate groups of buttons from each other.

4. Drag a **Space** from the window and drop it just to the left of **Dimensions**.

   Now, let's take advantage of toolbar customizability and rearrange the buttons in the following order:

   - **From Scratch**
   - **Smoove**
   - **Outer Shell**

- **3D Text**
- **Section Plane**
- **Dimensions**
- **Position Camera**

5. Click and drag the buttons on the custom toolbar until it looks like this:

Figure 8.7 – Your custom toolbar buttons

You can see the advantage of customizing this toolbar is getting access to the tools you need. There is a limit, however. You can only add enough buttons to the toolbar to fill the width of the screen. Once the width of the screen is full of icons, you cannot add any more.

For this reason, depending on your display, you may want to turn on **Use Small Size** (at the bottom of the toolbar customization window). This will display smaller icons and allow more space for customization.

If you ever want to remove a button from the toolbar (including the ones that are there by default), you can simply drag them off and drop them on the modeling screen. Let's get rid of that **Dimensions** button.

6. Click and drag the **Dimensions** icon from the toolbar to anywhere on the modeling screen and release it. It will disappear in a small puff of white smoke.

   At this point, you can remove any of the icons you do not think you will be using and add some that you think you will be using more in the future. Once you are done, you can save the edits to the toolbar by closing the customization window.

7. Click on the **Done** button to close the toolbar customization window.

With that, you have learned how to customize the top toolbar in SketchUp for macOS. Now that we have seen how to organize toolbars, let's think for just a second about what we want to use the toolbars for.

Picking your toolbars

The next step in leveling up is speeding up your modeling process by displaying the toolbars or buttons for the commands that you don't access with shortcuts. What commands are those?

Looking back at the lists of commands you are using and are planning to start using from *Chapter 6, Knowing What You Need Out of SketchUp*, and knowing the shortcuts you created after reading *Chapter 7, Creating Custom Shortcuts*, what commands remain that do not have shortcuts? What commands are you using that have you going to menus each time you use them?

As you work through your next model, keep an eye out for the times that you hunt through a menu to find a command, especially those that are two levels deep. The time it takes for you to navigate through menus is time that could be used to create geometry! When you identify those commands from menus, you need to decide whether they deserve shortcuts or they should be in buttons on the screen. The commands that you are using that don't make the cut for shortcuts (that don't make the shortcut?) should be considered for toolbar status.

If you are on Windows, you will need to decide whether you want to show an entire toolbar to get to that one command, or whether it makes sense to create a custom toolbar to just get to the commands that you need. There is no point in displaying the entire **Drawing** toolbar just to get the **2 Point Arc** button on your screen.

If you are a macOS user, the decision is a little easier. Commands that need an onscreen button just need to get added to the toolbar.

Deciding what toolbars or buttons to have onscreen can be an evolving process. As you work more in SketchUp with an eye toward improving your workflow, you may come back and change your UI many times. The important takeaway here is that commands you use regularly should get keyboard shortcuts, while commands that are used less often should go into toolbars. Once you figure out what those commands are, you will find that you can model faster when you are spending less time hunting for the commands you need to use.

Another way to have information available to you as you model is by setting up your trays or dialogs, which is exactly what we will cover in the next section.

Setting up trays and dialogs

Let's start by quickly defining some terms. **Trays** are the stacking windows that snap to the side of your display and present information such as entity information, or shadows if you are running SketchUp on Windows. **Dialogs** are the floating, stackable windows that serve the same function in SketchUp for macOS.

Regardless of the operating system, there are 11 items that can be displayed:

- **Entity Info**
- **Components**
- **Styles**
- **Tags**
- **Outliner**
- **Scenes**
- **Shadow Settings**
- **Fog Settings**
- **Match Photo**

- **Soften/Smooth Edges**
- **Instructor**

If you are running SketchUp on Windows, you will also have a **Materials** panel available while SketchUp for macOS will have a separate, non-stackable, floating **Colors** window.

Since the terminology and function of these tools are so different on different operating systems, we will dive into each one separately, starting with trays in SketchUp for Windows.

SketchUp for Windows trays

Trays are containers that allow you to organize **panels**. In SketchUp for Windows, a panel is the name of the groups of controls listed in the previous section. Trays give you control over how and where these panels are displayed. The following screenshot shows a custom tray with four panels:

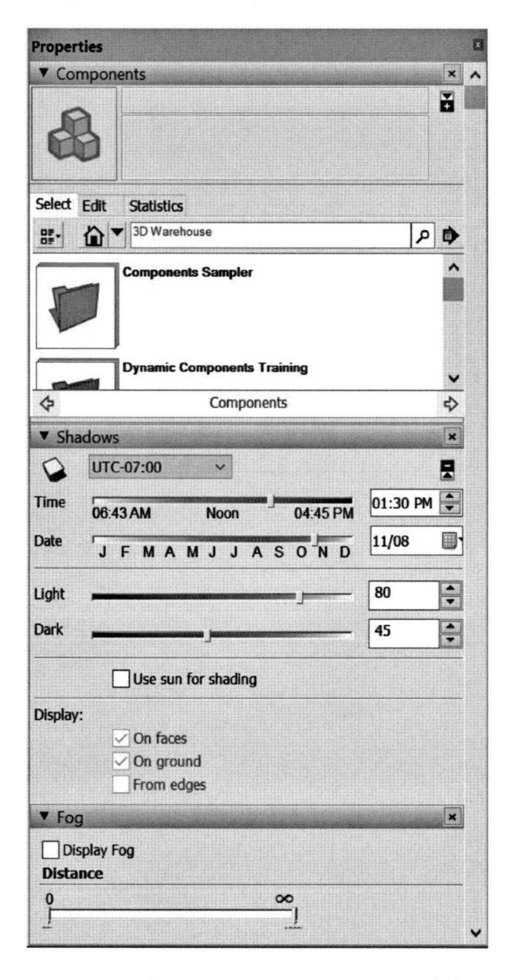

Figure 8.8 – A custom tray with four panels

Regardless of how you use your trays, you can rearrange your panels within the tray. Each panel can be individually collapsed or expanded by simply clicking on its title bar. Likewise, you can rearrange panels by clicking and dragging them into whatever order you need.

Now that we understand how panels work, let's look at how you can control the trays. When you first install SketchUp, you get a single tray (called the **Default Tray**) docked to the right side of the screen.

Many users just leave the Default Tray where it is without realizing the options they have to change where the tray can be. A tray can float or dock to any side of the screen. Docking a tray is as simple as dragging its title bar to the side of the screen where you want it docked and dropping it on the docking icon. You can create a floating tray by dropping the title bar into the middle of the screen. In the following screenshot, the workspace includes both a floating tray and a docked tray:

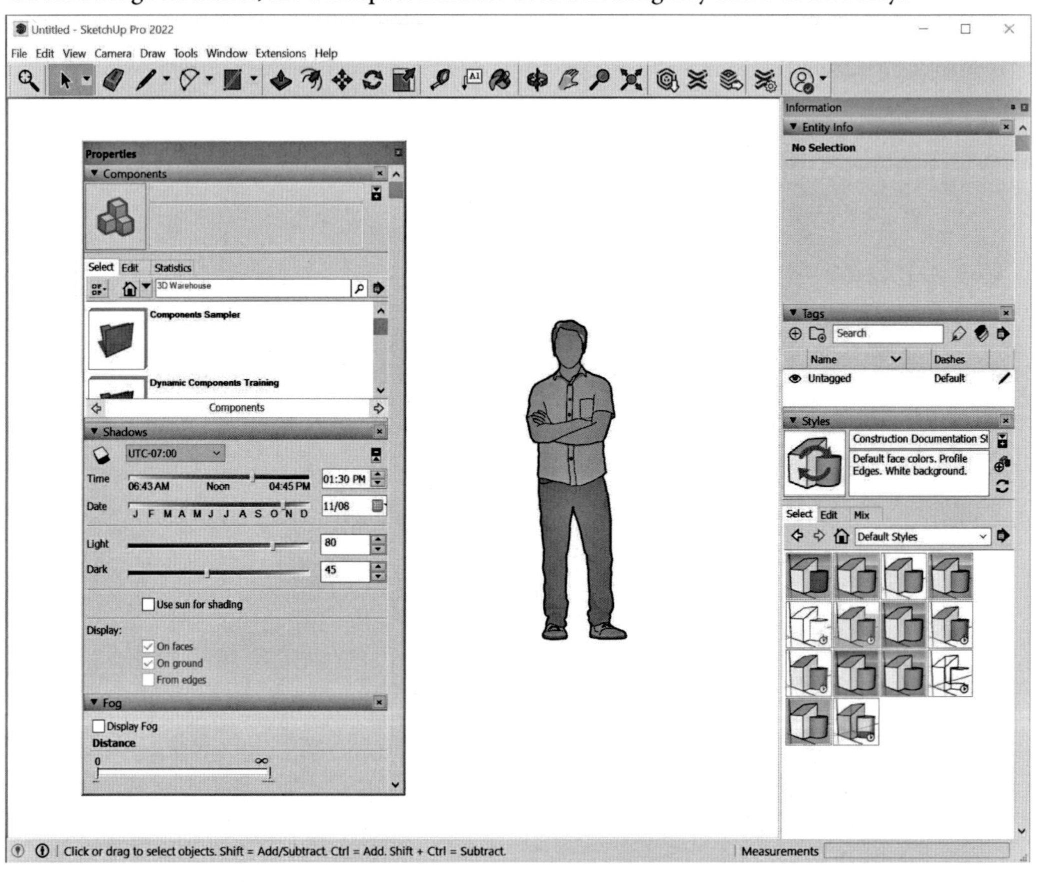

Figure 8.9 – The tray on the left is floating while the tray on the right is docked to the screen

You Will Probably Never Do This

Trays can also be docked to the top or bottom of the screen. Since the information displayed in the panels was originally designed to be displayed in a vertical bar shape, docking to the top or bottom and creating horizontal panels looks odd and wastes a lot of space. The option to dock anywhere is there, but I would recommend against using it to dock to the top or bottom.

Clicking the small pin icon at the top of a docked tray will cause it to collapse into a small tab on the side of the screen. To display a collapsed tray, just move your mouse on the tab. Clicking the pin of a collapsed tray will permanently display the tray again.

That's quite a bit of flexibility for displaying a tray! Plus, you have even more options once you have more than one tray. Let's create a few custom trays, then take a look at how multiple trays can be displayed:

1. Click on the **Windows** menu.

   From here, you have the option to explore current trays, manage trays, or create a new tray. Let's start by looking at what is in a default tray. The top part of the menu will list all the trays we have created. By default, there is only one tray in the list: **Default Tray**.

2. Click on **Default Tray**.

   This will display a flyout menu that gives you commands to hide or rename the tray, as well as displaying what panels are in the tray, currently. Panels with a checkmark next to them are displayed in the tray. From here, you can turn panels on or off and change what is included in the tray. All we want to do right now is make sure that the Default Tray is not visible.

3. If **Default Tray** is visible on your screen, click on **Hide Tray**.

   Now let's create a new tray that will display just a few of the panels we want to be displayed.

4. Click on **Manage Trays...**.
5. Click on the **New** button.

   This new tray will include the panels that display information about the model, so let's name it `Information`.

6. Type `Information` into the **Name** field.
7. Turn on the following panels:

   - **Entity Info**
   - **Styles**
   - **Tags**
   - **Scenes**
   - **Outliner**

8. Click on the **Add** button.

   Notice that when you turn on a tray, it automatically shows up in the default location; that is, docked to the right side of your primary display.

   Now let's create a second tray that will contain the panels that we use to set properties as we model. We will call this tray `Properties`.

9. Click on the **New** button again and type `Properties` into the **Name** field.

   Turn on the following panels:

   - **Components**
   - **Shadows**
   - **Fog**
   - **Soften Edges**

10. Click on the **Add** button.

    Notice that the new tray shows up at the same place that the previous tray did, but now there is a small tab bar at the bottom of the tray displaying the names of our two trays. Clicking either tab will allow us to switch between trays, displaying all of the panels within.

> **Panels Can Only Be Displayed Once**
> As you create trays, you will find that you can only have one panel active in one tray. If you were to go into our new **Properties** tray and turn on **Entity Info**, it would turn off in our **Information** tray.

So, why bother splitting the panels into two groups? Having your panels split between trays means you can display more of them at the same time. Since there is a limit to how much you can show in a tray (it can only display a stack of panels as tall as your display), you can only see so many panels at any given time. You can have more panels than you can display, but then you must scroll through the panels to see the one you need. With the panels in two trays, you can display more than one tray and see twice as many panels.

Right now, we have two trays, both docked onto the same side of the screen. This is nice if you are okay with clicking the little tabs to switch between trays. Let's see how we could actually display both trays at the same time:

1. Click and drag the **Properties** tab at the bottom of the tray into the middle of the screen.

   This will create a floating **Properties** tray. Now we need to dock the tray next to the **Information** tray. To do this, we will drag the floating tray into the docking widget in the middle of the screen. In this case, we will drag it and drop it onto the right side of the docking widget.

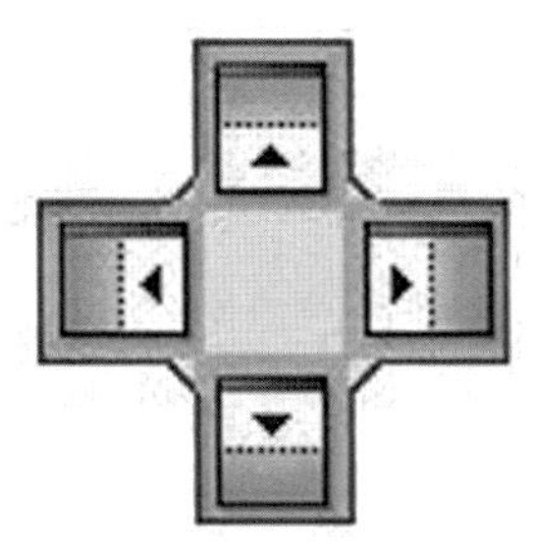

Figure 8.10 – Floating docking widget

2. Drag the floating **Properties** tray (by the title bar) onto the right icon of the docking widget.

   We now have stacked trays on the right side of the screen! I know, doing this on many displays means sacrificing a whole lot of your drawing area, but the important thing to know is that you can display more than just the Default Tray, docked to the right side of the screen.

   Since this current setup will not work well unless you have a super-wide 40" monitor, let's go ahead and collapse these trays and reclaim some drawing area.

3. Click the pin icon for both the **Information** and **Properties** trays.

With that, you have seen how you can customize the trays in SketchUp for Windows. Trays are incredibly flexible and give you a lot of options for displaying information.

SketchUp for macOS dialogs

The floating dialogs in SketchUp for macOS display the same information as the trays in SketchUp for Windows, but with less control. Using the dialogs is extremely straightforward. Toggle the visibility of each dialog by clicking on its name in the **Windows** menu.

When a dialog is toggled on, it will appear as a floating dialog. These dialogs (with the exception of the **Materials** dialog) can be stacked together in whatever order you like. This stack can be positioned anywhere you like on any display. Each dialog can be individually expanded or collapsed as you need it, as well.

> **Stack and Restack**
>
> Dialogs can be reordered at any point. To change the order of a stack of dialogs, pull them apart from the bottom. Dialogs can be disconnected from the dialog above them by clicking and dragging. They will also reattach by dragging a selected dialog below another.

Since the dialogs do not dock and will float anywhere you like, you have complete control over their placement. If you like, you can keep them all collapsed in a single stack on your primary display, as shown here:

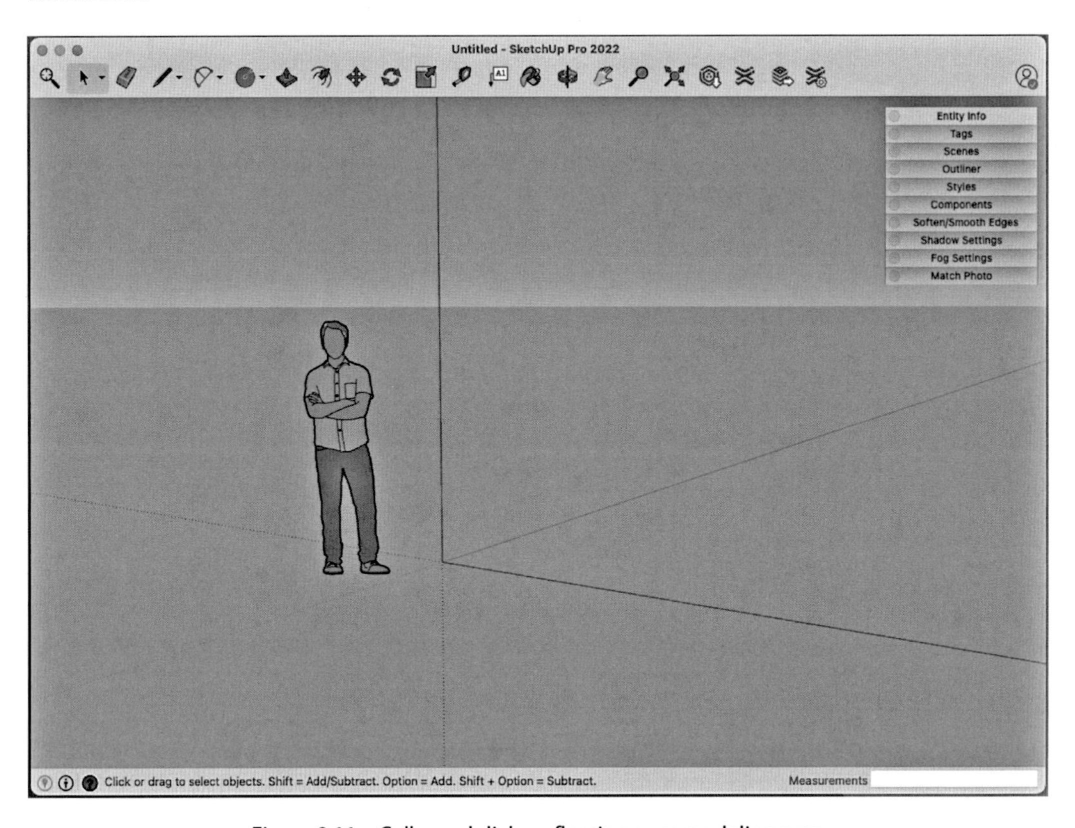

Figure 8.11 – Collapsed dialogs floating over modeling area

Or, if you prefer to see what is going on in them at all times, you could expand them into two or three stacks and let them float on a secondary monitor.

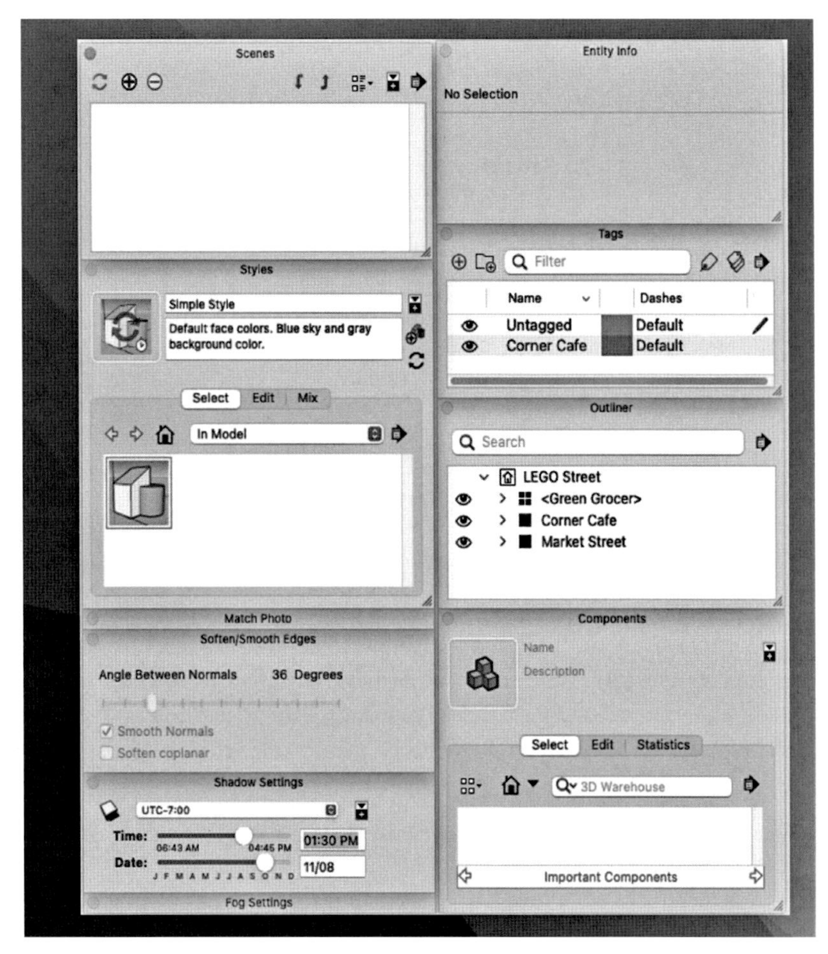

Figure 8.12 – Expanded dialogs in two stacks floating on a second display

Dialogs are simple to use, move, and rearrange. Take advantage of these elements by putting them wherever you need them, whenever you need them.

The ideal tray/dialog setup

As with other UI elements, the ideal setup of the trays or dialogs will be different for everyone. Try out different configurations and order of the elements as you model. See which dialogs or panels you need onscreen and which you like to have open versus which ones you open only when you need them. Do you like having them on the left or right? Since changing this UI is so easy, try a few different layouts and see what works best for you.

At this point, we have reviewed the default UI options. Up next, let's take a look at the UI elements that are made available with the installation of extensions.

Organizing a custom UI

Now that we have seen the options for organizing the default UI, we should probably touch on how the UI works for extensions. We will be spending plenty of time on extensions in *Chapter 13, Must-Have Extensions for Any Workflow*, but right now, let's discuss how we can add an extensions UI to our workspace.

In addition to adding menu options to the context menu or the **Extensions** or **Tools** menu, extensions creators have the option to add toolbars that can be accessed alongside the default toolbars or buttons or create brand-new custom UIs. Since we have already discussed how toolbars work, let's start by looking at how to work with toolbars for extensions. As with the other UI elements, this is different for SketchUp for Windows and SketchUp for macOS. We will start by exploring the options available in SketchUp for Windows.

Extension toolbars in SketchUp for Windows

Many extensions will make their own toolbar available for use via the **Toolbars...** option in the **View** menu. They will function very similarly to standard toolbars. They can be toggled on and off and docked to the sides of your modeling screen or can be allowed to float on any of your displays.

You cannot, however, drag buttons from the **Extension** toolbars, nor can you drop other buttons into them. This means if you want access to the **Extension** toolbars, you will have to display the whole toolbar.

> **Shortcuts Work for Extensions**
> If you dislike the idea of showing an entire toolbar just to get access to one or two commands from an extension, remember that you can assign shortcut keys to extension commands, just like native commands.

Extension toolbars in SketchUp for macOS

On SketchUp for macOS, you can activate toolbars for extensions, just like in other toolbars. They will, just like other toolbars, float wherever you like on any display. Additionally, most commands from extensions can also be added to the toolbar at the top of the screen.

When you install an extension, your buttons from the extension will automatically show up in the toolbar customization window. You can drag these buttons and drop them into your toolbar, just like the buttons for native commands!

Custom extensions UI

In addition to the toolbars and menu entries added to SketchUp by extensions, many extensions will display a custom UI as you use them. These UIs are generally displayed onscreen without many options for customization, regardless of the operating system. For the most part, extensions will display their custom UI in the same place, or honor where you have moved them. So, while you cannot control exactly how this UI works, you can plan on where it will show up when the extension is used.

Each extension developer has the option to create the UI that is appropriate for the function of their extension. In some cases, extension developers have a standard UI that they will use across all of their extensions, such as **Fredo6**'s colorful information bar that pops up when using one of his incredible extensions, as shown here:

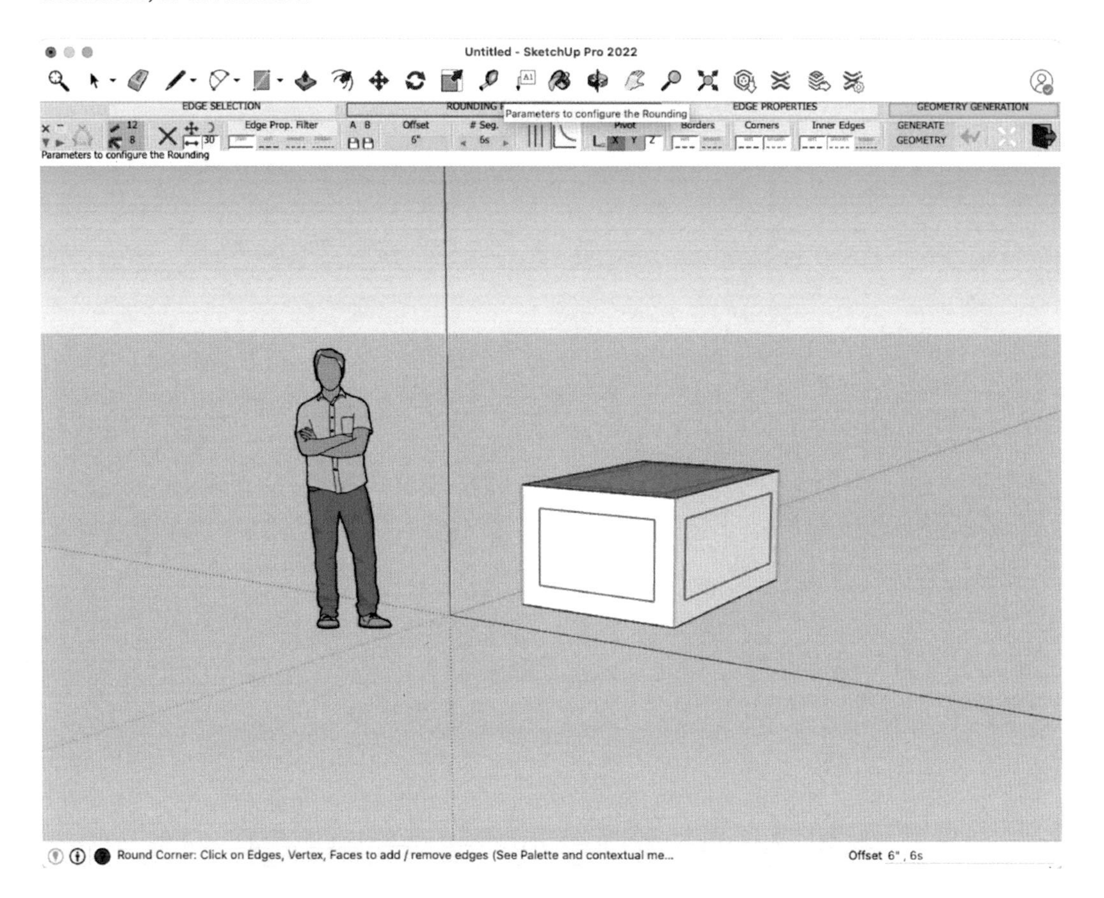

Figure 8.13 – Fredo6's RoundCorner extension shows Fredo6's standard information bar at the top of the screen

While Fredo6's UI always shows up at the top of the screen, some UI can be customized. **Thomthom**'s Vertex Tools UI appears as a custom tray-like element docked to the left side of the screen. The panels of the UI can be toggled on and off and the extension remembers the state of the UI after you close and reopen it. The Vertex Tools UI is shown here on the left side of the screen:

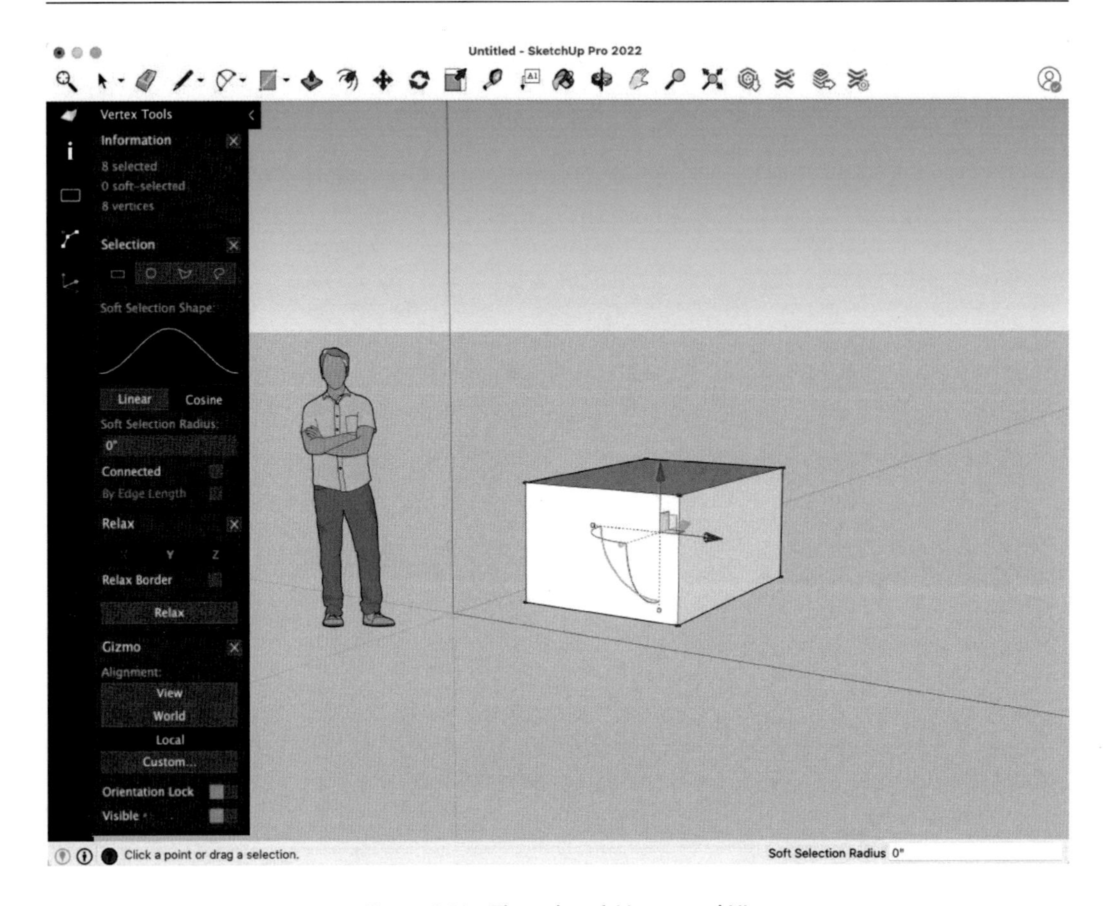

Figure 8.14 – Thomthom's Vertex tool UI

In both of these examples (and many other extensions), the UI only shows up onscreen while you are using the extension. This can make it a little tricky to tie the extension UI into your standard workspace. The best solution is to be conscious of the UI that will appear for the extensions that you are using. If you are using a Fredo6 extension and you know that the bar will be showing up at the top of the screen, don't put any floating toolbars up there. If you are using Vertex Tools from Thomthom, put your floating toolbars somewhere other than the left side of the screen. You cannot stop the custom UI elements from showing up; you can, however, make room for them when they appear.

Extensions are great for adding to your capability in SketchUp, but they can add a whole lot of UI to your screen if you are not careful. The key to working with an extensions UI, or any UI, for that matter, is figuring out exactly what you need onscreen and when, which is exactly what we will cover in the final section of this chapter.

Creating a next-level workspace

At this point, we have looked at all of the parts that make up your workspace: displays, toolbars, trays/dialogs, and the extensions UI. The looming question is, what is the best way to set this all up so that you have a workspace that will help you level up your SketchUp skills?

You have probably already guessed that there is no one answer to this question. Everyone will have a slightly different workspace that is best for them. Let's walk through some of the big factors in creating your ideal workspace. The first thing we can think about is where you will be working.

Where is your workspace?

Do you sit at the same desk every time you use SketchUp and have access to multiple monitors? Do you design on the go or out in your shop and rely on just your single laptop screen? Do you switch back and forth? I know we discussed single versus multiple-monitor setups earlier in this chapter, but it is time to think about the practical application of your setup. No theory, no dream setup, no *ideal situation*. Right now, where do you use SketchUp? This is the place where we will build your workspace.

What is the most important UI for your workflow?

This is a question that you should be asking yourself every time you model. What commands are you using on a daily basis? Which commands are used occasionally? What information are you referring to regularly? The easier these items are for you to get to, the quicker you can create 3D models. Let's do a quick math experiment and see what we are looking at, as far as time savings go.

Let's say you need to move your mouse from the middle of the screen to the **Draw** menu, then click on the **Line** command every time you need to draw an edge. Let's say that on average, it takes you about 3 seconds to do so. Now, let's say you grab the **Line** command 20 times or so every hour of modeling. That is a full minute spent every hour, just starting up the **Line** command!

I know, it may not seem like a whole lot of time, but it is a full minute of time that does not need to be spent. To get the full weight of how long a minute is, think about stopping in the middle of your productive modeling time and just sitting there, doing nothing for 60 seconds. That is what going up and clicking on menus is doing to your modeling time.

"*But it is just 1 minute!*", you say. One minute, for the **Line** command. But you are using more than just the **Line** command, right? Every time you run off to the menu instead of tapping a shortcut key or clicking on a button, you are adding to the time that you are not modeling. With this in mind, think about what items you have onscreen and how they help you be a better modeler.

Time to refer back to your lists from *Chapter 6, Knowing What You Need Out of SketchUp*, again. Look through the list of new workflows that you are developing. Look at the commands that you have identified as tools you need to start using. We talked about assigning keyboard shortcuts to many of these commands in *Chapter 7, Creating Custom Shortcuts*, but what about the commands that are part of your workflow that you did not assign shortcuts to? Should you be adding buttons for those to a custom toolbar?

To level up, you need to have the tools you need immediately at hand. Turn on or create toolbars that put your tools right where you need them and not stuck in a menu.

One workspace or two? Or three?

I honestly do not believe that there is one be-all and end-all solution for your workspace. I would guess that, for most of you, there are two or three. When you were looking at the kinds of models you currently use or the ones you want to start working on, back in *Chapter 6, Knowing What You Need Out of SketchUp*, you probably had more than one kind of 3D model. In most cases, different 3D models require different tools. This may mean modifying your workspace to display different tools for different models. Let me show you a couple of the workspaces that I use when I model.

I spend a lot of my time at work creating learning content for SketchUp users. When I do this, I try to keep the UI as simple as possible. I try not to have any extra toolbars onscreen and collapse the dialogs down, as shown in the following screenshot:

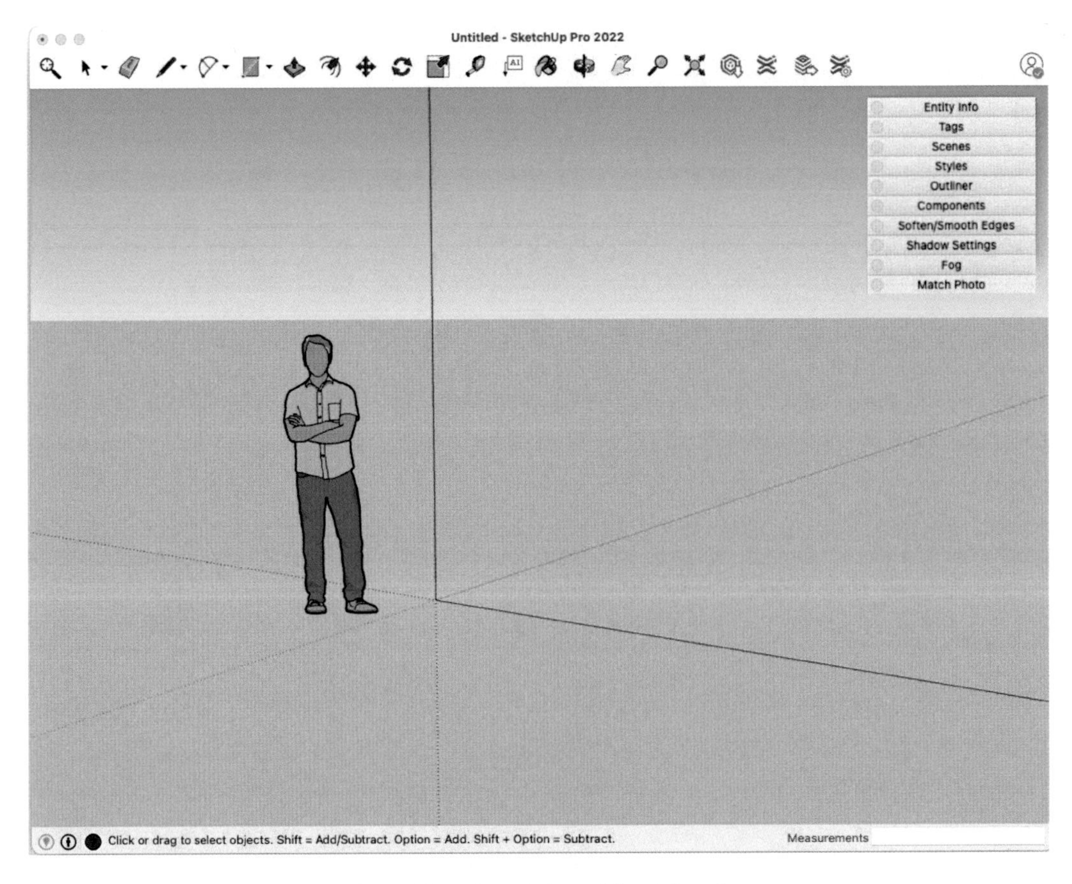

Figure 8.15 – My stripped-down "simple" workspace

Sometimes, I end up doing some organic or subdivision modeling using a handful of extensions. In this case, I turn on a handful of toolbars and make space for the extensions UI that I know will come up when I use the Vertex Tools extension from Thomthom. The following screenshot exemplifies this:

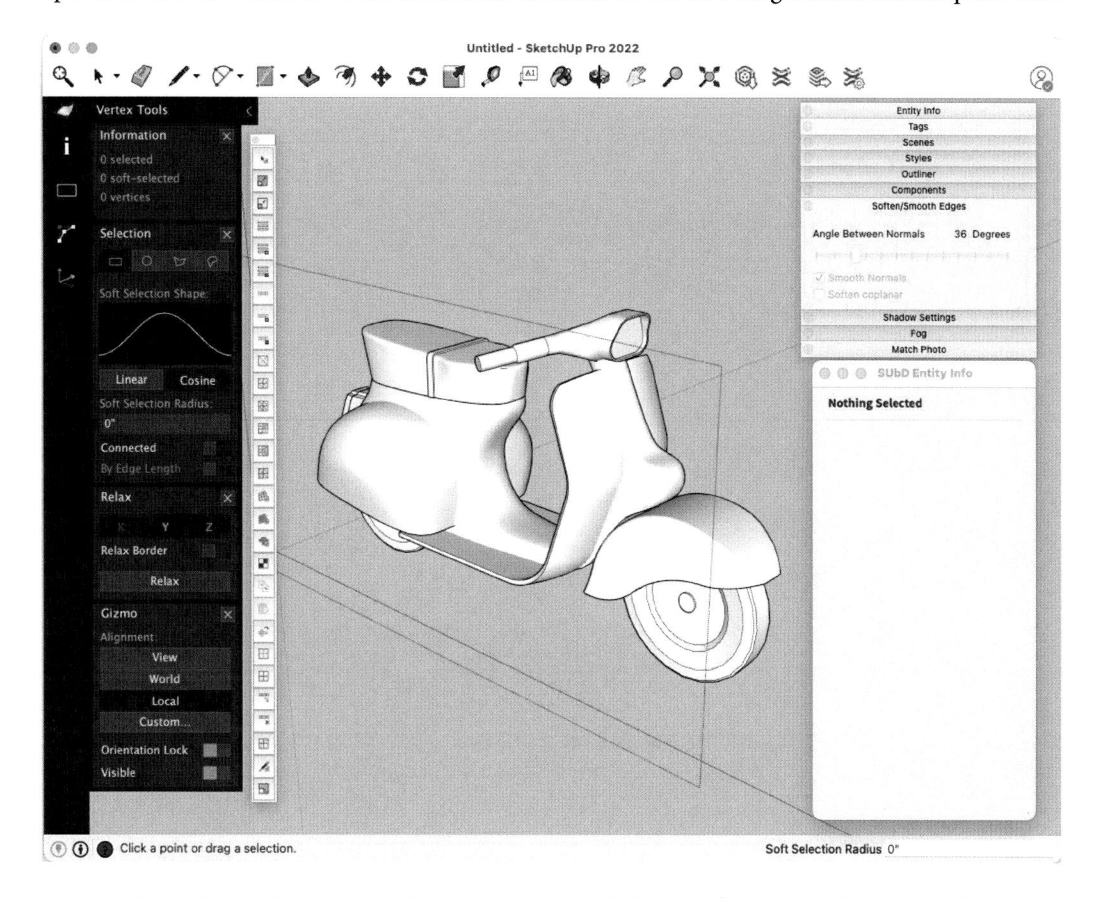

Figure 8.16 – My organic modeling workspace with the Vertex Tools UI visible on the left

Still, other times, when I am doing a live model on the SketchUp YouTube channel, I will create a multipurpose workspace. In this, I will often open one or more dialogs that I will be using regularly and turn on a couple of different extension toolbars with tools that I plan to use. This is a great general-purpose modeling workspace, and likely the one I would use most often if it were not too much UI for the teaching modeling that I do so often. My workspace for one of these sessions might look something like this:

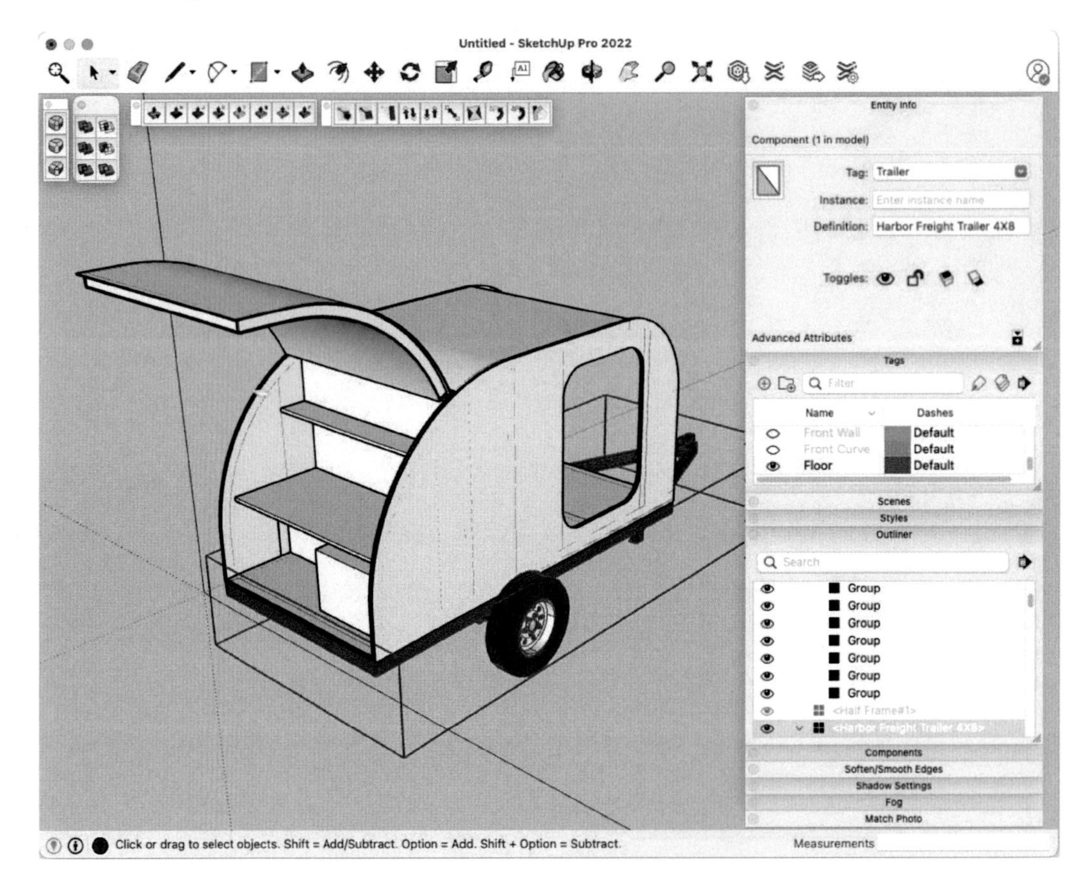

Figure 8.17 – Multipurpose modeling workspace

Why Not Show It All at Once?

There are plenty out there who would prefer to take advantage of multiple monitors or be willing to dedicate a good portion of their modeling screen to toolbars/trays/dialogs. In my case, because I do so much modeling intended to help beginners learn SketchUp, I cannot load up the UI. This does not, however, mean that it is a bad idea, or that you should not try it!

The downside to the multiple-workspace approach is, of course, that you have to spend time preparing your workspace for the type of modeling that you will be doing. Before you hop in and start modeling, you need to think about the tools that you want to use and get them onscreen.

Does that seem like it is, in itself, time-consuming? Does this defeat the purpose of having a time-saving workspace setup? Think about it this way. It takes the same amount of time to click into a menu to activate a toolbar as it does to click into a menu to activate a command. If you are going to use commands on the toolbar more than once in your modeling process, you come out ahead by turning on the toolbar.

Summary

The hope of this chapter was to give you the tools that you need to create a workspace that will help you to level up your 3D modeling skills in SketchUp. We have seen what UI elements you have access to as well as how to use and customize them. We have looked at the advantages of multiple-monitor setups and how to work with extension UIs.

Additionally, this chapter should have left you thinking about how you use the tools and how you can use them better. I am sorry if I suggested that there was one perfect way to organize your screen in SketchUp that would make you a better, faster modeler. There probably is a better workspace for you, but it is a one-of-a-kind solution that only you can build.

Now that you have the tools to create your perfect workspace, don't be afraid to dive in, make changes, and try out new screen layouts. Remember, the worst thing that can happen when you turn on a toolbar or move a tray is that you don't like it, and you can change it or go back to how it was. You literally cannot hurt yourself or SketchUp with anything we covered in this chapter, so do not be afraid to make changes to your workspace!

At this point on the leveling-up journey, we have talked about a lot of tools that can give you a leg up on the modeling process. Up next, in the next chapter, we will talk about how to take a huge step ahead at the start of any model you work on. We will also be looking at templates.

9
Taking Advantage of Templates

Templates are an amazing tool that can help you get a head start when you begin a new model. Most SketchUp users never get past using the default templates presented by SketchUp and do not realize how much time and energy can be saved by using the proper template. Understanding what is included in a template, how to edit templates, and when and why to use custom templates is a great way to level up your SketchUp skills. That is exactly what we will do in this chapter.

Specifically, we will be looking at the following:

- Understanding what makes up templates
- Exploring the default templates
- Deciding what templates to create
- Creating and maintaining custom templates

Technical requirements

In this chapter, you will just need access to SketchUp Pro.

Understanding what makes up templates

When you start a new model, you never really start with a *new* file. Any time you click **New** in the **File** menu or click on the **Simple** template in the **Welcome to SketchUp** window, you are opening a template. In fact, it would be fair to say that there is no such thing as an empty SketchUp model. If you have the SketchUp modeling screen up on your computer, even if there is no geometry, axis, or background colors, you are looking at the contents of a SketchUp file.

To dive right into this, that is all a template really is: a SketchUp file. When you start a new file, you are telling SketchUp to open up an existing SketchUp file. The difference between starting a new file via a template and using the **Open** command to open a file is that when you use a template, SketchUp knows that you do not want to save anything back to the template and prompts you to save it into a new file.

While clicking **New** from the **File** menu may seem to land you in a brand-new, empty SketchUp file (you did not tell it to use a template, right?) it is actually opening up your default template and presenting it to you. We will get into setting a default template later in the chapter, but for now, the important thing to understand is that you are always using templates when you start a new file.

Contents of a template

It's easy and fair to say that a template includes *everything in a SketchUp model*, but we can probably drill down a little bit and look at the important parts of a model that you may want control over when starting a new model.

Simply put, a template contains the following information:

- **Model info**
- **Styles**
- **Tags**
- **Scenes**
- **Components**
- **Geo-location**
- **Model contents**

There are a few additional pieces of a model that a template will control, but they fall under one of the other items on the list (for example, **Fog** and **Shadows** are controlled by **Styles**). Let's spend a little time looking at each of these items, how they affect a new model, and when you might want to use each in a template.

Model info

Simply put, everything that is in the **Model info** screen from the Windows menu is saved in a template. This is one of the pieces that make templates so powerful. Just look at all the tabs on the left side… so many settings!

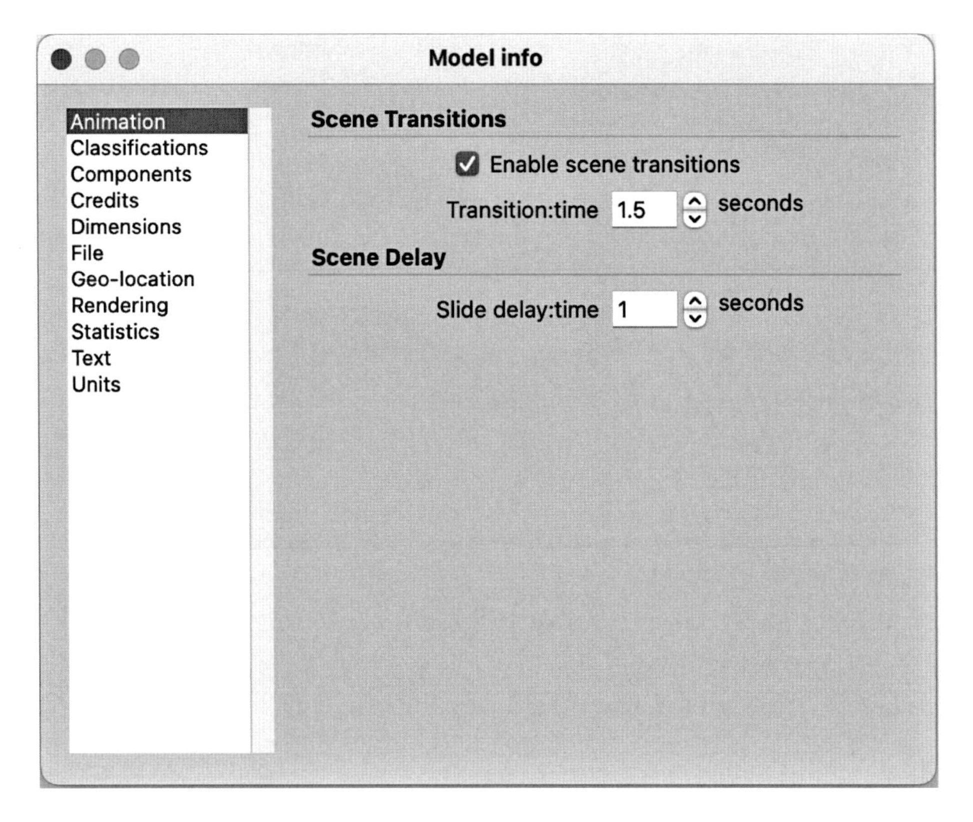

Figure 9.1 – Model info as seen in SketchUp for macOS

Some of the tabs in **Model info** contain settings that you may use infrequently (or maybe almost never), while others contain settings that can seriously change how you create a model! Here is a list and quick overview of the information on each tab, as well as the impact it can have on your model:

- **Animation** – The settings in this tab are used when running animations and also when changing from one scene to another. It is worth modifying these settings before saving a template if you do not regularly create animations or want a quicker (or no) wait time when jumping between scenes.
- **Classifications** – This tab allows you to import, export, or delete data schemes used for transferring data between programs. The default data schemes are **Industry Foundation Class** (**IFC**). Unless you have a workflow that sometimes has you export IFC3 and other times IFC4, you will likely not create a template because of these settings.
- **Components** – This tab has a few settings that change how objects are displayed in SketchUp. You can control how much the rest of the model is faded when you edit an object and whether the axis should be shown on each one. These are useful controls and something that you will probably set once, based on personal preference, before saving your first template.

- **Credits** – This tab allows you to mark your model with your information. This is great if you plan to share this model via 3D Warehouse (more on 3D Warehouse in *Chapter 12, Using 3D Warehouse and Extension Warehouse*) or any other file sharing process. Once again, a setting that you should set before you create a template.
- **Dimensions** – This tab has control of on-screen dimensions. This is a great tool that will allow you to make dimensions appear the way you like and should be modified before you save your template.
- **File** – This tab contains basic information about the model. If you do fill this information in, then you will probably want to do so on a file-by-file basis, and not as a part of a template.
- **Geo-location** – This tab includes information about the real-world location of this model. Setting this information and then saving it into a model will cause the model to be placed in a specific location before you begin modeling. If you are creating models that reside in the same location regularly, then you may want to save this into your template. In truth, most models, if geo-located, exist in a unique location, so this is something you probably will not set before saving a template.
- **Rendering** – This tab has controls for how the model is displayed on the screen. Again, something that you can use to make your model look the way you want, before saving your template.
- **Statistics** – This is an incredibly useful tab that will tell you information about your current model. Since this screen will only display information about your current model, there is nothing here that will be saved into your template.
- **Text** – Like the information displayed on the **Dimensions** tab, this info controls how on-screen text will be displayed. Set the settings as you like before saving your template.
- **Units** – This will control what units you will use to input display while modeling. If you work in a region that requires you to work in different units, this is a time-saver, as you can set your units and save them into your template.

So, looking back, you can see that there are a lot of settings that you would not want to go into and change every time you start a new model. Setting these values once and then saving them into a template means you do not have to change them every time you start a new job.

There are a few settings that may cause you to create more than one template, as well. Units, for example, might be a reason to have multiple templates. Even with all of the other settings the same, you may end up saving two templates, depending on the work you do. Perhaps you do most of your design in feet and inches when you do building design, but you also enjoy creating smaller models of 3D printing. In this case, you may end up with one template with units set to **Architectural** and the other with **Decimal** and **Millimeters** selected.

Styles

When you save a file as a template, it not only saves the current, active style but any other styles saved into the file. This means you can switch between multiple styles right at the start of modeling without having to find or create them. Every style displayed in your **In Model** style list will be saved as a part of the template. If this was my **Styles** window when I saved a new template, five different styles would be available when I start a new model, as shown here:

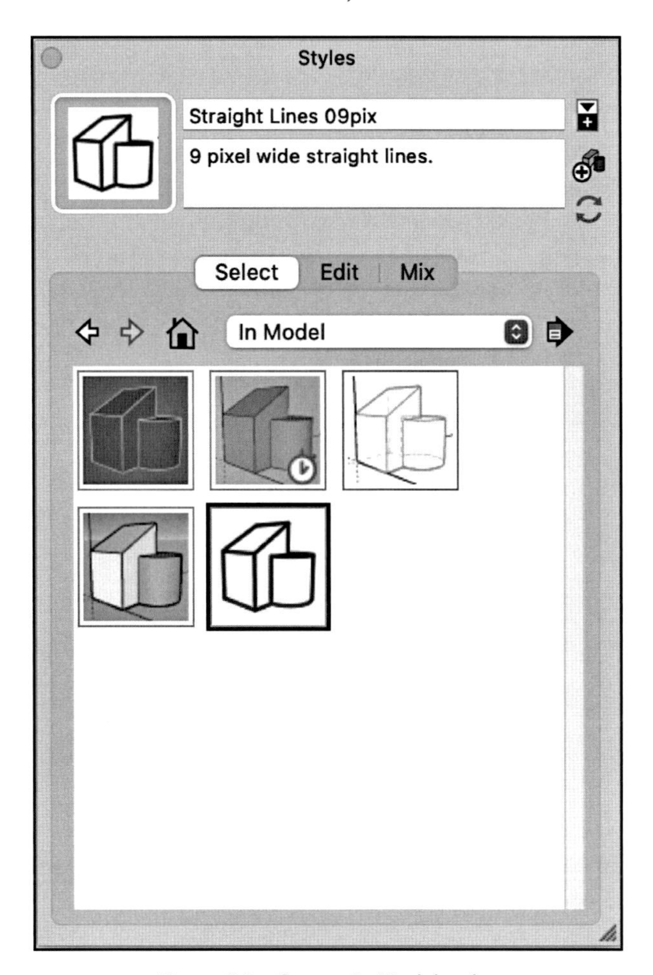

Figure 9.2 – Current In Model styles

Remember that styles not only control how your models are drawn on the screen but also the backgrounds, watermarks, and colors that are used to show highlighted or locked objects.

Tags

Many users end up developing a workflow that depends on certain tag names. Often, users that work collaboratively will establish a standard for tag use that can be used across a large, multi-user workflow. Regardless of the reason, creating your tags can be done once and saved into your templates, as shown here:

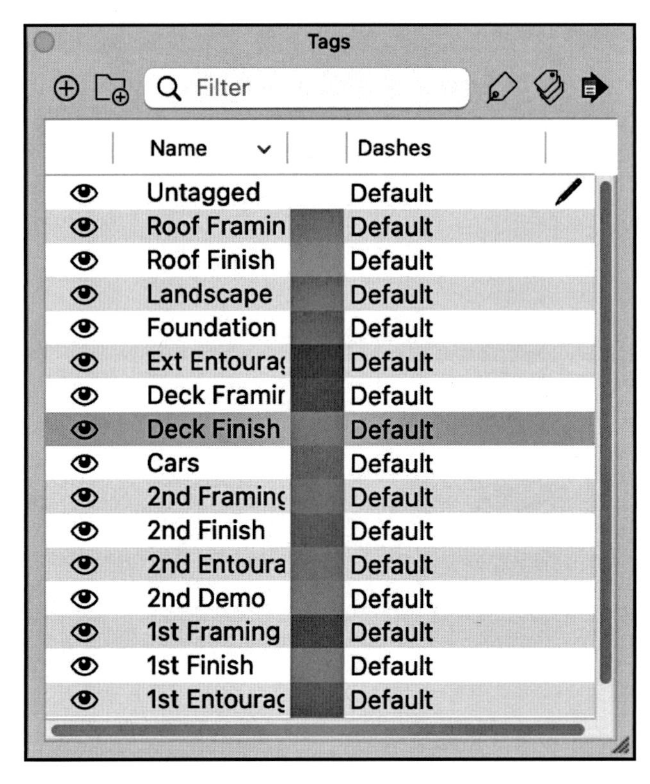

Figure 9.3 – Tag names created ahead of time and saved into a template

Even if you do not have a system that requires specific tag names, or a co-worker expecting items with specific tags, if you end up creating tags when you model, then creating them once and saving them into a template is a time saver.

Scenes

Many times, scenes are created specifically for a model. The view you need of a 1,500-square-foot home will be different from a 30,000-square-foot office building. Despite this, you can still benefit from creating and saving scenes in your template. Created scenes such as an overhead view with the camera set to **Parallel Projection**, or views of the model from the north, south, east, and west, can be saved into the model and updated once you have created your model, as illustrated in the following screenshot:

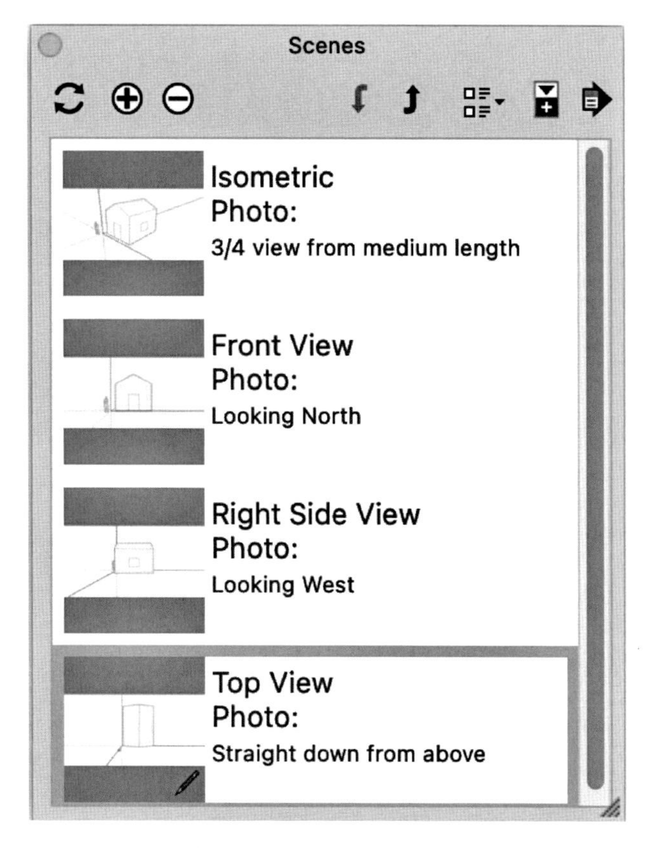

Figure 9.4 – Commonly used scenes created and saved into a template

Additionally, remember that scenes do not need to be tied to only a camera view. You can create working scenes that change styles, or toggle tags on or off and save them as a template, giving you a head start on working on your model.

Components

Every component created or imported into your model is saved in that file. This means it is saved into the template, as well. Remember, components do not have to be on screen to be saved. This can give you a huge leg up on modeling items that you regularly include in your work. If you model kitchens and have a set of cabinets that you like to start with, they can be saved as components in your template and dropped into a new model right at the start!

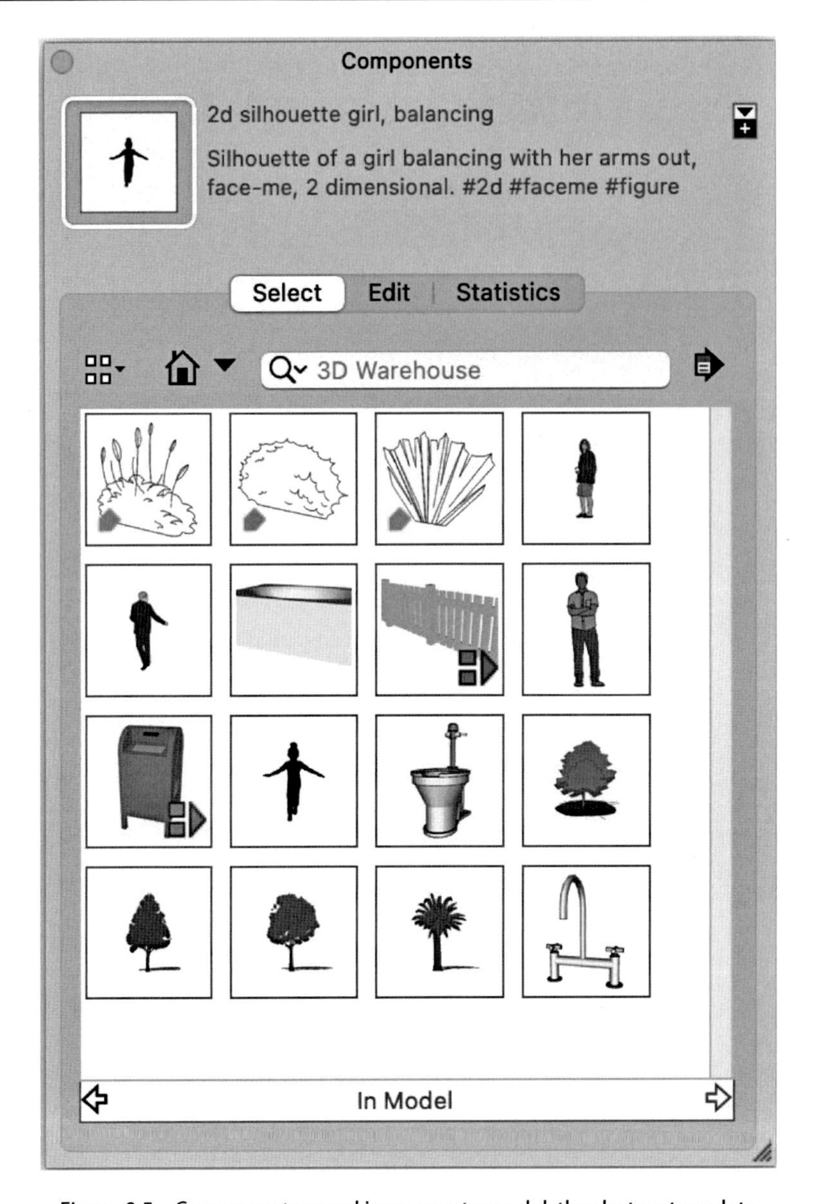

Figure 9.5 – Components saved in an empty model, thanks to a template

It is worth mentioning that since they are saved into the file, too many components can cause your template's file size to grow. It may be worth only including the components that you know will be used in your model and not every useful component that you have come across. Too big of a template file can cause long save and load times. This is not worth it to include a bunch of components that you will not use in most of your models.

Geo-location information

Geo-locating models is an amazing function in SketchUp that allows you to place your model in a real-world location. This can mean a quick and easy reference to surrounding geometry and accurate shadows. So, why would you want to save geo-location as a part of a template?

In some cases, you may be creating multiple models that will reside close together. It is also possible that you will be creating multiple models to pitch for a project, and they all need to be placed in the same location. Regardless of the reason, should you ever need geo-location to be the same for multiple models, know that setting the geo-location before saving a template will save it as a part of the model.

> **Keeping Geo-Location out of Your Model**
> Remember (and this holds true for everything in this list) that if you do not want the geo-location to be included in your template, you will want to clear that information before you create the template. Once added to the template, there is no quick or easy way to remove it.

Model contents

You know that little two-dimensional person standing next to the origin whenever you start a new model in SketchUp? Those are called scale figures, and they exist to let you know how big something is on your modeling screen. Without them, it would be impossible to know whether a rectangle drawn on the screen was the size of a laptop or a football field.

Figure 9.6 – The default template for SketchUp 2022 includes Niraj as the scale figure

In a new model, the template usually only contains the scale figure. It does not have to be limited to that though. If you want your new model to include a grid, a specific set of entourage, or even a template for a building, you just have to save it into the template that you will be starting from.

Even if nothing is on screen when a template is created, the camera location and settings are saved. This means you can have a totally empty screen, but SketchUp will remember where the camera was pointed at the time that the template was created.

Understanding everything that is saved into a template is extremely important. Once you know what is in a template, you can start to explore how to use the templates. Let's start by looking at the default templates and see how and when we might use some of them.

Exploring the default templates

Every time you start SketchUp Pro, you have the option to start a new model using a template. On the **Welcome to SketchUp** window, the top section lists your most recently used templates. Clicking the **More templates** link to the right of the template tiles will launch a list of all the default templates:

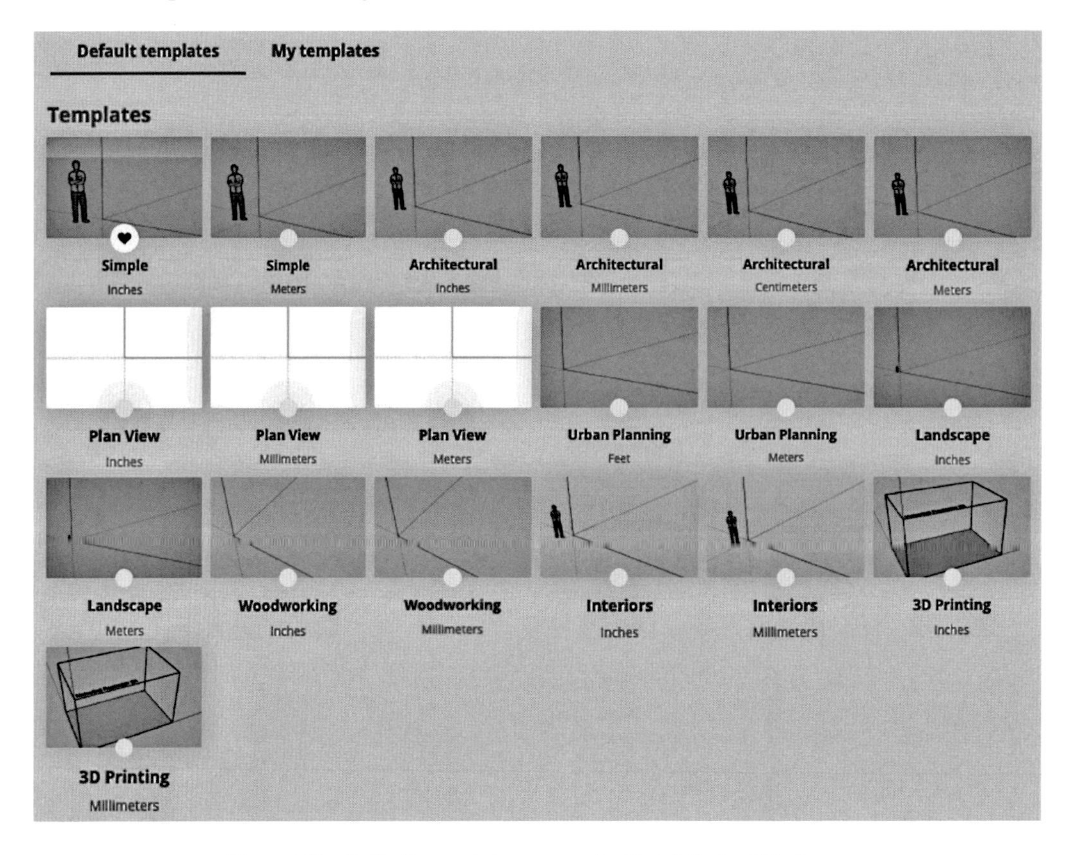

Figure 9.7 – Default templates as seen in SketchUp for Windows

As you can see in *Figure 9.7*, from here, you can start a new model using any of these templates.

Getting to Templates from the File Menu

If you are already in SketchUp (past the **Welcome to SketchUp** window) you can view templates by selecting the **New from Template...** option in the **File** menu.

SketchUp Pro 2022 comes with 19 different default templates. These are really eight different templates, offered in different default units:

- **Simple** – **Inches** or **Meters**
- **Architectural** – **Inches**, **Millimeters**, **Centimeters**, or **Meters**
- **Plan View** – **Inches**, **Millimeters**, or **Meters**
- **Urban Planning** – **Feet** or **Meters**
- **Landscape** – **Inches** or **Meters**
- **Woodworking** – **Inches** or **Millimeters**
- **Interiors** – **Inches** or **Millimeters**
- **3D Printing** – **Inches** or **Millimeters**

The settings of each of these templates are the same, except for the units, so what you would see in the **Architectural – Inches** template will be exactly the same as the **Architectural – Meters** template except for the default units.

In general, these templates are set up to start you from a position where you can start modeling different model types with little to no adjustments. The titles give you an idea of the types of models intended for each template. This does not mean that you cannot start with the Architectural template and then model a planter, for example, but using the template that most closely aligns with your intended modeling can save you time and energy. Knowing what each template is set up for can help you to pick a template when you start a new model or choose a template to customize when creating your own custom templates. Let's take an in-depth look at the templates.

Simple template

The Simple template is intended to be an all-around template that could be used for any modeling session. This template is the default template when a new version of SketchUp is installed, as shown in the following screenshot:

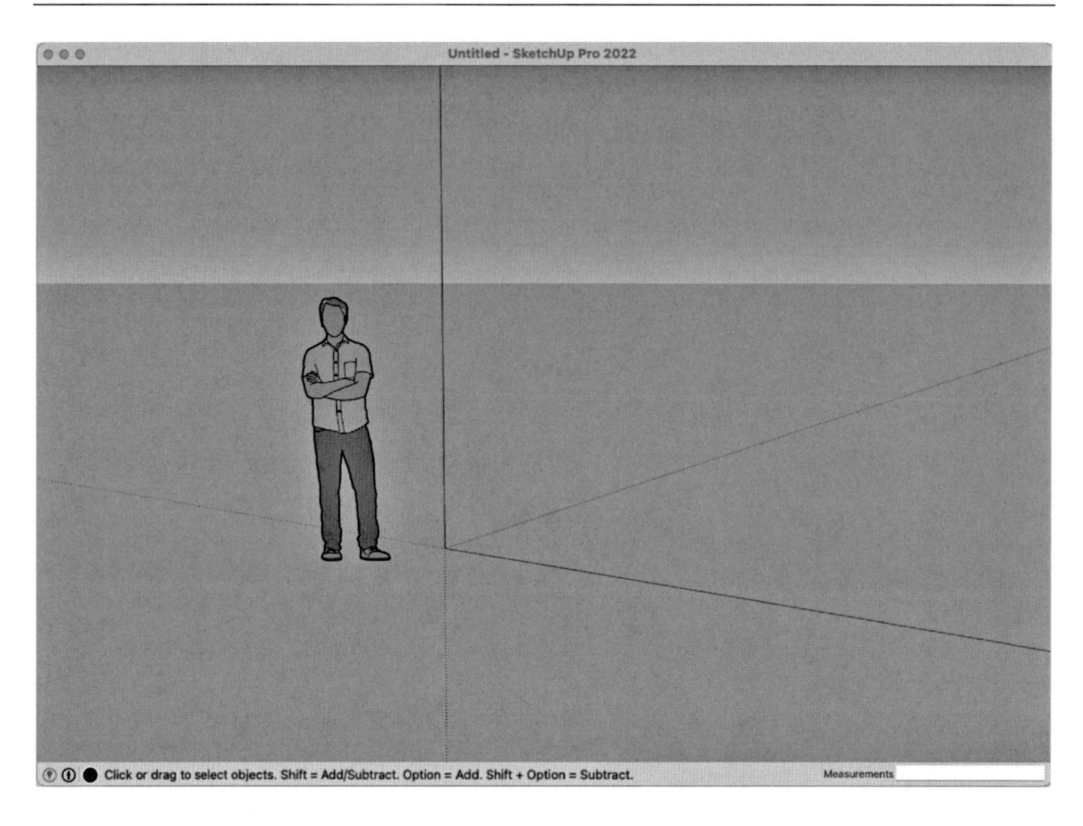

Figure 9.8 – Default template

This template was designed to be a good starting point for anyone using SketchUp. It has a neutral gray ground plane and a blue sky in the back. The camera is set to **Perspective**. This view is pretty easy to customize into any type of template you might want, especially if none of the other templates align with the kind of modeling you do.

Architectural template

The Architectural template was designed for anyone modeling structures in SketchUp. Here is a brand-new empty architectural file, ready for a building model!

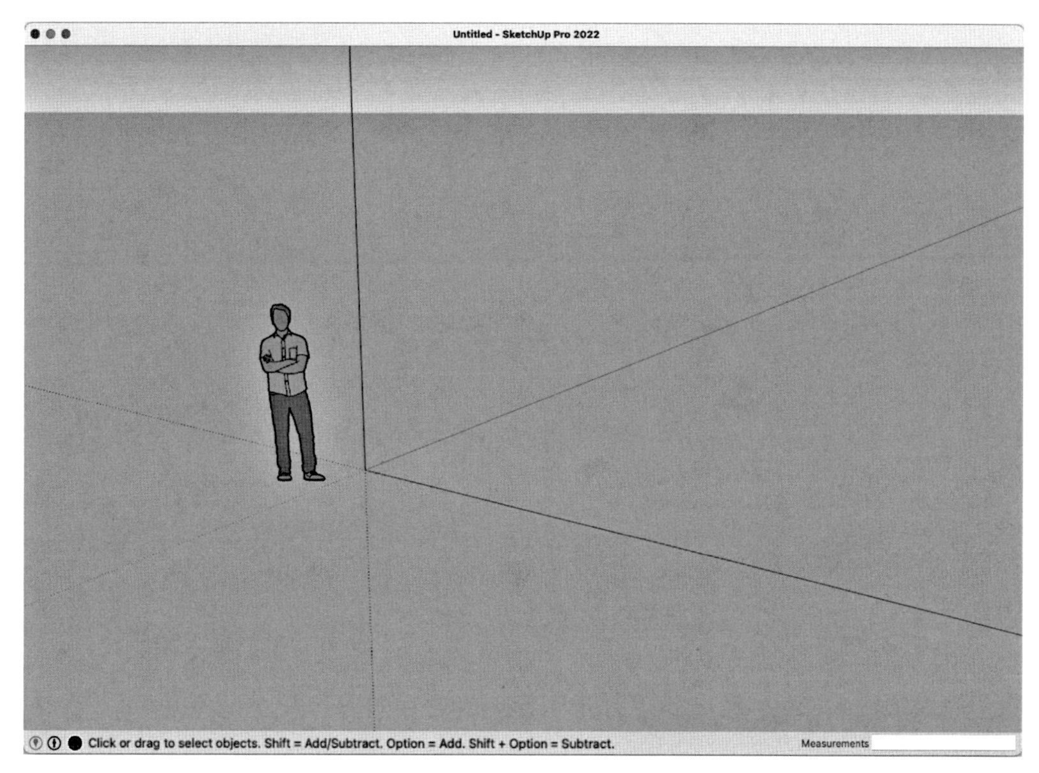

Figure 9.9 – Architectural template

The Architectural template is intended for modeling buildings, so the camera is pulled back slightly farther than in the Simple template but is still set to **Perspective**. The background and sky are more muted colors, drawing attention to the building being modeled rather than a vibrant background. This is the ideal template to customize if you are using SketchUp as a building design tool.

Plan View template

The Plan View template is intended for someone planning to draw flat on the ground (potentially in 2D). Take a look at this empty Plan View template:

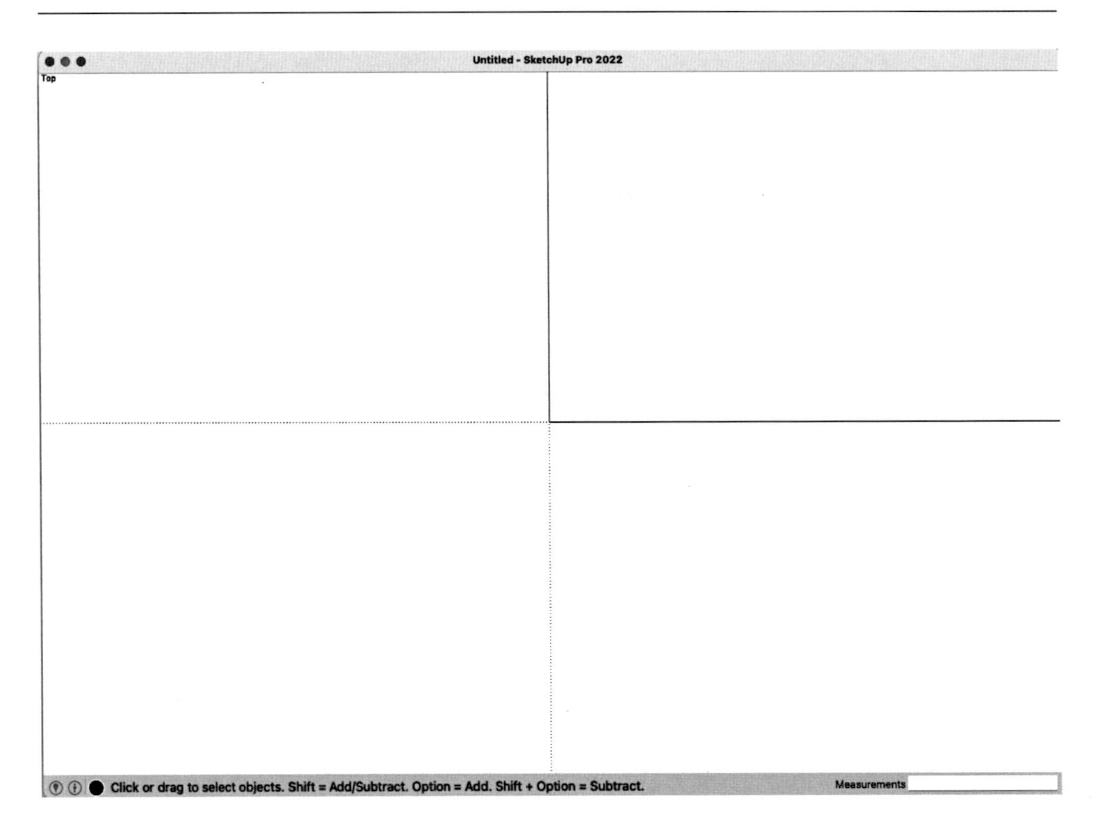

Figure 9.10 – Plan View template

This template starts with the camera directly above the axis, looking straight down. The sky is disabled, and the background is set to pure white. Unlike the other templates, this template has nothing in the drawing area to start with (no scale figure). This can make this template a little hard to start drawing in as there is no reference for how big anything is. This is a template that you might use if you are planning to sketch a rather large 2D model. Since SketchUp is not really intended for drawing in 2D, and there is nothing about this template that makes it any better for doing so, it is not a template that I would even recommend.

Urban Planning template

The Urban Planning template is designed for an urban planner or anyone planning to model large, sprawling models. See how this empty Urban Planning template is pulled way back, ready for a large model:

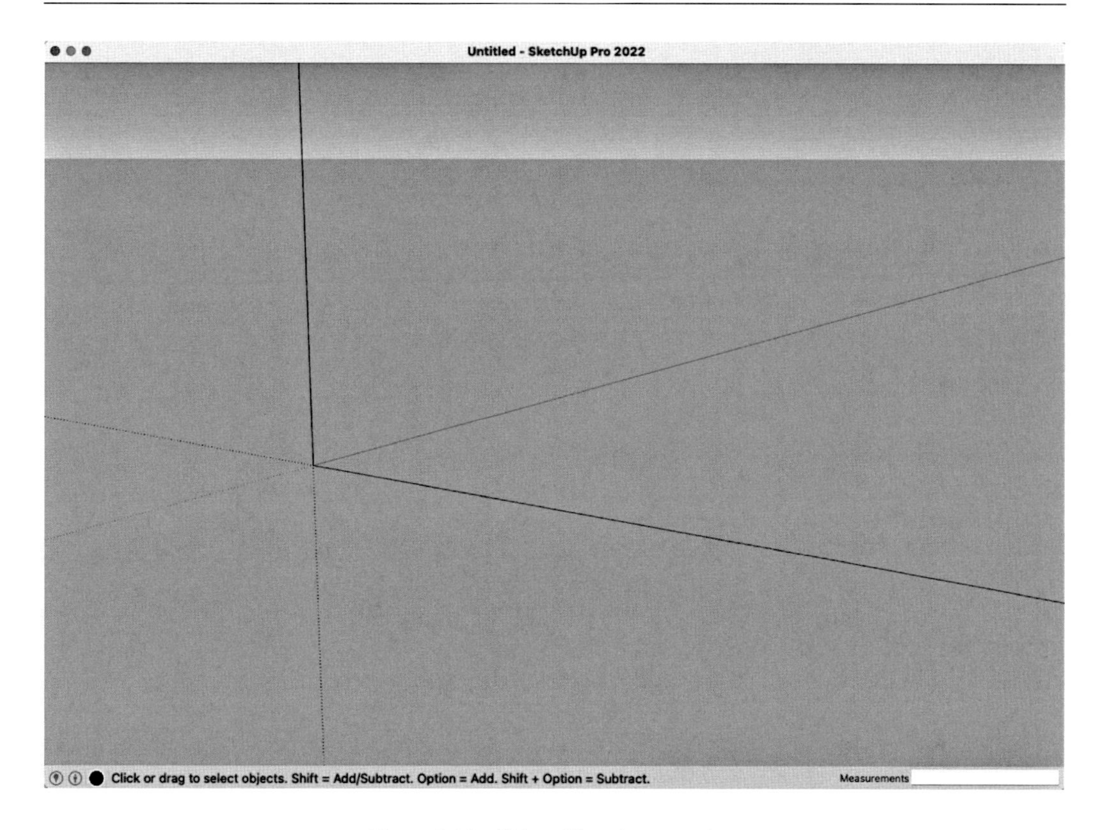

Figure 9.11 – Urban Planning template

This template has a **Perspective** camera at a three-quarter view pulled back quite far from the origin. The ground is set to the same neutral color as the Architectural template but with the bright blue of the Simple template. There is no scale figure in this template, and the units are only available in larger values (feet or meters). As the name implies, this template is a great starting point for someone who is doing urban planning in SketchUp.

Landscape template

The Landscape template is very similar to the Urban Planning template with a few differences. See the similarities to the Urban Planning template? Since we are talking about landscape, this template starts with a nice green background:

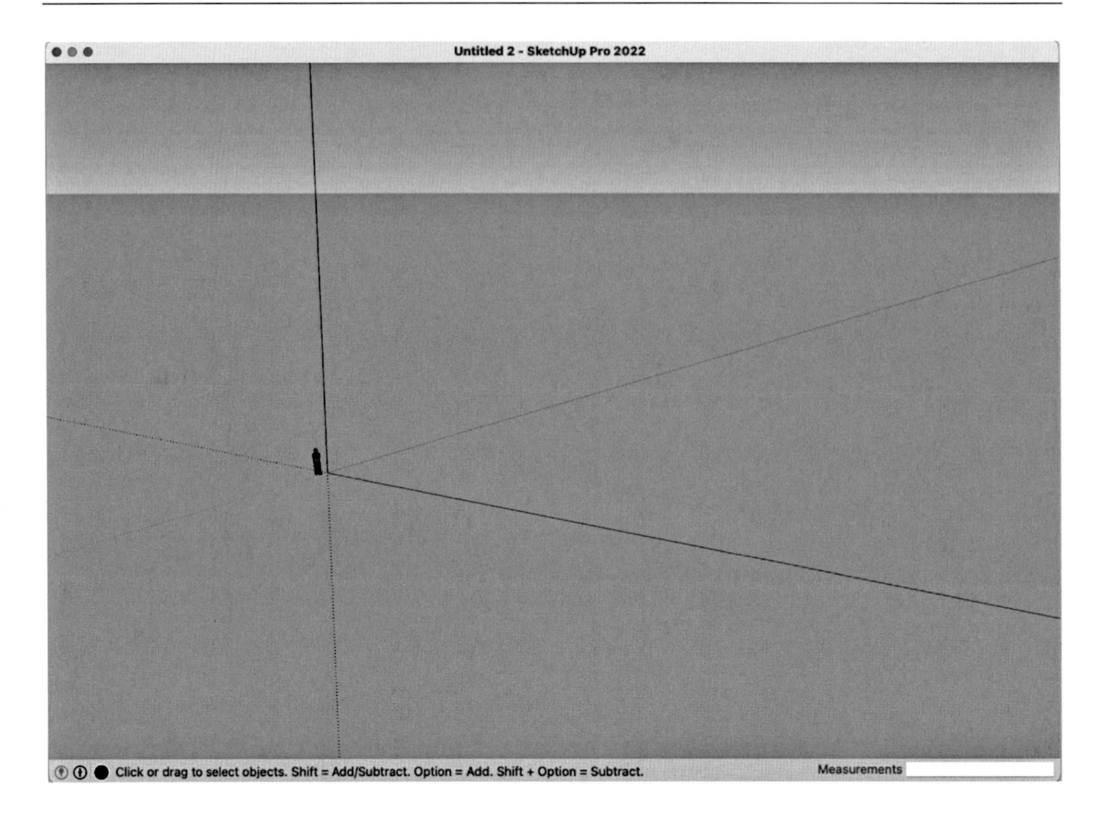

Figure 9.12 – Landscape template

This template has a camera setup like the Landscape camera but includes a green ground plane and a scale figure near the origin. This template is intended for use or customization by those planning to do landscape design in SketchUp.

Woodworking template

The Woodworking template was designed with woodworkers in mind. See how the empty Woodworking template is zoomed in, ready for a smaller model:

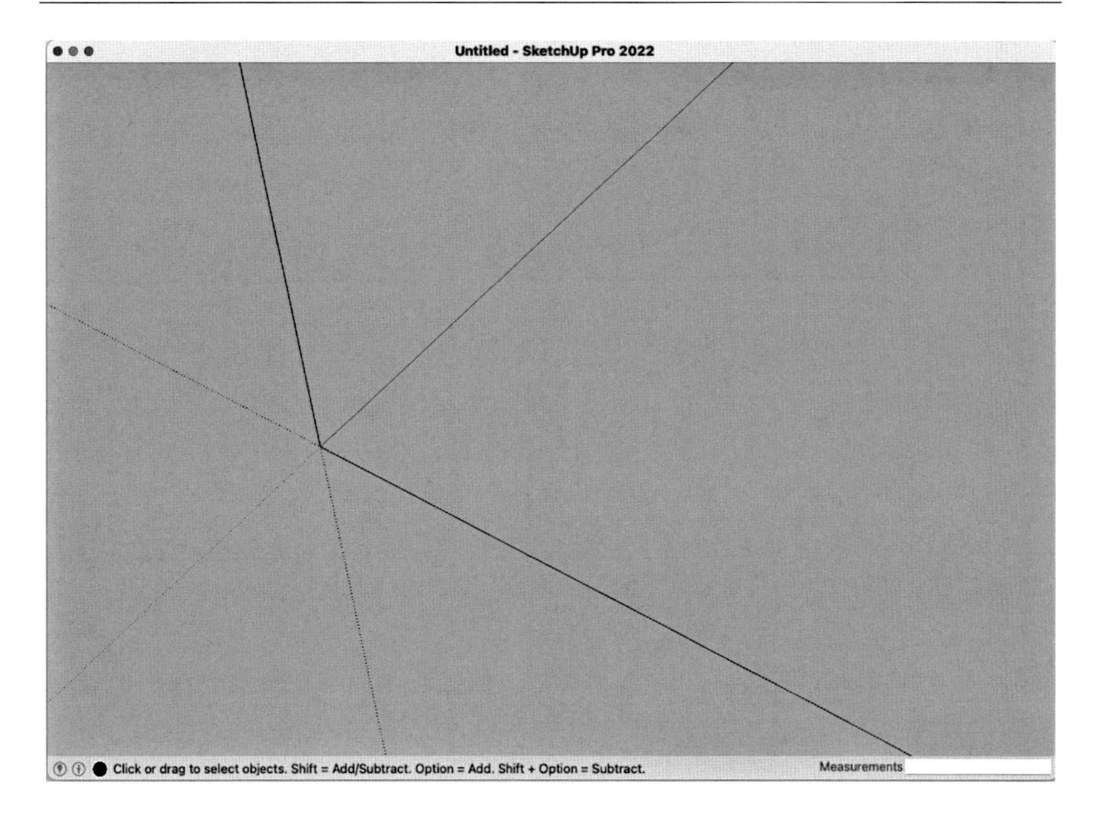

Figure 9.13 – Woodworking template

This template has a camera near the origin and no sky. The background is a neutral tan color, and it is available in units often used by woodworkers (inches and millimeters). This template is intended for use or customization by those creating woodworking or other smaller-scale models in SketchUp.

Interiors template

The Interiors template was designed for anyone designing the inside of a room or building, as shown here:

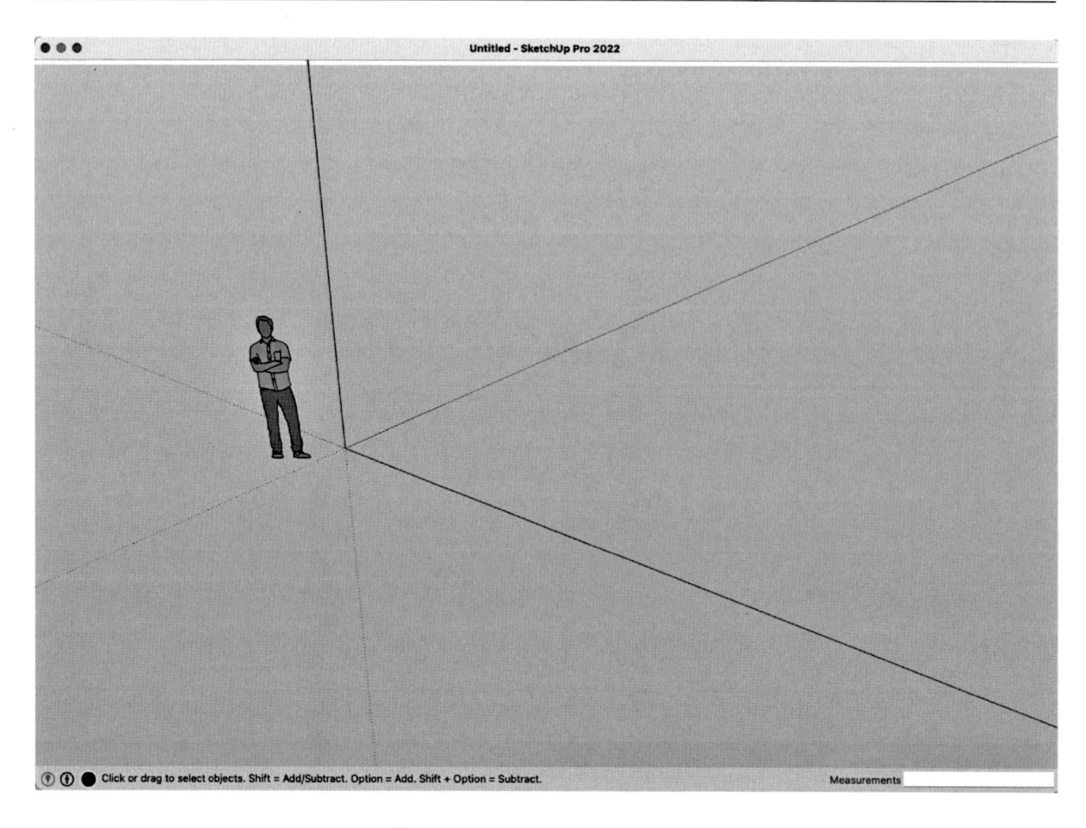

Figure 9.14 – Interiors template

This template has a camera setup similar to the Architectural template. The background is a neutral tan color, and it has a pale sky. This template is the one you may want to use or customize if you do interior or kitchen/bath design work.

3D Printing template

The 3D Printing template was designed for someone modeling with the intention of exporting to a 3D printer. Notice that the 3D Printing template is the only template that includes an item other than a scale figure:

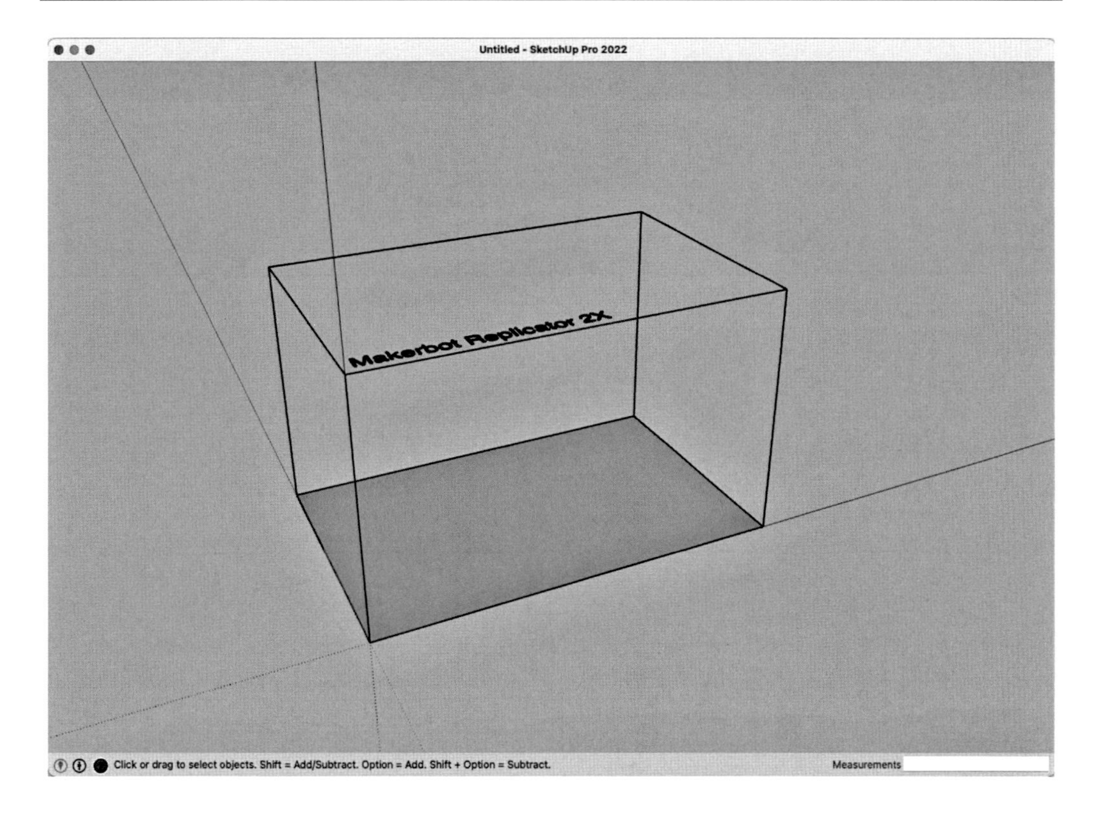

Figure 9.15 – 3D Printing template

This template has the camera zoomed in close to the origin and has a light gray ground plane and bright blue sky. The big differentiator for this template is the **3D Printer Build Volume** dynamic component at the origin. This dynamic component can be changed (right-click, **Dynamic Component | Component Options**) to show the exact volume of the 3D printer you are using. The aim of this template is to help 3D printers know whether their model would fit into their printer's printable volume.

> **My Personal Recommendation**
>
> While this sounds great, I recommend not using this template, even if you are 3D printing. Unfortunately, the dynamic component adds a lot to the file size and can cause issues with your output if you accidentally export it along with your 3D print.

Those are your options when it comes to default templates. If there is a specific template in this list you want to use each time you start a new model, you can click the little heart icon below the thumbnail in the **Templates** window. Doing so will use that template when you click **New** from the **File** menu.

Now that you understand where we are starting from, let's talk about what sort of templates you will want to create.

Deciding what templates to create

At this point in the book, you have probably picked up on the pattern and have a good idea of what this section will include. No, I am not going to tell you exactly what templates you should create and when to use them because everyone has different requirements for using SketchUp. Just like before, I am also going to ask you to look back at the work we did in *Chapter 6, Knowing What You Need Out of SketchUp*.

As we have seen, the whole point of templates is to give you a leg up as you start a new SketchUp model. To truly take advantage of them, we need to continue looking at what you hope to get done with SketchUp. If we look back to the lists of workflows you are interested in pursuing, you may be able to create templates based on each of these workflows. Alternatively, it is entirely possible that the default templates work for you but just need a slight change to be the ideal place for you to start a new model. Let's dive into both ideas and see how we might go about setting up our custom templates.

Modifying default templates

There is a good chance that one of the default templates will come close to satisfying your needs. Often, new users will get in the habit of opening one of the default templates, making one or two changes, then proceeding with their modeling process. Then, the next model they start, they do the same thing!

Now, if you are okay with this practice, please go ahead and keep doing it, but if you are serious about leveling up your skills, then this is an easy place to make a change. If you find yourself doing this, take a look at what it is that you are changing. Some people erase the scale figure from the template or change it to their own. Others immediately change the background color. Others go right away into **Model Info** to change their units. If this is you, here are some simple steps to never have to run through this routine again!

1. Start a new file as you normally would (using your preferred default template).
2. Make the changes you normally make.

   This is important! Run through and make all the modifications to the starting file you would throughout the entire model. If you swap out the figure and change the units at the beginning of the modeling process and usually make changes to your styles after you have done a little modeling, make those style changes right now, upfront. Make every change you normally would in this step.
3. Click on **Save as Template...**.
4. Name your template and check the **Set as Default Template** checkbox.

That's all there is to it! The next time you start a new file (via the **Welcome to SketchUp** window that pops up when you open SketchUp or by clicking **New** in the **File** menu), you will be greeted with a new file that already has all of your pre-modeling routine completed! This seems like a simple process (because it is) but it can add up to a lot of saved time and set you down the path of leveling up.

This is a great way to account for the little tweaks you may do when using a default template, but what about the times when you need a whole new template? When is it worth starting from scratch and making something that is very different from what the default templates offer? Let's dive into that right now!

Customizing your templates

Before you dive in and start clicking **Save as Template...** over and over, I recommend looking back to your list of workflows (from *Chapter 6, Knowing What You Need Out of SketchUp*). Think about each one of these workflows and the sort of modeling space you would want to be in for each. As an example, I will grab a few workflows from my list and think about how I would create a template for each. My workflows will be as follows:

- Modeling door and window components
- Creating a library of Face Me components
- Modeling Star Wars spaceships

These three workflows will be very different from one another and require a different starting model. In some cases, it may help to start into one of these workflows and see what sort of changes you make to your starting template before you save anything.

> **When to Template**
> It may seem odd to consider going through the process of saving a custom template for certain workflows. In some cases, you may be creating a one-off model that you will not need to create ever again. In the case of the examples created previously, I would imagine that I will create at least a dozen or so models in each workflow. This more than merits the creation of templates. However, as we have seen, template creation is such an easy process, and templates are so easy to maintain, that there is not really a downside to spending the time to create a template for anything that you plan to model more than once.

Let's take a look at how we would create templates for each of these workflows. For these examples, I will present a *brief* explanation of what I think my modeling needs will be for the template. I will also start every template from the Simple template. As you create templates, you might want to use a different template or another of your custom templates as the starting point to save some time.

Creating a template for modeling door and window components

A template for this type of modeling would be fairly close to a default template. It would be something that was used in 3D, and I would prefer not to show a sky. I am going to be building these components off manufacturer drawings that are in decimal inches, so I need those units as default. I also like to model with the default material, so to have the models stand out, I would like the background to be a color other than white. I also like to put dimensions on my door and windows as I create them, so it would be good to bump up the font of dimensions from the default to make them easier to read. Ideally, this will be as efficient a template as possible, so I do not want shadows or profiles, which can slow down modeling.

Based on this list of needs, I will start with the Simple template from the defaults and make the following changes:

- Disable **Sky** in **Styles**.
- Set units to **Decimal** and choose **Inches** in the **Units** tab of **Model Info**.
- Set the **Background** color to a pale blue color.
- Modify **Dimensions** in **Model Info** and increase the font to **18**.
- Disable **Profiles** in **Styles**.

This will create a model that looks like the following:

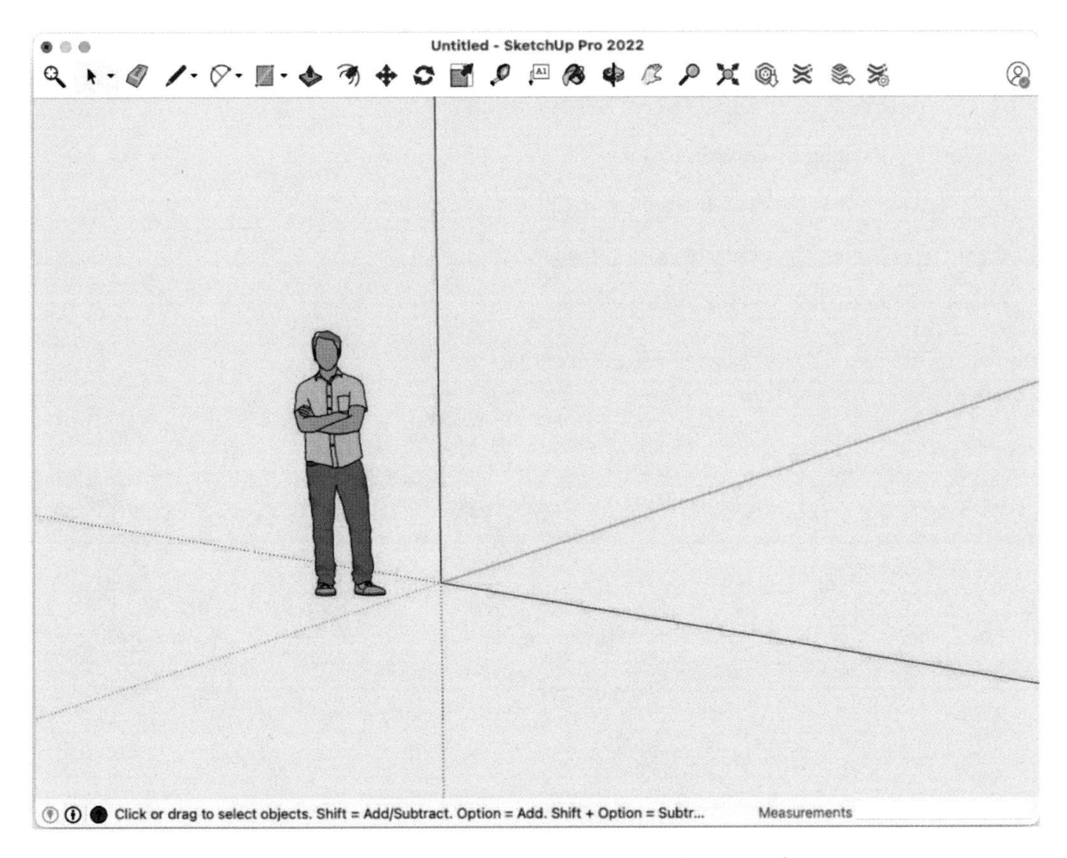

Figure 9.16 – Newly created Doors and Windows template

Once these changes have been made, I can save this model as a template, call it `Doors and Windows`, and I am all set to start my new workflow!

Creating a template for modeling Face Me components

This template will be used to import and then trace full-size images of humans. Since Face Me components are modeled in 2D, this template would not start with the camera in a three-quarter view but looking straight on at the origin. The camera should even be in **Parallel Projection**. Since I will be drawing figures that are around human size, I can zoom in so that a 6-foot-tall image roughly fills the modeling screen. The ideal view for tracing images is in X-ray mode (so that new faces do not cover up the reference image), so that should be turned on. I also know that it is easier to trace an image with edges that stand out against the image, so the edge color should be set to yellow, rather than black.

Again, I will start with the Simple template from the defaults and make the following changes:

- Use the **Front** view from **Standard Views** in the **View** menu.
- Turn on **Parallel Projection** in **View**.
- Zoom in so that the default scale figure fills the screen, vertically.
- Turn on **X-Ray** from **Face Style** in the **View** menu.
- Set the **Edge** color to yellow in **Styles**.

This will create a model that looks like the following:

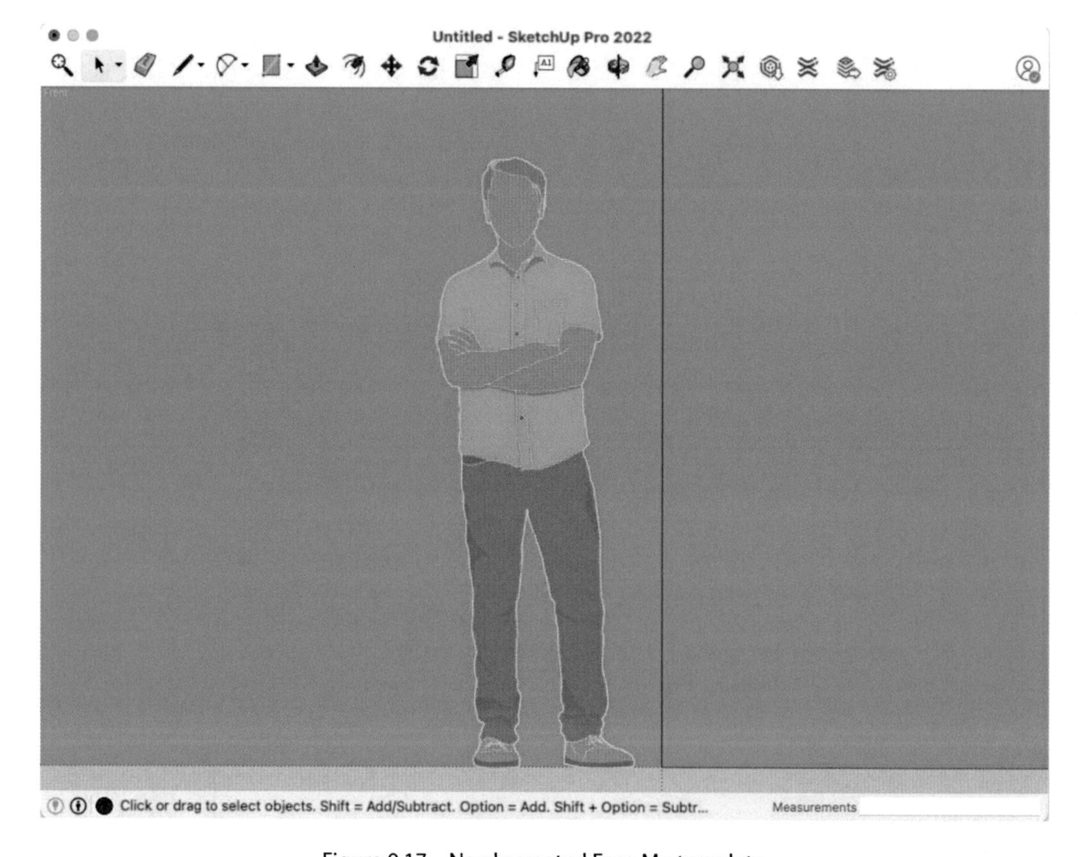

Figure 9.17 – Newly created Face Me template

Once these changes have been made, I can save this model as a template, call it `Face Me`, and I am all set to start creating some 2D entourage components!

Creating a template for modeling spaceships

In this workflow, I will be referencing imagery of spaceships from Star Wars. While I may end up adding textures later, I want to do most of the modeling in plain white. I need to be able to see all of the larger models, so I would like to start zoomed out. To get the feel of spaceship modeling, I would like to model against a gray background. Most of the images that I am modeling off of have dimensions referenced in decimal millimeters.

Based on this list of needs, I should probably start with the Simple template from the defaults and make the following changes:

- Set **Face Type** in the **View** menu to **Monochrome**.
- Zoom out so that the default scale figure is small in the lower-left corner.
- Set the **Background** color to a light gray color and disable **Sky** in **Styles**.
- Set units to **Decimal** and choose **Millimeters** in the **Units** tab of **Model Info**.

This will create a model that looks like the following:

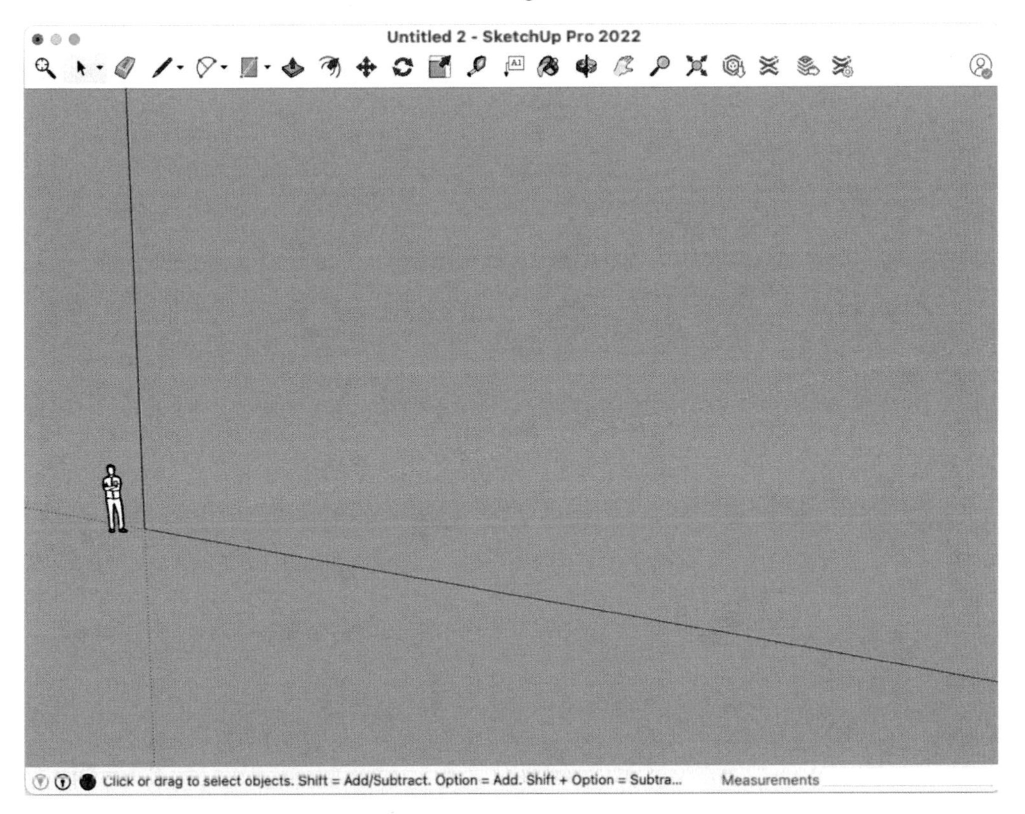

Figure 9.18 – Newly created Spaceship template

Once these changes have been made, I can save this model as a template, call it `Spaceships`, and I am all set to start my new workflow!

I know we did not go through these step by step, but I am hoping that the lists of changes for each template got you thinking of the steps you might take to create your own templates. The important takeaway for this section is not how I made each of these templates specifically, but the process used to break down the needs of a template for a specific modeling workflow and the changes that should be made in order to make a useful template. You can use this process to make any template. Once done, try using the template and see whether there are any changes or tweaks you need to make. Template creation can be seen as an iterative process. As your modeling workflow changes, so will your templates.

Creating templates is not a difficult process. The bigger challenge can be maintaining your templates once you start making them. It can seem like a good idea to save every model into a template as you get started, but eventually, this will leave you drowning in templates. To keep that from happening, let's take a look at how to maintain your template collection.

Creating and maintaining custom templates

We have covered the basics of saving a template, but let's dive just a little bit deeper and look at every field on the **Save As Template** screen and how and when to use them. The process for creating a template is extremely easy and is the same on SketchUp for Windows or macOS. Let's assume that you have done everything you need to set up your template. At this point, it is time to save your file as a new template. Let's run through the steps:

1. Set up everything in the current file as you want it saved.

   This is probably the most important part of creating a custom template. We already discussed how to figure out exactly what to change to create a template in the *Deciding what templates to create* section.

2. Click **File**, then **Save as Template…**.

   This will bring up the **Save As Template** window:

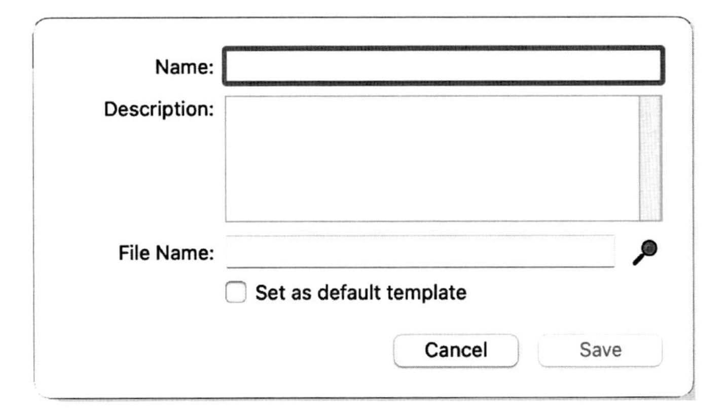

Figure 9.19 – All information for a custom template is entered in this window

3. Enter a name and description of your custom template.

   Notice that the name of the template file is displayed in the **File Name** field at the bottom of the window. By default, the filename will match the template name, but you can change it if you like.

The macOS Magnifying Glass

In the **Save As Template** window in SketchUp for macOS, there is a magnifying glass icon to the right of the **File Name** field. Clicking it will allow you to navigate folders but not change where the template is saved.

4. If you want SketchUp to use this template every time you start a new file, check the **Set as default template** checkbox.
5. Click the **Save** button.
6. You have created a custom template!

Creating a custom template is incredibly easy, and when a user first learns the steps to create one, they often wonder, "Why don't I have a bunch of templates?" The simple fact is that a well-thought-out template can get you started so much quicker than modifying a default template every time you start a new model. There are concerns, of course. The more templates you create, the more files you have to maintain.

On the plus side, template maintenance is not a difficult business. As you create custom templates, you will be able to see them in the same window that you open default templates from. At the top of the screen, there is an option to view **Default templates** or **My templates**. Clicking on **My templates** will show all of your templates:

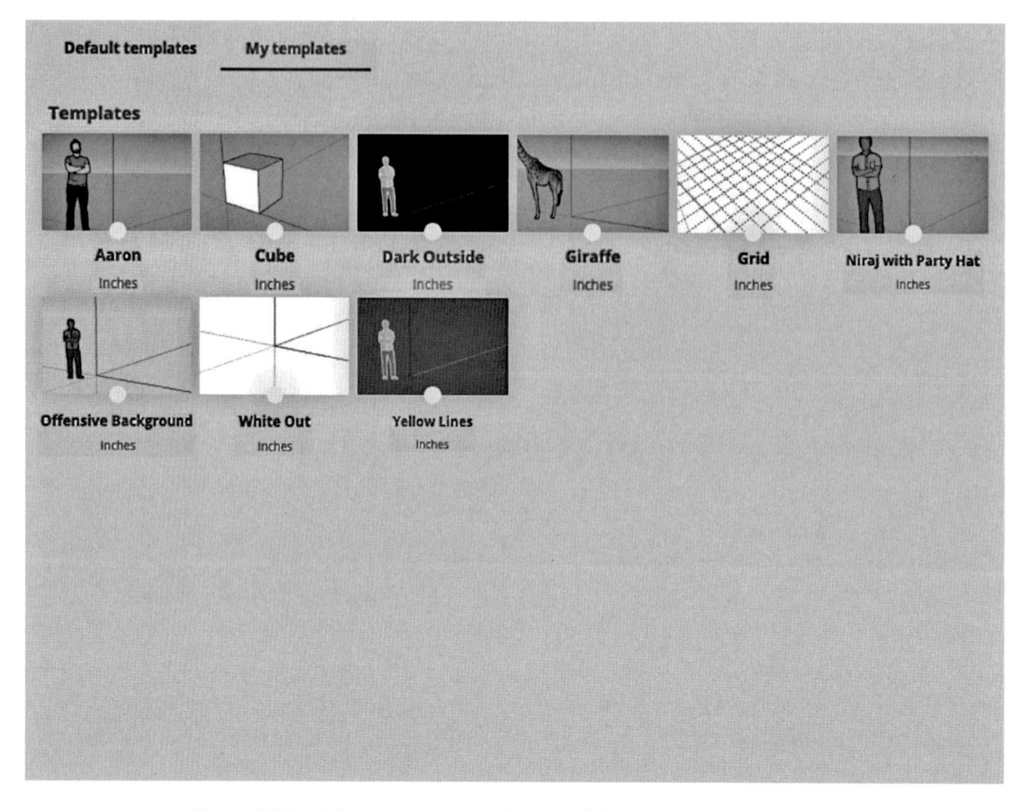

Figure 9.20 – A few custom templates visible through My templates

> **Viewing My templates**
>
> In order to view your custom templates, you have to have at least one custom template created and saved. The **My templates** option at the top of the page will only appear after you have saved at least one template.

Seeing all of the templates is great, but there are no options for modifying these templates in this window. To edit or remove any of your templates, you will have to get hands-on with your files.

Deleting and renaming templates

Choosing to use a template is as simple as clicking one of the template thumbnails when you start a new model. To delete a template from the list, you will have to remove the SketchUp file (`.skp`) from the template folder. To find where templates are saved, check the **Files** tab in the **Preferences** window. **Preferences** can be found in the **Windows** menu in SketchUp for Windows, and in the **SketchUp** menu in SketchUp for macOS. Do you notice anything different about the **Templates** section of the **Preferences** window?

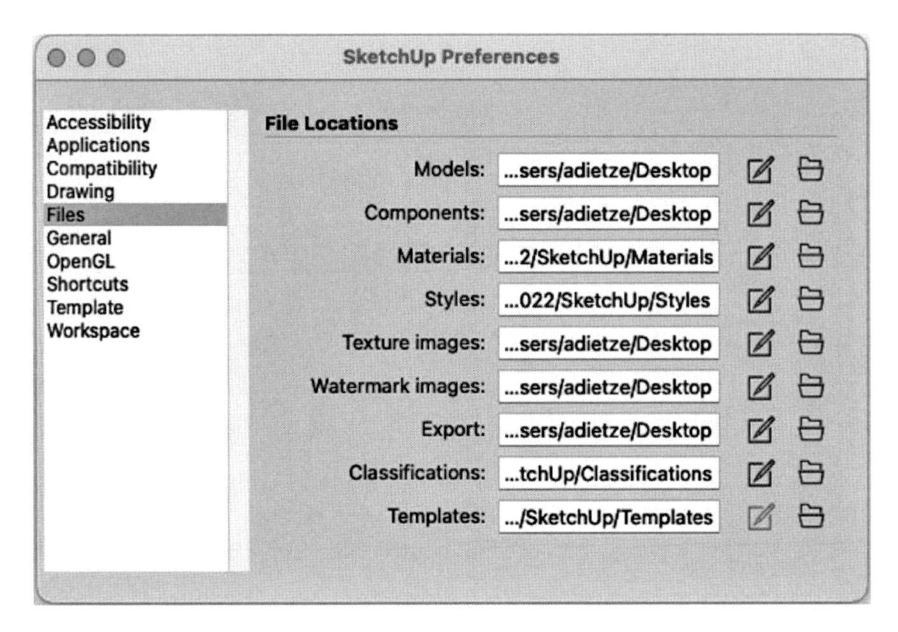

Figure 9.21 – Templates path shown at the bottom of the Files tab as seen in SketchUp for macOS

The last file path displayed is the location of your templates. As of the time of writing this book (SketchUp version 2022), this path can be viewed but not edited on either operating system.

Clicking the folder icon to the right of the **Templates** file path will open a window displaying the files saved there. From here, you can rename or delete templates just as you would any other file on your system. Changes made to the files in this folder will be reflected in **My templates** the next time you view that screen.

This brings us to the end of the chapter and should have you thinking about the templates that would benefit your workflows and help to level up your SketchUp skills!

Summary

With this, we have fully explored the power and time-saving nature of templates. Not only do you understand what each of the default templates might be used for, but you should also have an idea of how you might use them as starting points for your own custom templates. You have seen how to create templates and know how easy it is to maintain your list of custom templates. Finally, we set up a basic process for thinking about what sort of templates you might make and how to go about breaking down the needs of your workflow so that a template will save you time every time you start a new model.

Up next, we will continue to look into ways to level up your modeling speed with a dive into additional hardware options in *Chapter 10, Hardware to Make You a More Efficient Modeler*.

10

Hardware to Make You a More Efficient Modeler

Up until now we have spent time looking at how to use SketchUp to improve your 3D modeling skills. Practice and planning are the best possible way to get better at something. There is, however, something to be said about the tools you use, as well. I know, there is the old idiom that says, "*A bad carpenter blames his tools*," and I totally agree. If you practice and learn how to do something, you should be able to do it with even the simplest of tools. However, I will make up my own idiom and say, "*Level up your tools, level up your skills!*" Now this may not be true in all situations, but in the case of 3D modeling in SketchUp, the right set of tools can help increase modeling speed, cut down on errors, and make you more efficient.

It is worth noting right now that nothing in this chapter is a requirement for running SketchUp. SketchUp can be run on a laptop with *no* peripherals at all. People do that every day. The point of this chapter is to explore the hardware options that are out there, available to you and how they may help you to become a better SketchUp user.

Just My Personal Opinions

I should also point out that I have no tie to any of the manufacturers of any of this hardware. I do not benefit from you purchasing anything recommended in this chapter. While in some cases I will relay information regarding hardware that I have heard about, I will also make recommendations for the hardware that I have used and believe will help you in your quest to level up.

In this chapter, we will be exploring five different pieces of hardware that you may want to consider:

- Mice
- Monitors
- Keyboards

- Tablets
- 3D Mice

Technical requirements

All you need for this chapter is the ability to read! You may want to look up some of this hardware, so access to a browser may be a good idea.

Mice

A good mouse is not just an option, but pretty close to being a requirement, when talking about SketchUp. Back at the beginning of the book (*Chapter 1, Reviewing the Basics*) we talked about why a 3-button mouse was so important when navigating in SketchUp. Yes, it is totally possible to use SketchUp with a 1-button or 2-button mouse and, yes, you can also get around with a touchpad or trackpad. Heck, if you have to, you could probably struggle through using no mouse if your computer has a touch screen, but this book is not about getting through it. This book is about getting better at using SketchUp, so I will be talking about the mice that I have seen and heard about, that make modelers better.

Basic 3-button mice

Before we dive into the multi-button, advanced-feature mice that are out there, let's spend a few minutes talking about a basic 3-button mouse. For anyone using a touchpad or 1-button or 2-button mouse, this will be an upgrade that could make SketchUp far more usable.

You know the mouse I am talking about. These are the ones that come with a lot of computers. If you have to purchase a new one, they usually can be found for under $15 USD at stores like Best Buy or Target or even cheaper at Amazon.

Figure 10.1 – No-frills, wired, 3-button mouse

The nice thing about these mice is the middle -button/scroll wheel. Using that to do all the orbiting and zooming means never having to run to the toolbar or rely on shortcut keys to change your view.

While we are talking about straight forward 3-button mice here, there are options. If you do any amount of travel, it's nice not to have to deal with a USB cable on your mouse. Nowadays, even the less expensive models can be found wired or wireless. In fact, it's pretty easy to find mice that uses Bluetooth rather than having to rely on a USB dongle. Where you can get a wired 3-button mouse for under $10 USD, you can generally find a 3-button wireless mouse for under $20 USD.

Programmable mice

If you are being serious about levelling up your SketchUp skills, it's worth looking into a mouse with more functionality and more buttons than the basic models will offer you. While these models generally do cost more and will require additional setup, it is worth it if you are serious about becoming a better, faster, more efficient modeler.

In the world of programmable mice, there is no shortage of choice. Most of these mice offer things like additional buttons, higher-fidelity tracking, or gesture support. They may be wired, have a USB dongle or use Bluetooth. Some models designed for gaming will offer things like customizable weight, lights, or even a little fan inside intended to keep your hand cool during marathon design sessions.

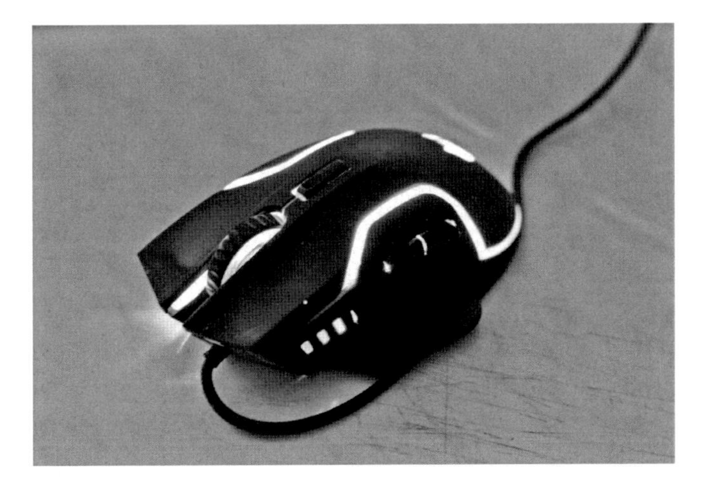

Figure 10.2 – A 7-button gaming mouse A+ Kuafu from SketchUp forum user Mihai.s

While there are a lot of options in this sort of hardware that I don't really see as a benefit (lights, weights, and more), having additional buttons on a mouse that can be programmed to activate commands is a huge advantage. Just like it is a time and energy saver to be able to use the scroll wheel to zoom and orbit, it's a huge benefit to be able to activate your most commonly used commands by pressing a button without having to move a finger.

Some of these mice looks and feel like a standard mouse while some get creative with their button placement. Logitech offers a line of mice with increasing capabilities that looks and feel very much the same, one to the next, but offer more and more customizability. Other mice, like the Razer Naga mice, offer up to a dozen programmable buttons on the side where your thumb sits. If you are diving into the world of customizable mice, I recommend reading online reviews or stopping by a bricks-and-mortar store where you can at least put your hand on one of these devices before you buy. It is pretty easy to spend $100 USD on a mouse, and it is a good idea to *try before you buy*!

Speaking of Price

To some, it can seem weird to spend $100 USD on something like a computer mouse. If you are a full-time designer, the mouse is a thing that is in your hand all day long. Aside from the computer that it is connected to, it is likely the tool that you use more than any other. If spending a little more means having the functionality that will help save you time as you design, it is worth it! If spending a few dollars more means you get a namebrand mouse with a warranty and a track record for lasting years, it is worth it!

Mouse gestures

Gestures are movements you make with your mouse to which you can assign functionality. With a certain mouse or software, holding down a button on the mouse and moving it to the right may activate the **Move** commands in SketchUp. Gestures are another way to save your time and energy by keeping your hand on the mouse when switching commands.

While many of the programmable mice out there will allow you to assign shortcuts to gestures through their own proprietary software, there is software out there that can be used with any mouse. In fact, if you type `mouse gesture software` into Google, you will get several pages that list the top options available!

Gestures are an amazing way to add customized functionality to your workflow. Think of gestures as shortcuts to your mouse! In general, the way they work is by assigning a keystroke to a movement of the mouse. So, if you were to hold down the right mouse button and slide your mouse to the right, it would be the same as hitting the *R* key on your keyboard. In SketchUp, hitting *R* will activate the **Rectangle** command. Things like gestures do not replace shortcut keys but make them accessible without having to move a hand to the keyboard!

While I recommend the use of mouse gestures, I have only used them as a part of the driver for my preferred mouse, so I cannot recommend a specific software for use. There are, however, more than a few options on the market. The good news is that these programs are available for free or very cheap ($10 USD). The bad news is that most of them that I came across were Windows only, so Mac users may not be able to try them.

OnScreen Keys

While speaking of software that adds to what a mouse can do, it makes sense to mention that there are software tools out there that allow you to bring up a series of on-screen keys or shortcuts via your mouse. This may be as simple as calling the default on-screen keyboard via Windows, or using the on-screen radial menus in 3Dconnexions's driver. Again, having access to programmable keys without taking your hand off your mouse can be a huge time-saver.

Author's recommendation

At the end of every section, I figured it would make sense to tell you what hardware I have been using or preferred. Please understand that, while I have tried out several options, my search has not been exhaustive. I do use SketchUp a lot and have found a set of peripherals that helps me and makes me an efficient modeler.

My preferred mouse is one of the MX Master mice from Logitech. I have had four mice from this line (at the time of this writing I have a MX Master 3). I have used a lot of Logitech equipment over the past few decades and have found their hardware to be well made and their software (the *Logi* app) as one of the better tools for customizing the functionality of a mouse.

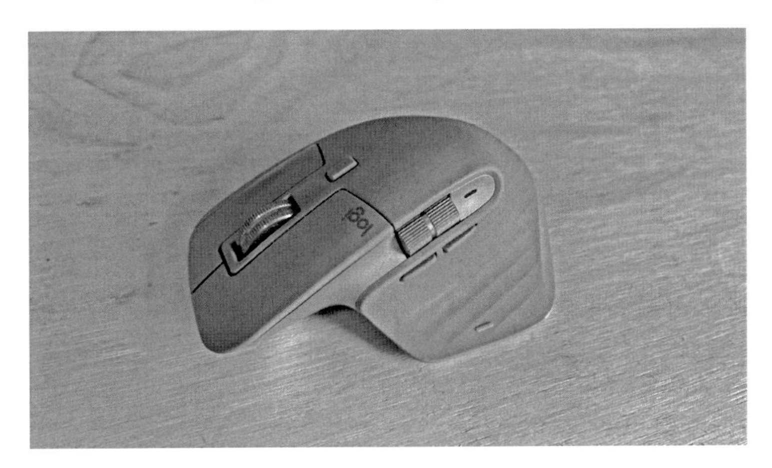

Figure 10.3 – My grey Logitech MX Master 3

I like that it is a larger mouse and fills my hand. I also like that it is Bluetooth (no wires or dongle needed), and it can be synced to three different devices (I often run my MacBook, a Windows laptop, and use it on my iPad). The thing that drew me to it, however, is the number of shortcuts I can assign to it.

I currently have seven shortcut keys assigned to the mouse. Two (**Line** and **Eraser**) are assigned to the *Forward* and *Back* buttons that are right above the thumb rest (did I mention that there is a thumb rest? How fancy!). If I press down on the thumb rest, I enter **Select** mode. If I hold down on the thumb rest and gesture up, down, left, or right, I activate **Push/Pull**, **Move**, **Rectangle**, or **Rotate**. Having these, my most commonly used commands, available to me without ever lifting a finger off my mouse is a time saver that really adds up. More than that, it means that I can keep modeling without breaking my flow to switch commands.

It is one of the more expensive options out there. It retails for around $99 USD (but it can usually be found on sale). To be honest, I have never let that bother me when I consider that this mouse will last me years. In the end, it probably costs me a few cents per day to use a piece of hardware that I trust and that saved me so much time while modeling!

The mouse is probably the number one peripheral that you should consider when upgrading your equipment. As far as SketchUp goes, it is the peripheral that will be used the most. Another important peripheral to consider, is your monitor!

Monitors

Back when we were looking at setting up your workspace in *Chapter 8, Customizing You User Interface*, we talked a little bit about monitors. While I cannot say that I can dive quite as deep into the world of displays as I can with mice, I do have a few thoughts to share that I think are important, as far as SketchUp is concerned.

Video cards

I know, we are here to talk about monitors, but we should spend a moment or two talking about the hardware that drives the monitors. When running SketchUp, you do need a video card that is capable of fully supporting OpenGL. Most video card from Nvidia and AMD will support OpenGL in a way that keep SketchUp running smoothly. In some cases, though, on-board, or integrated video cards fall short of full support. At the same time, some newer video cards designed for high-end graphics for video games also lack in OpenGL support.

To be honest, it would probably be possible to dedicate an entire chapter to video card selection and how OpenGL works, but it would not be in the spirit of this book. I really just wanted to drop this section in to let you know that if you are planning to purchase a new computer, or add a video card to your machine, just make sure that the manufacturer has OpenGL support in the specifications. If that is the case, then you should be able to run SketchUp on additional monitors without any issues!

Laptop displays

I can already hear the arguments starting. I know that there are plenty people out there who only do 3D modeling on their big, beefy desktop computer. If that's you, that is great and you can skip to the next section! If you are someone like me who primarily uses SketchUp on a laptop, I wanted to mention a few things about this setup.

Using the integrated laptop display as your primary display may work great for you in many cases (assuming that you have OpenGL support as mentioned in the *Video cards* section). However, you may, in this case, want to invest in a secondary monitor for serious modeling sessions. While you may have a beautiful, high-resolution laptop display, and may even be proud of the fact that you are running a 16" notebook, the fact is that a second display, even an inexpensive one, may give you access to a SketchUp workspace that is twice the size!

Laptops are great and getting better every day. Being able to pick up SketchUp and take it wherever you need to go to work is amazing. To be honest, compared to other 3D drawing programs, SketchUp is far less demanding and runs great on most laptops (I have seen some other drawing programs that slow to a crawl on an average laptop). Do not take anything said here to suggest that you should change machines. If you are serious about leveling up your SketchUp skills, however, you may want to consider adding an additional display to your current laptop only setup.

Picking out a monitor

When it comes to monitors, we really are spoiled for choice. In fact, monitor manufactures seem to crank out new, higher resolution monitors so quickly, it is hard to look at specific examples (the monitor on my desk is only a few years old and already discontinued. While I cannot recommend a specific monitor, I will present a few things to think about as far as monitors go:

- **Resolution** – SketchUp is really pretty basic as far as resolution requirements go. There is some basic anti-aliasing (the process by which a line is made to look smooth on the screen), and you have the option to control how X-ray views work, but other than that, SketchUp displays some pretty simple graphics. When it comes to the resolution that you need to display, it is possible that other programs you are running may be more important to satisfy than SketchUp. Many SketchUp users end up doing video work with the assets they create or end up in a rendering software. In either of these cases, picking out a monitor based on how it will display these programs may trump what is needed to display SketchUp.
- **Size** – Is bigger always better? Yeah, probably. I know that entire articles have been written about the ideal monitor size and formulas have been created to help figure out what size a monitor should be based on viewing angle and distance from screen and phase of the moon. In the case of a monitor for working in SketchUp, a larger monitor means more space for UI and a larger modeling area. This does not mean that you need a 49" curved monitor just to run SketchUp, but it doesn't mean that monitor would not help, if you had it.
- **Quantity** – Here's the big question. With modern computer and operating systems coming out of the box with the ability to drive 4+ monitors, just how many monitors should you have? Do you keep it simple and work with a single monitor, or do you go full *Minority Report* and cover the wall of your office with displays? Personally, I have been a fan of a dual monitor setup for as long as I can remember. I like dedicating one, primary display to the program I am using (after all, SketchUp only runs in a single window) and rely on the second monitor to display everything else. However, if I was getting into something that would benefit from running on an

additional display, I may consider adding another monitor. If you do any amount of rendering, for example, you may see the benefit of adding an additional display where you could keep an eye on something like the status of renders. Really, with the how affordable monitors are recently, the number of monitors you need is most likely limited by your desk size more than anything else!

Figure 10.4 – My friend Nick Sonder shared an image of his amazing 3 monitor setup

As you consider monitors for your SketchUp setup, I would be sure to think about the other software you are running and the physical limitations of your workspace. Having a quality monitor will give you a good-looking workspace and last for a long time. It is up to you to decide how much money and desktop space you want to dedicate to monitors.

A Note About Touchscreens

I debated over including touchscreens in this section. While touchscreen technology has come a long way and is now even standard on many laptop computers, the fact is that SketchUp does not support this functionality very well. Yes, basic functionality like clicking on a menu or icon can be performed using a touchscreen, SketchUp does not leverage anything else that a touch screen offers (gestures, pinching, and likewise). Since this whole book is about what you can do to improve your SketchUp skills, it seemed like an option that did not need more than this note.

Author's recommendation

To be honest, I keep it pretty simple in this department. I tend to work in one of three locations (home office, desk at work, and in the recording studio). Each of these locations has a secondary monitor for me to plug into. Most of these monitors are high-resolution and 24+ inches. Most, if not all my actual SketchUp design work is done on the built-in display of my MacBook. The secondary monitor is always being used and displays things like email, Messenger, reference images, or recording/streaming software. In a pinch, I can run everything with just my laptop, but I will admit that I feel less productive that way. I end up switching windows in the middle of a modeling session or running things in a smaller screen, so I can see other programs at the same time. If I really want to be productive and model the best I possibly can, I do need a secondary monitor to the side of my laptop.

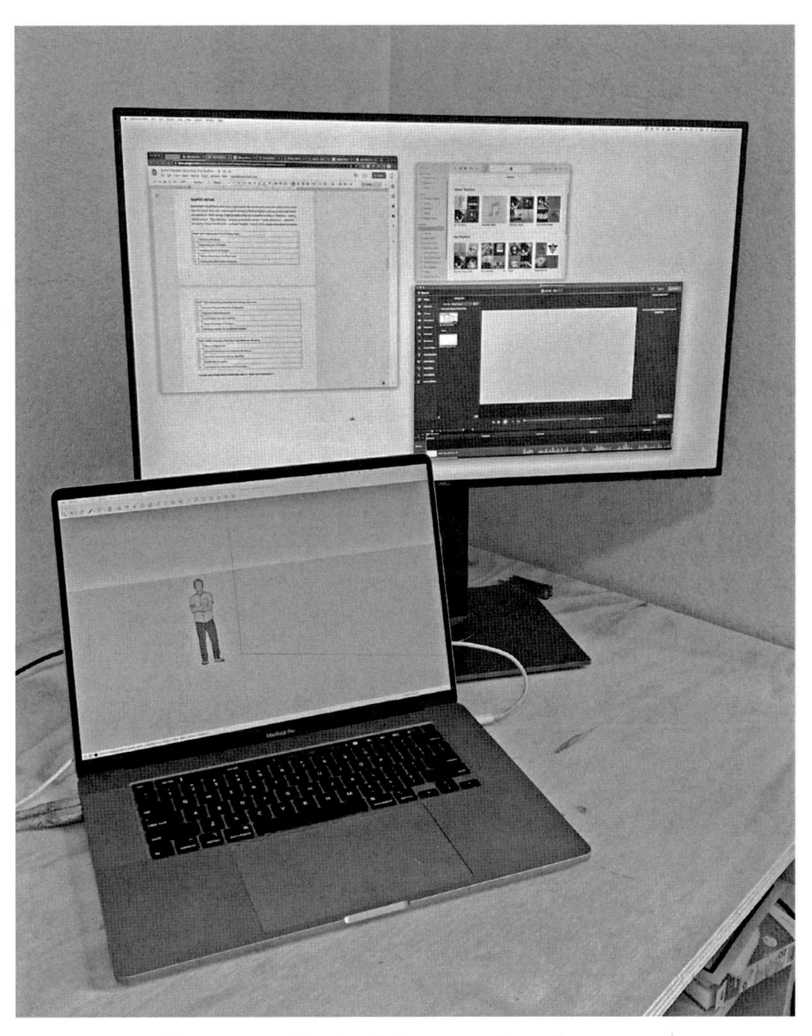

Figure 10.5 – My simple laptop and monitor setup

Displays are, of course, required to run SketchUp or any computer software. Likewise, you will, sooner or later need to have a keyboard. Are all keyboard created equal? Of course not! Let's find out what to look for in the next section.

Keyboards

As with almost all computer programs, SketchUp Pro will require you, at some point, to use a keyboard. If you are keyboard averse, you could do a lot of your input and editing using only your mouse by using commands from the menus, but sooner or later you are going to need to name a component or enter the dimensions of a line and for that, you will need a keyboard.

For most users, the keyboard is a fairly vanilla peripheral that they do not think about upgrading while others will spend time researching the perfect keyboard or even build their own! If we really wanted to dive down the rabbit hole, we could get into things like interchangeable keycaps and mechanical versus membrane keyboards. For this book, though, we are exploring things that will help us to level up our SketchUp skills. The main area of focus will be around taking advantage of the shortcuts that we set up in *Chapter 7*, *Creating Custom Shortcuts*, and thinking about what keyboards are out there that can speed up our input.

To this end, there are three keyboard options I would like to talk about: Ten key keyboards, custom keypads, and virtual keyboards. I am assuming that whether you are using a laptop or desktop computer, you have access to some sort of keyboard. From here, let's look at how you might be able to customize your setup with an effort to level up.

Tenkey keyboards

If you are already using a full-size keyboard, you may already have a ten key keypad on the right side of your keyboard. If you are using a laptop keyboard as your primary computer or a compact keyboard, you may not have a ten key keypad. Just in case anyone is not familiar, the ten key keypad is a rectangle of numerical keys set to the side (usually the right) of the main keyboard. These keys are setup in the same configuration as the keys on an adding machine. While these are often leveraged by people using accounting or spreadsheet input, they have their place in 3D modeling, as well.

Figure 10.6 – Standard ten-key keypad on a full-size keyboard

While the rest of the keyboard is useful for using shortcuts, the ten key keypad is useful for entering dimensions. As you use SketchUp, you will be entering dimensions. This may be as you are drawing (specifying line lengths or polygon sizes) or as you are editing (specifying move lengths or scale dimensions). Depending on your setup, the ten-key keypad can save you quite a bit of time for this input. If you are right-handed, this might mean sliding your hand off the mouse to the edge of the keyboard to type numbers (while left-handers will already have their right hand on that side of the keyboard). I know, in the ideal workflow you would not have to take a hand off the mouse but think about how much quicker it is to move your hand to one spot where all of the number keys are clustered, rather than running your hand back and forth across the full length of the keyboard to find the numbers that you need.

I know, we are talking about seconds or fractions of seconds, but we are also talking about doing this dozens or hundreds of times each day. Just like when we talked about using shortcuts keys rather than menu commands, the time it takes to move your hand from your mouse over to the *1* key at the top left of your keyboard will add up, if you have to do it 30 times in a single model.

Plus, it is much easier to remember where the keys on a ten key keypad are located. Where you may have to look down at your keyboard to find the *4* key each time, there is a good chance that, with a little practice, you might be able to jump your finger right to the *4* key just by feeling the keys. If you have ever seen an accountant using an adding machine, you will know that they can input numbers for hours without once having to look at the keypad.

OK, so let's say you are bought into the idea of trying a ten key keypad but use a laptop without one. Maybe you have a compact keyboard like an Apple Magic Keyboard or a Logitech MX Keys Mini. You can't just add keys to your existing keyboard, right? Well, in this case, you kind of can do that. If you search `ten-key keypad` on Amazon, you will get a list of hundreds of keypads, both wired and wireless, that will sit right next to your existing keyboard. As a plus, they are fairly inexpensive (many options for under $30 USD), so it is something that you could try and, if you do not love it, you are not out a huge amount.

Figure 10.7 – SketchUp forum user Endlessfix sent me this image of his custom ten-key keypad

Next time you do some serious, heads down geometry modeling, keep an eye on how often you are moving your hands across the keyboard. If you have a ten-key keypad on your current keyboard and do not use it, give it a try to see if it helps. If you notice that you spend a lot of time hunting for number keys, consider picking up a cheap external ten key keypad. It may be a small effort that ends up speeding up your input and helps you to level up!

Customizable keypads

Another option to consider is a customizable keypad. With these keypads, you can program whatever keystrokes you want onto any of the keys on the device. This can mean having all of your most important shortcut keys clustered together right next to your non-mouse hand. Imagine having access to 32 of your most used commands on something like this:

Figure 10.8 – Paul Mcalenan's Stream Deck with SketchUp shortcuts and icons

Products like the P.I. Engineering's X-Keys keypad or Elegato's Stream Decks give you the ability to program any keystroke or, in some cases, macros to a single key. Many of these products even allow you to place an image on each key either through removable key caps or using small LCD screens in the keys! These keypads are amazing tools that can save you a lot of time, especially once you get used to using them. The downside is they can be expensive. A quick search on Amazon for a 24-key X-Keys keypad shows $149 USD as the starting price.

If that pill is a little too expensive to swallow, there are software options to get you a similar result. There is several keyboard-remapping software out there that will allow you to change the keys associated with a specific keyboard (to find examples of such software, you need to only type `key remapping software` into Google). This means you can pick up an inexpensive 10-key keypad and run a remapping software to change what keystrokes it passes to the computer. This means that the *1* key on the keypad could pass an *R* to your computer, turning it into the **Rectangle** shortcut!

If you are a tablet user, there are even tools out there that will allow you to program shortcuts keys on your tablet. If you already own a tablet, this can mean adding an app and using the hardware that you already own! Think of how cool it would be to have pictures of all your shortcuts on your existing hardware like this:

Figure 10.9 – My friend Box sent me this pic of his Android tablet running Touch Portal

This all may seem like a lot but finding the right options for a keyboard can seriously speed up your input and help you to level up. Remember back in *Chapter 7*, *Creating Custom Shortcuts*, when we did a little math to figure out how much time can be saved if you program the proper keyboard shortcuts? Imagine shaving another 30 – 50% off those numbers by not having to move your hand at all to get to your shortcuts!

Virtual keyboards

These options are amazing, but all require you to move a hand at some point. Imagine if you could do all of your numeric input without taking your hand off your mouse! This is what on-screen keyboards facilitate. If I am working on my laptop and want access to a ten-key keypad, I could always pull up something like this:

Figure 10.10 – 3Dconnexion's on-screen Number Pad

On-screen keyboards are nothing new. In fact, Windows has been shipping with an on-screen keyboard capability since Windows 7! Both Mac OS and Windows offer some sort of virtual keyboard, natively, but there are additional options out there as well. If you do a search on Google, you will find many options out there, many of which can be downloaded for free.

Author's recommendation

I have to be honest and say that my current keyboard game is unimpressive, at this point. Right now, I just use the keyboard on my MacBook. In the past, I have used a number of different keypads for shortcuts and for years used a Logitech ten-key keypad to do most of my input. While I cannot see myself changing my keyboard setup drastically (I travel and have a hybrid work schedule that splits my work week between home and the office), so I doubt I will add a full-size keyboard with a ten-key keypad or a separate programmable keypad. I have been thinking about adding a Bluetooth ten-key keypad that I can throw in my bag when I am on the go, but that has not been something that I have moved on, yet. Maybe it is time for me to think about leveling up!

Figure 10.11 – My current lackluster keyboard setup

This was probably a lot more than you were expecting from the keyboard section for the book. I know it was for me! It was pretty cool to learn about all of the options out there for getting quick access to the keys that you use most. As an added bonus, most of these solutions can be leveraged in any other software you use, as well!

For the final two sections, we will be looking at a few hardware options that may not be something that you use at all right now. These go beyond the default mouse, keyboard, monitor setup that many of us have used for years.

Tablets

Tablets, or graphic tablets, have been around since the 1950s but available to home computer users beginning in the 1980s. The basic premise is that you can control the cursor on your computer with a pen like device resting on a flat tablet, rather than with the mouse. Over the past several decades, the graphic tablet has evolved to include the ability to sense pressure, the tilt of the pen, and some have automatic handwriting recognition. While they are incredible tools with so many benefits and options to a designer, we are, once again, going to try to narrow our focus to how they are helpful to a SketchUp user.

Advantages of a tablet

The aim of this section is not to talk you into using a graphic tablet over a mouse. It is, however, aimed at helping you to look at the options that are out there and, maybe, show you a solution that you do not know about. Despite the wide use of mice throughout the computer using world, there are a few significant advantages that graphic tablets have over them. Here are a few of them.

A more natural feel

For many humans, writing or drawing with a pen or pencil is much more natural than trying to make precise movements with a mouse. Yes, it is easy to argue that this is a matter of practice, and, given enough time, most people would be able to be just as precise with a mouse as they would be using a pen. The fact is, though, most of us learn to draw and write using some sort of drawing tool like a pencil. Think of a child learning their alphabet. More likely than not, they will be using a crayon or pencil, no mouse in sight. Given that this could be the tool of choice for the next dozen or so years, it makes sense that we are more comfortable with something like a pen on a tablet than we are with a mouse.

Ergonomics

Another thing that makes tablets very popular with people who spend a lot of time designing is the ergonomic advantage that they have over standard mice. Many people who spend a lot of time using a standard mouse (myself included) have, at some point or another, had to deal with pain caused from repetitive motion. Experts say that these ill effects are made worse by having your palm facing down (slightly twisting your forearm). Using the pen of a tablet helps to relieve some of the stress associated with this sort of repetitive strain. I know a handful of designers who started with mice but have move to a tablet for this specific reason.

You can actually test this idea fairly easily right now! Sit upright with your dominant hand (the one that runs the mouse) resting on the desk in front of you, arm bent at a 90-degree angle, palm facing down. This is the standard posture one would take to use a mouse. If you lift your hand just off the desk and move your fingers around a little bit, there is a chance you can feel some tightness in your forearm. Now, if you rotate your hand so that your thumb is facing directly up and wiggle your fingers, you may notice some relief in your forearm. This is the difference between working with a mouse and a tablet.

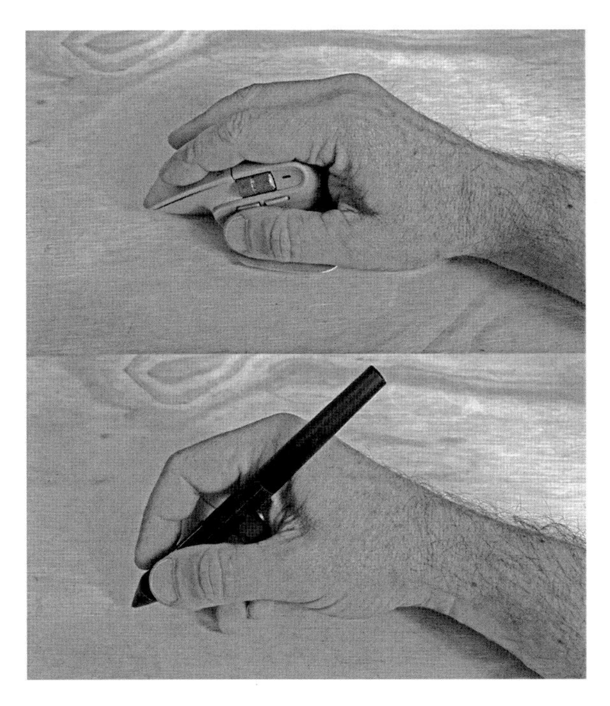

Figure 10.12 – Natural mouse hand versus natural graphic tablet pen hand

The difference is subtle, but most can feel the change. This is the slight relief that working with a tablet over a pen can offer. The difference may not be world-changing for you right now, but if you suffer from any tendonitis or repetitive strain injury in your forearm, it may be a relief!

Vertical Mice

It is worth calling out at this point that there are mice that allow you to keep your hand in this vertical position as you use them. If keeping your hand in this upright position helps, but you don't want to make the leap to using a tablet, you may want to take a look at a mouse like the Logitech MX Vertical.

Customizability

While we did talk about mice that you can program and customize in the *Mice* section of this chapter, mapping custom functions is just something that comes along as a part of using a graphic tablet. Every graphic tablet that I have used has had at least three buttons that can be customized in the included driver software.

Graphic tablets are designed with the understanding that the user will have one or even two hands on the tablet. So, many compensate for that fact that the user will not have quick access to the keyboard by adding multiple programmable buttons to the tablet itself. This may be as simple as a button or two along the side of the tablet, or the tablet may have a dozen or more programmable buttons available. This is a great option for activating SketchUp shortcuts without having to take a hand off the tablet! Add to that, many pens have two or three buttons on them (most of which are also customizable) and you have a lot of options for setting up your tablet!

Disadvantages of tablets

Of course, there are downsides to using a graphic tablet (if everything was perfect, we would all be using them, right?). As with the other peripherals, I don't want to spend a ton of time arguing the merits of each peripheral against the other, but I figured it was worth mentioning at least two issues that play against using a tablet.

Size

If you are someone who is mobile, perhaps with a laptop as your primary computer, and you need a solution that allows you to pick up and move from one workspace to another, tablets can be an issue. Now this is not a world-breaking problem. Tablets are light, generally not larger than a laptop and can even be wireless (more on the features and tablet options later in this section), but they are more work to pack up than a mouse.

If you are a "*work in one place all of the time*" type, a tablet will still need some consideration, as it can tend to take up more space than what a mouse will require. Again, not a huge issue, but something to consider.

Price

Just like with mice, there is a range of tablets out there. The big difference is the size of the range! While mice can be picked up for $10 USD and run up to $150, graphic tablets start at $35 USD or so, and can cost you $2500 USD or more. This can be nice for someone who wants to give a tablet a try (spending $35 USD on a tablet, using it for a month, and finding out that it is not for you, is not a huge loss), and gives you plenty of room to grow if you do like the tablet life. It is, however, across the board, a more expensive peripheral than a mouse.

Picking a tablet

There was a day not too long ago when, if you wanted a graphic tablet, your options were the line of hardware from Wacom. They did (and still do) offer a range of simple, compact graphic tablets that connect to your computer via USB up to enormous 24" graphic display tablets with touch features and dozens of programmable buttons via an external programmable control deck. Today, however, there are other manufacturers like Huion that offer a range of tablets at a more affordable price. In fact, if you search for graphic tablets, you are likely to get overwhelmed by the number of options available to you! To that end, let's take a quick look at a few of the options you may want to consider when picking out a tablet.

Size

Any of the tablet manufacturers you see will offer a range of different size tablets. This is probably the first (and maybe most important) thing to think about when you start looking. While you may be looking for a smaller size because you travel and want something that is easy to slip into your backpack, it is important to consider the way that you use the tablet as well. Remember, you will move the pen to move the mouse, so if you want to move the mouse from one side of the screen to the other, you will need to move the pen from one side of the tablet to the other. The larger the tablet, the larger this motion will be. I personally have found that I prefer a more compact tablet that allows more cursor movement with smaller gestures.

Wired or wireless

This may be something to consider, again, if you travel with your computer. While most tablets connect to your computer via USB cable, there are more than a few available that connect wirelessly via dongle or Bluetooth. Even if you do not have a road-warrior lifestyle, you may like the simplicity of working with a wireless device. Unlike a mouse, tablets can be used in multiple positions. Many tablet users will pull the tablet onto their lap (like a drawing pad) or rotate it on their desktop to get a more natural position. This is all easier to do when you do not have to worry about a cable sticking out of the side of the device.

Display

Standard graphic or graphic display tablet? The difference is that the graphic display tablet acts as a monitor, as well as an input device. This means you are drawing directly onto the screen, rather than moving the cursor with the pen. Graphic display tablets really are amazing tools that bring another level of interactivity to design. Right now, I would say it is the way to get closest to your work possible. The downside, however, is that display tablets cost hundreds or thousands more than their non-display siblings. While I have met quite a few SketchUp designers that have purchased and use graphic display tablets, none of them use them solely for SketchUp. These devices were primarily created for use in drawing, painting, and photo-editing software, but can be used for SketchUp, as well. If you are a multitalented designer who has workflows that lap into graphic editing, then something like a graphic display tablet may be a worthwhile investment.

Pens

Many tablets have a single option for their pen. Usually a one or two-button device that may or may not need to be charged or might have batteries in it. There are, however, options out there that will offer more control or buttons. The way that a tablet works with SketchUp is fairly straightforward (SketchUp is looking for where the cursor is moved and left or right mouse clicks), but if you end up using other programs alongside SketchUp, then things like tilt-sensitivity or an airbrush-like grip may be something to consider when picking out your tablet setup.

Learning to Use a New Peripheral

Whenever you need to learn to use a new tool, it is going to take some time. If you are thinking about trying a tablet, you should plan to give yourself time to learn and get used to using it. When you first start using it, you will most like feel like you are struggling and that you are less effective than you were before. That is normal. That is learning! I personally have found that I do best when I do not give myself the option to use the "old way" and force myself to adopt the new solution. When I started using a tablet, I put my mouse away and allowed myself to only use the tablet for a full week. This was enough time to get used to using the tablet as an input device and finetune the settings so that it worked for my specific needs.

Once again, I am not trying to sell you on graphic tablets, but I have seen so many users changing to using them that I figured it was worth looking into. Tablets really are amazing input devices that are very usable and customizable.

Author's recommendation

I do own a tablet but to be honest I do not use it that often. While I do spend a lot of time in SketchUp, I do use other programs, as well. I find that I am more likely to use my tablet in drawing and painting programs than I use in SketchUp. It's not that I don't like using a tablet in SketchUp, in fact, I have made a few videos about using a tablet with SketchUp and really got to the point where it felt good. I think that the reason that I don't use the tablet more in SketchUp is the amount of time and energy I have put into customizing my mouse setup.

Having said all of this, I do use a Wacom Intuos Pro with a Wacom Pro Pen 3D. It is an amazing device that can be use wired or wireless, has eight programmable buttons and a touch ring. The pen has three buttons and feels very natural in my hand.

Figure 10.13 – My Wacom Intuos Pro with Pro Pan 3D

Given the chance, and if I did more graphic work, I would love to dive into the world of graphic display tablets. I have long dreamed of working on a 24-inch Wacom Cintiq. At this point, however, with my need to change workspaces and travel, and my focus on SketchUp, my Inutos Pro is the perfect tablet.

Tablets are great devices that can change the way that you interact with your computer. Another device that can do that for you is the 3D mouse and we will see how they work right now.

3D Mice

Unlike traditional mice which are used to move your cursor on the screen, a 3D mouse will move your camera through 3D space. This device is not a replacement for your regular mouse, but a piece of equipment that can work alongside it. A 3D mouse is not technically a "must have" device to run SketchUp, but more and more professional users are adding them to their standard gear for good reason. 3D mice come in different shapes and sizes, but all center around the basic functionality of a puck that can move in 6 different axes. These movements are translated into camera or model movements inside of SketchUp, allowing you to spin around a model as if you were holding it right in your hand, or flying around a model like you would in a helicopter.

Before we dive into the advantages that a 3D mouse can bring to your SketchUp game, I want to again say that I am just giving you an idea of what is available when it comes to these peripherals. I have no association with any hardware manufacturers and am only relaying my experiences with these devices. Now that is out of the way, let's start this section by looking at the biggest advantages of using a 3D mouse in SketchUp.

Multitasking

A 3D mouse separates movement if the camera inside the SketchUp window from the input happening with the keyboard and mouse. This means freeing up the mouse in your dominant hand to just do the modeling. Rather than using the middle wheel to orbit, then zoom to get to a specific view, and then move the cursor to a toolbar to select the next command, a 3D mouse takes care of moving the camera where it is needed while you pick the next command. By taking the orbit/zoom/pan functions off the standard mouse, you are freeing up the primary mouse to just perform modeling tasks.

Smooth camera moves

I have said it before and I will say it again, SketchUp has one of the easiest solutions to move through 3D space of any 3D modeling software. Using the middle button/scroll wheel to zoom and orbit is easy to learn and incredibly efficient. it does not always, however, show very well to others. When we jump through a SketchUp model, quickly zooming into an area then orbiting around the model, we see it as a quick and efficient way to move though 3D space. To someone watching from the outside, it can be jarring and possibly a little disorienting. To the one in control of the scroll-wheel, every movement makes sense because we are expecting it. If we are presenting to someone else, it can be seen as unexpected jumps that can take away from the experience.

When presenting with a 3D mouse, the moves through the model look smooth and even. You can zoom, pan, and rotate the model all at the same time. You are spending a lot of time and energy making a model look as good as possible in SketchUp. Once it's time to take that model and present it to someone else, doing so with a 3D mouse will allow you to smoothly move through it, making the experience as enjoyable as possible.

Customization

Just like many of the other peripherals in this chapter, 3D mice can be customized to do what you need them to do. This may mean changing the way that the puck interprets movements. choosing to have the 3D mouse, move the model or move the camera around the model. Many models also come with one or more buttons. Much like the programmable keys on a graphic tablet, these buttons can be programmed to activate shortcuts or modifier keys.

As with other peripherals, 3D mice come with a driver program that will allow you to edit all of the settings of the device. These functions can be general or tailored specifically for a single program.

Choosing a 3D mouse

While I seem to remember two or three companies that offered 3D mouse hardware a few years ago, a few minutes on Google verified that there is one manufacturer of 3D mice selling hardware as of the writing of this book: 3Dconnexion. 3Dconnexion offers a range of 3D mice of different sizes and capabilities.

At the low end (low is a relative term as the most basic 3D mouse from 3Dconnexion does cost $99 USD) is their basic SpaceMouse. This device is little more than a weighted base, a puck, and a USB cord. While 3Dconnexion does offer a wireless version as well (great for travel), the SpaceMouse is a great no-frills controller for moving in 3D space and not much else.

Figure 10.14 – Standard SpaceMouse

At the other end of the spectrum is the SpaceMouse Enterprise (this costing $499 USD). In addition to the puck to move through 3D space, this bigger model offers dozens of programmable shortcut keys. One of the disadvantages of using a SpaceMouse is that you have to dedicate your non-mouse hand to controlling the puck, which means taking your hand off of the keyboard. All those beautiful shortcuts you have created end up going unused unless you want to swap your hand back and forth from 3D mouse to keyboard, repeatedly.

The SpaceMouse Enterprise remedies this situation by giving you a slew of customizable buttons as well as dedicated modifier keys, and even short cuts to specific views! With a standard mouse in my right hand and a SpaceMouse Enterprise in my left, I rarely have to touch the keyboard for anything other than typing in component names or when I first save a model.

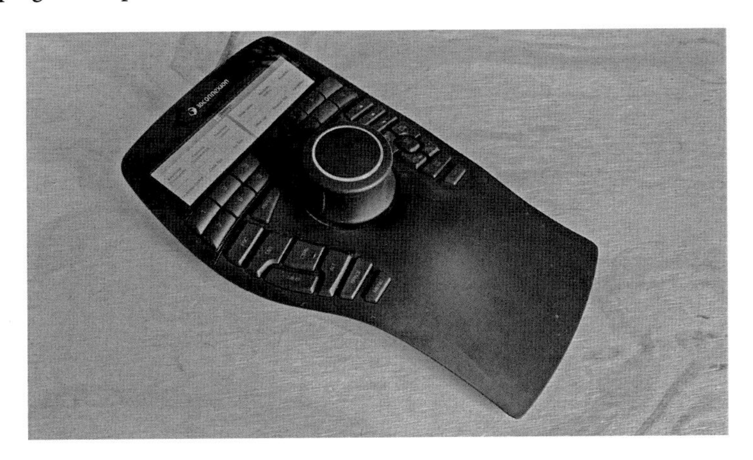

Figure 10.15 – The SpaceMouse Enterprise

3D mice are a great way to level up your SketchUp skills. They speed up design by putting the load of model navigation on your non-dominant hand and will give a serious leg up to your model presentations with smooth navigation. If you have the means to purchase such a piece of hardware, you will likely become one of the many 3D modelers out there who swear by this piece of hardware.

Committing to a 3D Mouse

If you do plan to enter the world of 3D mouse use, be forewarned that there is a learning curve. The first time someone uses a 3D mouse, they end up spinning their model off into no-man's land or zooming right through it. It will take time to get used to using it and even more time before you become capable of smoothly swooping through your model. When I decided to learn to use a 3D mouse, I actually disabled the scroll wheel on my mouse for three days. This forced me to learn to use the 3D mouse to move through my models. At this point, I am so used to using the 3D mouse I actually feel a subtle pain any time I use SketchUp without it!

Author's recommendation

Once again, I want to say that you do not need a 3D mouse. They are optional peripherals, and you can be an amazing SketchUp modeler without one. Having said that, I can tell you that I am a quicker modeler when I am using a 3D mouse than when I am not. When I model with my 3D mouse in front of people, they always ask how I move so smoothly though the model. My personal experience is that they do help and did bring me to a new level of ability in SketchUp.

Having said that, I actually use two different 3D mice. When I am at my desk, I use a SpaceMouse Enterprise. I love all of the programmable buttons and how I can keep my hand in one place and get to every important key while moving through the model with the puck. I also have a standard SpaceMouse that I keep in my backpack so I can plug it in if I am modeling anywhere outside the office.

Figure 10.16 – My two 3D mice

If you feel like you have plateaued with your SketchUp growth and are looking for a way to get further than you can with a standard mouse setup, or if you find yourself presenting a lot of SketchUp models to others, a 3D mouse may be worth considering.

Summary

With that, we have looked at all of the peripherals that can be directly used with SketchUp. It is important to say again, hardware alone will not make you a better 3D modeler. Setting up your software the way you need it, and practice are the key to being a stronger 3D modeler in SketchUp. Alongside that, however, the right set of peripherals can help get you just that much more efficient than you currently are. Whether it is upgrading your standard gear like your keyboard, monitor, or mouse, or adding something like a tablet or 3D mouse, there is a possibility that the right peripheral may help you to level up your SketchUp skills.

With this, we have looked at how hardware can add to SketchUp. Up next in *Chapter 11, What Are Extensions?*, we will be looking at how we can add software to SketchUp that will increase your modeling capabilities and help you to level up!

Part 3: Extending SketchUp's Capabilities for Modeling

Even if you master SketchUp Pro, you are only working with a portion of the tools available to you. This section will encourage you to go beyond SketchUp and explore the entire Pro package.

This part contains the following chapters:

- *Chapter 11, What Are Extensions?*
- *Chapter 12, Using 3D Warehouse and Extension Warehouse*
- *Chapter 13, Must-Have Extensions for Any Workflow*
- *Chapter 14, Introduction to LayOut*
- *Chapter 15, Leveraging the SketchUp Ecosystem*

11
What Are Extensions?

When you talk about broadening your capabilities as a SketchUp modeler, using **extensions** must be a part of the plan. Extensions are little "mini-apps" that can be added to SketchUp to expand what you can model, how you model it, and what you can do with the model you create! Extensions extend the capabilities of SketchUp and are key to any 3D modeler's journey to level up!

When we dive into the world of extensions, there is a lot to talk about. In fact, we will be touching on extensions for the next few chapters. In this chapter, we will take a look at what exactly extensions are and how they work with SketchUp. Specifically, we will be discussing the following topics:

- Learning the basics of extensions
- Installing extensions
- Using extensions

Technical requirements

For this chapter, you will need SketchUp Pro and your favorite web browser. We will also be installing an extension together later in this chapter. We will be using `su_bezier_1.1.1.rbz`, which can be downloaded here: `https://extensions.sketchup.com/extension/8b58920d-0923-42f8-9c72-e09f2bba125e/bezier-curve-tool`.

Learning the basics of extensions

SketchUp is an amazingly easy-to-learn 3D modeling software. One of the main reasons it is so easy for so many people to learn is that it keeps things simple. All the tools work in a similar fashion and there are no superfluous commands or UIs.

The downside to this simplified approach is that SketchUp tends to stay somewhat generic in its toolset. Now, this is great if you want a 3D modeling tool used by architects, landscape designers, video game designers, and more. With this simplified approach to 3D modeling (everything you model is just a collection of edges and faces), anyone can model anything in SketchUp. On the flip side of that coin, however, is the fact that there are no tools to make the specific item you might need in your model.

SketchUp does not have a "wall" command. SketchUp does not have a "vegetation" tool. There is no tool to animate your models once they are completed. In many workflows, SketchUp serves as a starting point for these models. Smart designers take advantage of SketchUp's multi-purpose modeling tools and crank out conceptual models fast, then export their geometry to another software to continue their process. More and more often, however, SketchUp users are finding ways to stay in the software for longer or, in many cases, developing workflows that allow them to never have to leave SketchUp at all. The secret to these workflows? Extensions!

A brief history of extending SketchUp

There was a day (not too long ago) when there was no way of modifying SketchUp. There were no extensions, and no way to add to the stock capabilities. Version 4 of SketchUp changed everything with the inclusion of these things called **plugins**.

I know, we now call them extensions, but when they first came out, they were referred to as plugins. Plugins were these cool little mini-programs that added tools or macro-like functionality to SketchUp.

Plugin or Extension?

Today, there are still those who will refer to extensions by their original name of plugins. There are some out there who claim that a plugin is a smaller extension or an extension that is manually installed rather than through the Extension Manager. Still others say that a plugin is any extension not downloaded from the Extension Warehouse. The fact is, they are all the same thing. Call them what you like, they are the same bits of code that add, or extend, the capabilities of SketchUp.

The idea of adding third-party scripts to software was not a new idea (lots of software were opening up to external developers in the late 90s and early 2000s), and by 2004, SketchUp opened its doors and added the ability for plugins to be added as well.

Since that release, over 1,000 extensions have been created. Multiple online repositories and stores have been created. Millions of users have taken advantage of these added capabilities and downloaded free and paid extensions to expand their workflow and make SketchUp into a tool that reaches far beyond the abilities of the software alone.

Writing software that runs inside of SketchUp

This chapter will not be going into how to write an extension. In fact, that is information that will not be included anywhere in this book. The simple fact is that I am not a developer and do not have the skills to teach anyone how to write code for SketchUp extensions or anything else. I do, however, know enough to tell you how extension developers create code and can point you to a few online resources if you want to learn more.

In the world of software, there are open source and closed source software. Open source software shares its source code with the world, allowing anyone to get into the inner workings and make changes to the program or create brand-new software using its code. Closed source software does not release its source code. The software is released and can be used by users, but the code that is used to create the software is kept private. SketchUp is a closed source software.

There are, however, ways for developers to interact with closed source software. One way is through an **application programming interface** (**API**). An API is a way for third-party developers to write software that can directly interact with other software. In the case of SketchUp, developers can write software that can access certain functionality inside SketchUp and add capabilities, report on information in a model, or change the way that a user interacts with SketchUp. Generally speaking, APIs only give third-party developers access to a limited set of functionality (a third-party application cannot take over a software completely using an API), but enough to create new tools that will be useful to their users.

SketchUp uses a programming language called **Ruby** for its API. Ruby is a simple, yet powerful programming language that is easy to read and write. Many third-party developers have come into working with SketchUp without previous experience in Ruby, but have been able to pick it up quickly.

> **Ruby Can Be Just the API**
>
> I did want to drop a note to anyone who is interested in using the API but concerned about the limited capabilities of a simple language like Ruby. Ruby is the API language, but not necessarily the only language used in extensions. Third-party developers will regularly write code in other languages that will interact with their Ruby code. While some can get everything they need to be done directly in Ruby, it is possible to leverage other languages to create extensions.

If this overview has sparked an interest and you want to learn more about how extensions are created or are considering getting into creating extensions of your own, here are a few links that you should take a look at:

- **The SketchUp Developer home page** (`https://developer.sketchup.com`) – This is the official SketchUp page with information about the API, the development process, access to the SketchUp **Software Development Kit** (**SDK**), and an application to become an official SketchUp Extension Developer. This page hosts links to resources, examples, and information that can get you started writing extensions of SketchUp.
- **The Ruby programming language** (`https://ruby-lang.org`) – This is the official site for the Ruby programming language. From here, you can learn about Ruby specifically, as well as access examples and download links for Ruby programming tools.

- **SketchUp's API** (`https://ruby.sketchup.com`) – This page goes deep into the SketchUp Ruby API. This page includes a full class list and information about the API with examples, debug tools, and more. This page gets deep into the technical stuff, so only head over here if you are ready for some serious technical talk!

I want to mention once again that the goal of this section is not to teach you how to write code or even suggest that you should. This is really just here to set the stage for what extensions are and where they come from. If you are already considering writing your own extensions, I gave you a place to start, but the rest of that journey will be beyond the content held in this particular book. For now, let's take a look at the community that surrounds those who choose to learn about the SketchUp API!

A community of extension developers

Up until now, I have used the term "third-party developer" a few times. Just to clarify, this means an individual or group of people who are developing extensions for SketchUp that are not directly connected to Trimble or the SketchUp development team. Third-party developers are those users who take matters into their own hands and extend the capabilities of SketchUp and take it to new heights. And there are more than just a few of them out there!

One of the great things about extensions development is that anyone can do it. This means a SketchUp user out there may write an extension for their own in-house use. While many developers share or sell their creations, there are a large number of developers that create code for use for themselves or their company without ever distributing it to others. For this reason, it is difficult to say how many third-party developers there are in the world.

It is safe to say that there are hundreds of developers. I think it is even safe to say that over 100 developers have created extensions that are publicly available to purchase or download right now.

Despite many of these users developing their code alone, they are not by themselves in their efforts. SketchUp extension developers have created a worldwide support system. In addition to the support offered through the SketchUp Developer home page, sites such as SketchUcation (`https://www.sketchucation.com/`) or The Official SketchUp Forum (`https://forums.sketchup.com`) both offer sections dedicated to supporting the efforts of extension developers. The amazing thing about these platforms is that many of the more prolific developers actually hang out on these forums, ready to help newcomers develop their extension development skills!

Once again, I am not trying to push you into extension development, but know that if you are interested, you will not be alone! There is a whole world of developers out there ready to lend you a hand!

Are extensions safe?

Before we dive into getting a hold of some extensions and putting them on your computer, the question that should be in your head right now is, "Is it safe to install this software?" Any time you download and install software, you should be thinking about where you are downloading from and what it is you are putting on your computer. You should never download software unless you know the site you are downloading from is legitimate. Both Windows and macOS have in-built features that will check with you if you try to install software that is not from a registered development company.

As with other software, you should make sure that extensions that you install are not going to do anything bad to your computer. The best way to keep safe, as far as extensions go, is to download from a reputable location. Do not download extensions from sites unless you know that someone there is checking them to see that they are safe and will not cause problems. To that end, let's discuss the best places to find extensions.

Where to find extensions

So, we have talked a lot about what extensions are and who makes them. The big question is where do you find them? There are two big repositories online and there are a handful of developers with sites that contain multiple extensions. We will take a look at a few of these:

- **Extension Warehouse** (`https://extensions.sketchup.com`) – This is the official extension repository hosted by Trimble. The Extension Warehouse houses hundreds of extensions from dozens of developers. Every extension on the Extension Warehouse has been reviewed by SketchUp team members. This verification includes making sure that the extension does not contain malware and will not conflict with any other extensions. Extensions are not verified to work perfectly or do everything that the developer says they do in the best way possible, but you will at least know that an extension from here is safe.

  A listing page for an extension will include images or videos, a description of the extension, and information about an extension including important data, such as the versions of SketchUp it supports, the release date of the extensions, and what languages it supports. Here is an example listing page:

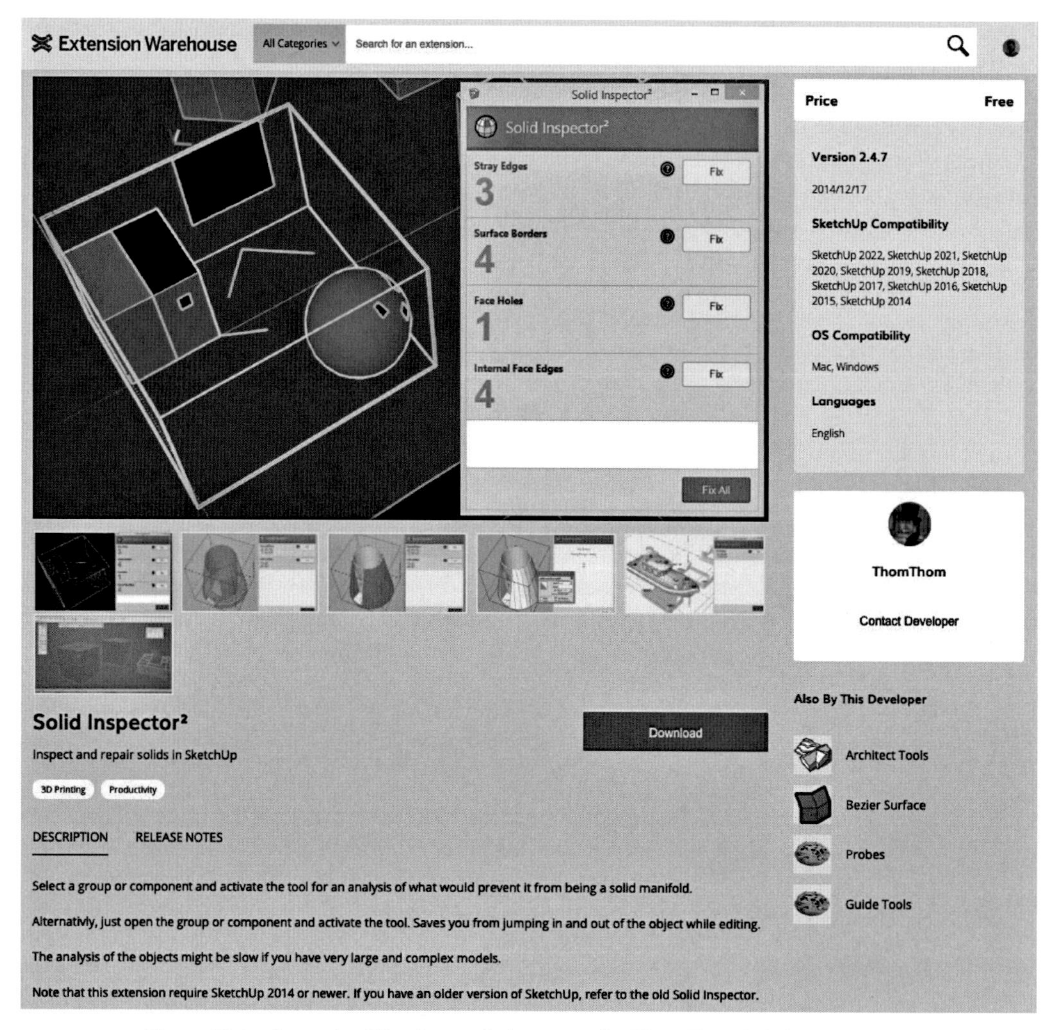

Figure 11.1 – Extension Warehouse listing page for ThomThom's Solid Inspector²

If the extension is free, you will have the ability to download it from this page. Likewise, paid extensions can be purchased before download. As a bonus, since the Extension Warehouse is created and maintained by Trimble, it can be accessed directly from inside the software by clicking on the **Extension Warehouse** button or selecting **Extension Warehouse** from the **Extensions** menu, as shown here:

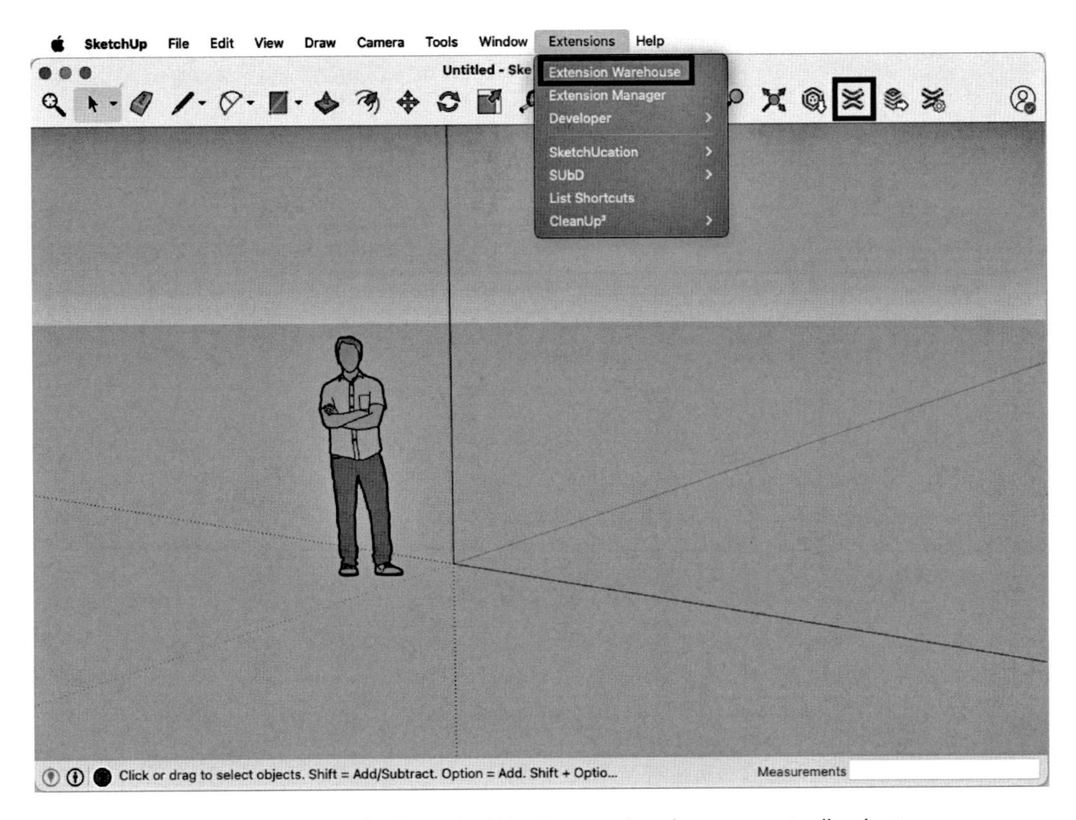

Figure 11.2 – Access the Extension Warehouse using the menu or toolbar button

When you access the Extension Warehouse from inside SketchUp, there is the added bonus of extensions downloading and installing automatically! More on that in the *Installing extensions* section, though. If you want to know more about using the Extension Warehouse, know that we will be diving in much deeper in *Chapter 12, Using 3D Warehouse and Extension Warehouse.*

- **SketchUcation Plugin Store** (`https://sketchucation.com/pluginstore`) – SketchUcation is not only a great user-created and user-moderated site with an awesome forum and great SketchUp-related resources, but it also includes a repository of over 800 free and paid extensions.

One of my favorite things about using the SketchUcation Plugin Store is that the listing pages for the extensions can link directly to a forum page for that extension. This means seeing feedback from users, updates from developers, and a good amount of information before downloading and using an extension. Here is a listing page from the SketchUcation Plugin Store:

Figure 11.3 – SketchUcation Plugin Store listing page for Fredo6's Animator

Another amazing feature of SketchUcation's Plugin Store is the ability for users to donate to developers. While there are paid extensions listed here, many are available to download and use for free. If you use an extension from a specific developer, the SketchUcation Plugin Store allows you to send them a little support in the form of a donation.

Ads

One thing that you may experience when using SketchUcation's Plugin Store is ads. Unlike the Extension Warehouse, which is paid for completely by a company, SketchUcation is run by a community of users and relies on ads to pay the bills. The nice thing is that they do tend to sell their ad space to companies advertising SketchUp-related products, so they are generally relevant. In fact, I have learned about a new extension or two by visiting and seeing a couple of the ads there!

Unlike the Extension Warehouse, SketchUcation's Plugin Store cannot be directly accessed from within SketchUp unless you download and install their own extension first. Their extension, Extensions Store, can be downloaded for free and once installed, will give you the ability to search, install, and even donate to developers, directly from SketchUp.

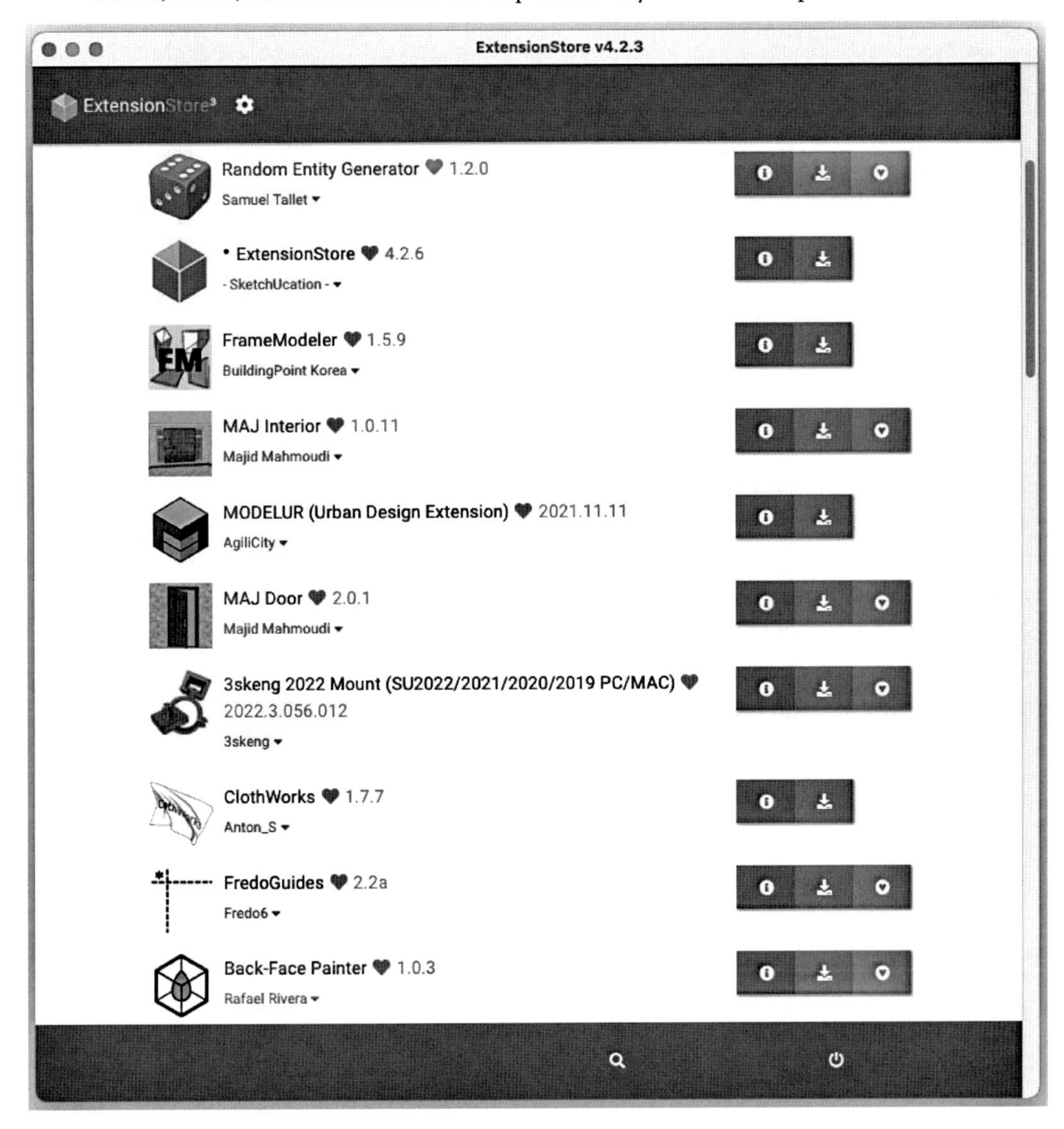

Figure 11.4 – SketchUcation's Extension Store

While there are many extension developers that list their work on both the Extension Warehouse and SketchUcation Plugin Store, there are a few who prefer the Plugin Store exclusively. For this reason, there are some great extensions that can only be found here.

- **Developer websites** – While almost all extension developers will list their work on one of the two sites just mentioned, a few will also maintain their own website to house their work. Here are just a few developer sites that are worth a look. I do want to say that, due to the vast size of the developer community, I cannot list every developer and their work. These sites are sites that contain multiple useful extensions from reputable professional developers:

  - **mind.sight.studio** (`https://mindsightstudio.com`) – This site creates extensions such as PlaceMaker (which generates complete urban landscapes with buildings), Skimp (a tool that allows you to decrease the amount of geometry in imported models), and Artisan (a full-feature organic modeling toolset). While most of these extensions are also available through the Extension Warehouse, this site has additional videos, documentation, customer testimonials, and, in some cases, different options for purchase.
  - **Estimator for SketchUp** (`https://estimator4sketchup.com`) – If you are using SketchUp for any level of building creation, you should check out Estimator for SketchUp. Not only is this site home to its namesake (Estimator is a great way to generate material lists in SketchUp), but it also houses an extension to automate framing in SketchUp and a utility for PDF import.
  - **Evil Software Empire** (`https://evilsoftwareempire.com`) – Don't let the name fool you, there is only good happening here. Evil Software Empire is the home of some great extensions such as SubD (an organic modeling tool) and Vertex Tools (which modifies how you use SketchUp by allowing you to modify the points in the model, rather than the edges and faces).

These are just a few examples of reputable locations where you can purchase and download extensions. Wherever you download extensions from, you will need to get them installed in SketchUp sooner or later, and that is just what we will be talking about next.

Installing extensions

Getting an extension installed is not a difficult process, but there are a few points of which you will need to be aware. The first is what kind of file you are trying to install. When you download an extension, you will download one of three different file types:

- `.exe`/`.dmg` files – In some cases, extension developers may choose to create their own installers for their extensions. In this case, they will distribute an installer specific to the supported operating system.
- `.rbz` files – These are the most commonly distributed extension files. These files are actually zipped or compressed files that include the Ruby code needed by SketchUp to make them run, and then whatever additional code or files are needed.
- `.rb` files – Files with the `.rb` extension are just single Ruby files. These extensions rely solely on the code created in Ruby to do what they do.

Since there are different files that can be distributed by extension developers, there are different ways to install the files. Let's take a look at how to get each type installed.

Installing an .exe or .dmg file

Since these installers can be customized to do whatever the developer needs, we cannot run through the exact steps of what they may show; however, the process of launching them will always be the same. If you download a file with the `.exe` extension, you can install it by simply running it (double-click on the file) on a computer running Windows. A file with the `.dmg` extension can be run the same way (simply double-clicking the file name) on a computer running macOS. In either of these cases, the installer should walk you through any additional steps that need to be performed to get the extension installed.

Installing a .rbz file

Extensions packages such as `.rbz` files are the most common and are the type of file that the extension installation system was created to support. In this example, it will be assumed that you are installing the file linked to at the start of the chapter (`su_bezier_1.1.1.rbz`) and that you have downloaded and saved the file into your `Downloads` folder.

> **Windows or macOS**
>
> Note that the SketchUp for Windows UI is being shown in this example, but the steps and UI are the same, regardless of OS.

Let's step through the process of installing this extension:

1. In SketchUp, open the **Extension Manager** window by clicking on the icon on the main toolbar or selecting **Extension Manager** from the **Extensions** menu:

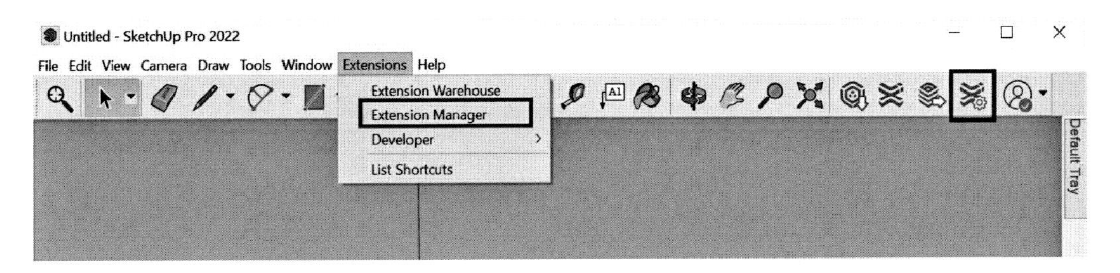

Figure 11.5 – Extension Manager can be found in two locations

2. Click the **Install Extension** button at the bottom of the window:

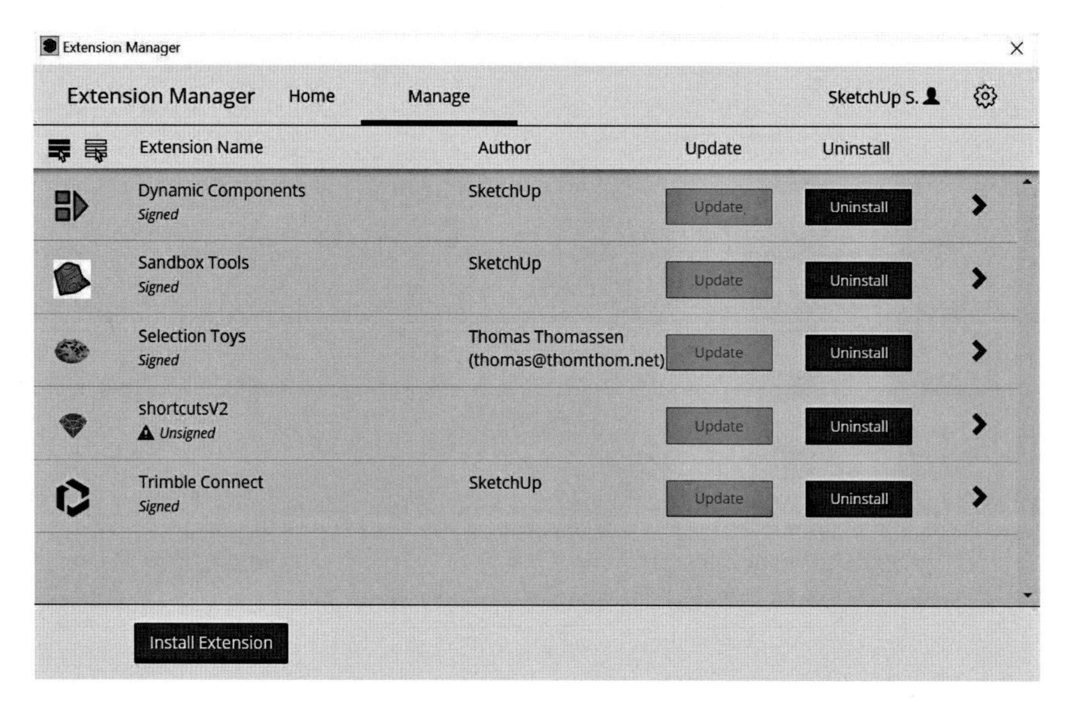

Figure 11.6 – The Extension Manager window in SketchUp for Windows

3. Navigate to your `Downloads` folder, select `su_bezier_1.1.1.rbz`, and then click the **Open** button:

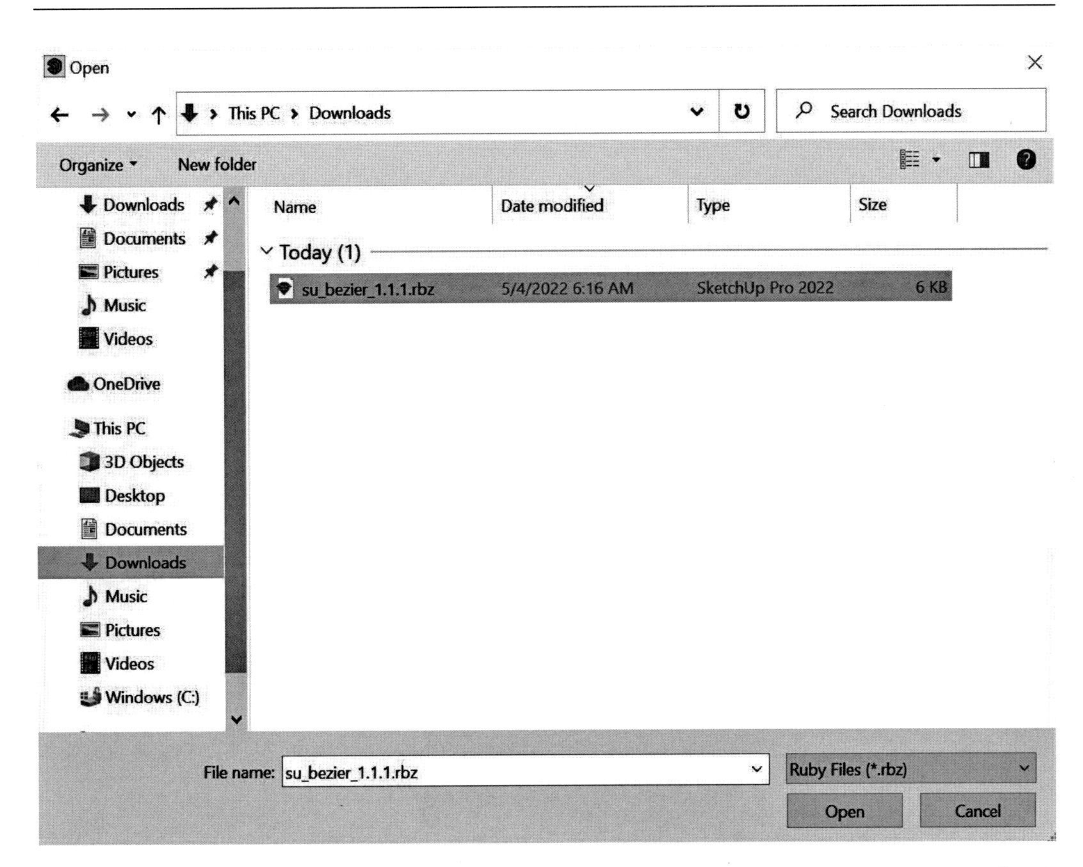

Figure 11.7 – File selection happens in the OS file navigation tool (in this example, Windows File Explorer)

4. After a few seconds, **Bezier Tool** will show up in your list of extensions!

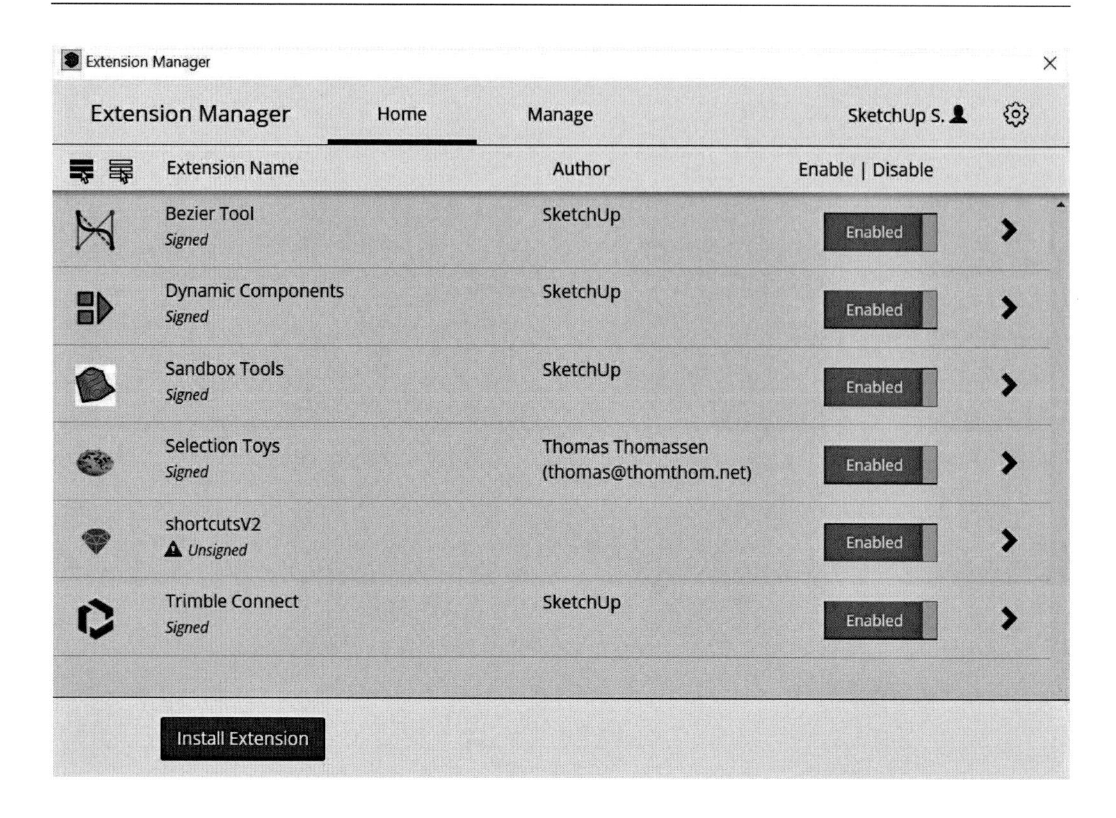

Figure 11.8 – Bezier Tool is now displayed as an installed extension

Some extensions may require you to restart SketchUp before you can use them. In this case, a message will pop up after the installation. Quitting and restarting SketchUp will allow the extension to load.

Installing a .rb file

Unlike installing a `.exe` or `.dmg` file, `.rb` files will not run when double-clicked. In fact, doing so may cause your computer to ask how you want to open the file. Since `.rb` files are just Ruby code, they need to be called by SketchUp through the API (which means you cannot just tell SketchUp to open them as they would a `.skp` file). `.rb` files have to be manually placed into the appropriate folder. If you are the adventurous, computer-savvy type, you can navigate to the following path and manually place the file where it needs to go:

- SketchUp 2022 for Windows extension folder:

  `/AppData/Roaming/SketchUp/SketchUp2022/SketchUp/Plugins/`

- SketchUp 2022 for macOS extension folder:

  `/Library/Application Support/SketchUp 2022/SketchUp/Plugins`

However, if you don't want to root around your computer trying to find the correct folder to drop these files into, there is an easier way. All you need is the original `.rb` file and a compression utility (a program that makes `.zip` files). Here are the steps, regardless of the operating system:

1. Use your preferred compression software to zip the `.rb` file.
2. Rename the new `.zip` file to `.rbz`.
3. Install via **Extension Manager** (which we covered in the *Installing a .rbz file* section).

This will mean changing the file extension of your extension. For example, if I was installing a file named `Drop.rb`, I would zip it and have `Drop.zip`. I would rename it to `Drop.rbz`, and it would be ready for installation!

You can see that installing extensions is not a difficult process. Assuming you have downloaded from a credible source, you should have no problem installing extension files. Once an extension is installed, there are a few points about using them. We will cover those points in the next section.

Using extensions

As a general rule, using a well-designed extension should be like using SketchUp itself. Extension developers are presented with tools to make the UI and workflow as close to SketchUp as possible. There are a handful of extensions that I use that I have completely forgotten are not core functionality because they are so well designed!

In this section, we will run through a few examples of how you might use extensions through the SketchUp UI. For each of these examples, I will mention a specific extension by name and share the Extension Warehouse link. You do not need to download and install these extensions, but the information is available to you if you want them.

Let's start by discussing where you can find your extension once you have installed it. Extensions developers have some flexibility with the placement of their extension UI. There are four main locations where extensions may appear once you have them successfully installed.

Extension toolbars

Most extensions can be used via a toolbar created by the developer. These toolbars are turned on just like stock toolbars are turned on and can be customized the same way (see *Chapter 8, Customizing Your User Interface*). Depending on the extension, you may have access to a single toolbar or a few different toolbars. Some extensions offer a toolbar with a single button while others do not give you access to a toolbar at all. Take this toolbar in **1001bit Tools** (from developer GOH C), for example, (`https://extensions.sketchup.com/extension/e5b1211a-8d1a-4813-bdc3-b321e5477d7b/1001bit-tools-freeware`):

Figure 11.9 – Extension toolbar for 1001bit Tools

If you are a button user, these toolbars are a great solution to getting access to the functionality of a bunch of different extensions at once. Whether you are on Windows and can dock these toolbars around your modeling space, or on macOS and end up with toolbars floating around your model, extension toolbars are a good way to make extensions a part of your modeling workflow.

Extensions in the Extensions menu

You might think that most, if not all, extensions would show up in the **Extensions** menu. Oddly enough, on my own installation of SketchUp, I only have a few extensions with a UI in that menu, despite having dozens of them installed. There is no rule saying that a developer must place their extension into this menu, despite its name. However, there are some that do end up in the **Extensions** menu, like the **Eneroth Scaled Tape Measure** extension:

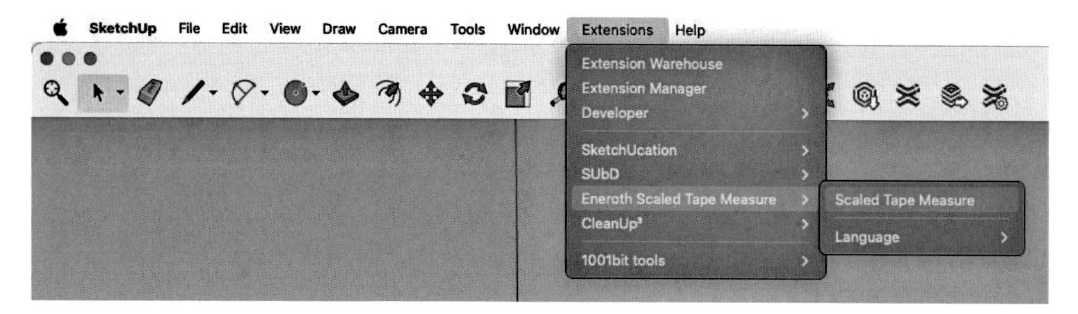

Figure 11.10 – Eneroth Scaled Tape Measure in the Extensions menu

The nice thing about extensions that do install into the **Extensions** menu is that they tend to have everything you need in one place. The commands for the extensions are generally listed along with the settings or other UI controls often existing in a single menu option.

Extensions in the Tools menu

Another place that makes sense for extensions to end up is in the **Tools** menu. Many extension developers seem to prefer to put their extensions here, rather than in the **Extensions** menu. Some extension developers have many extensions available. Fredo6, for example, has a whole suite of extensions available through the SketchUcation Plugin Store. All of his installed extensions can be found in a single submenu in the **Tools** menu:

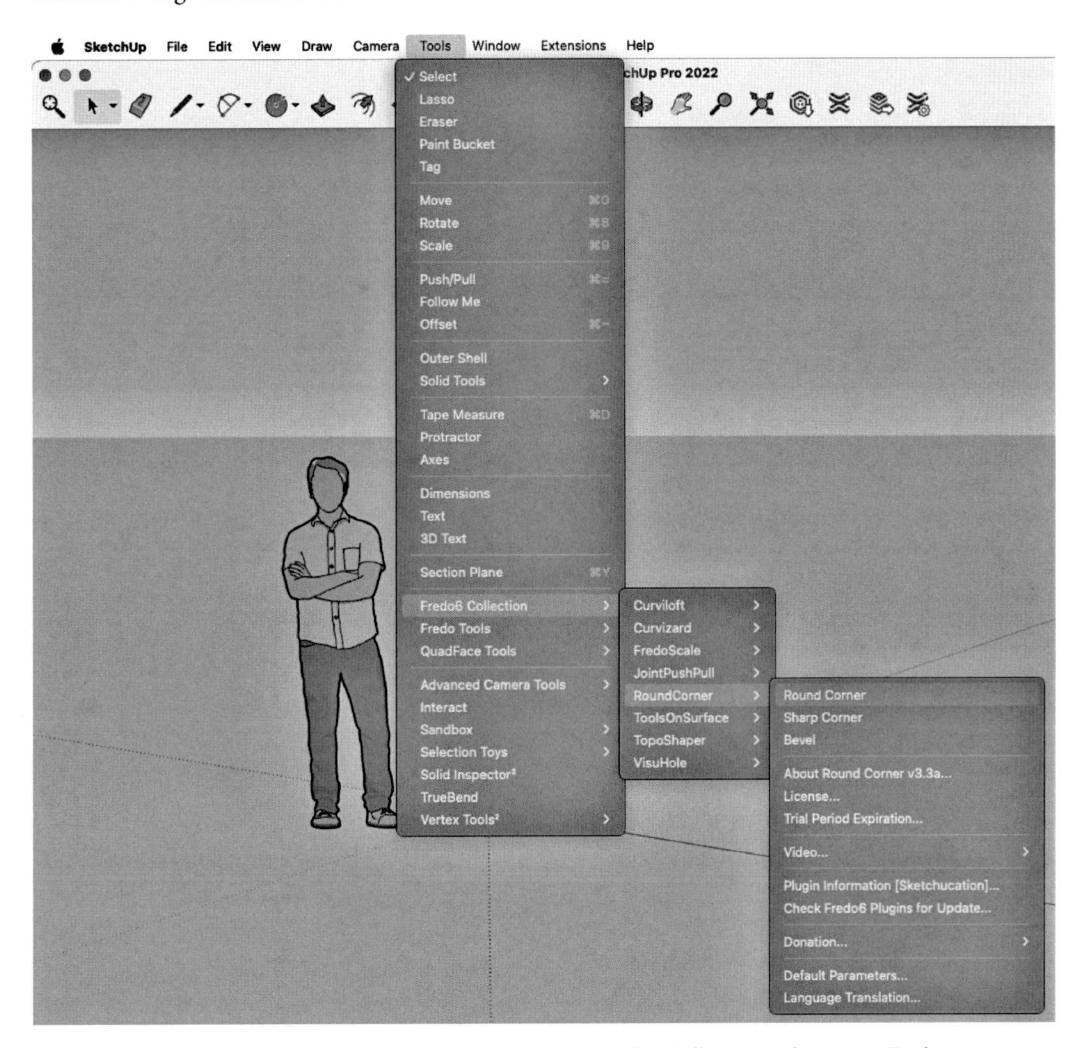

Figure 11.11 – Round Corner extension in the Fredo6 Collection submenu in Tools

If you rely heavily on menus to find extension tools, this can get to be a little bit daunting. Hunting down commands several menus deep is not quick, or easy. In fact, trying to use these commands regularly this way would be a step backward in productivity. If this were an extension that you used regularly, you should really be assigning it a shortcut or adding the toolbar to your UI.

Extensions in the context menu

Some extensions add options to your context menu (the context menu is the menu that pops up next to your cursor when you right-click on a selected entity). This can be a very nice feature, depending on the extension. If the extension performs a single action on selected geometry, the context menu is a great option, especially if this is not a command that you use often enough to assign a shortcut key to. An example of this is ThomThom's Selection Toys context submenu:

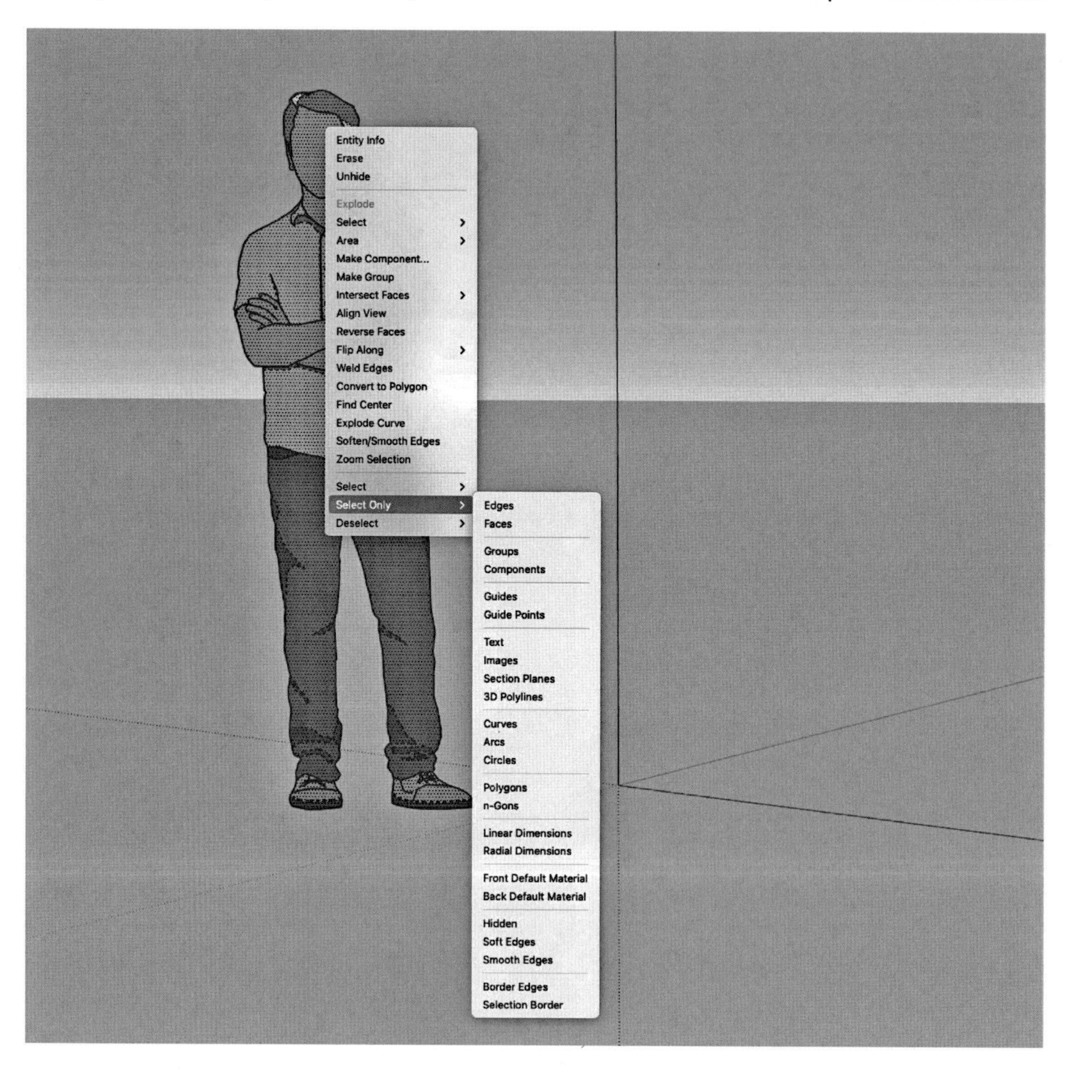

Figure 11.12 – The Select Only filter from ThomThom's Selection Toys in the context menu

Something like this example, a selection filter, contains a lot of options, and setting a shortcut for each would mean using a lot of keys. Having it available in the context command, however, means all of the filters are only a click away at any given time.

Finding extensions after installing them

Most extensions will end up in one of the locations we have just mentioned. If you ever install an extension and cannot track it down, or if you are looking for a specific extension and cannot remember where it lives, there are a couple of solutions. The first is the **Search SketchUp** command. The **Search SketchUp** command will add installed extensions to its search database, allowing you to search for commands by their name.

For example, if I cannot remember where **Round Corner** is installed, I can simply type `round` into the **Search** field and I get all of the commands, including extensions, that include the term "round":

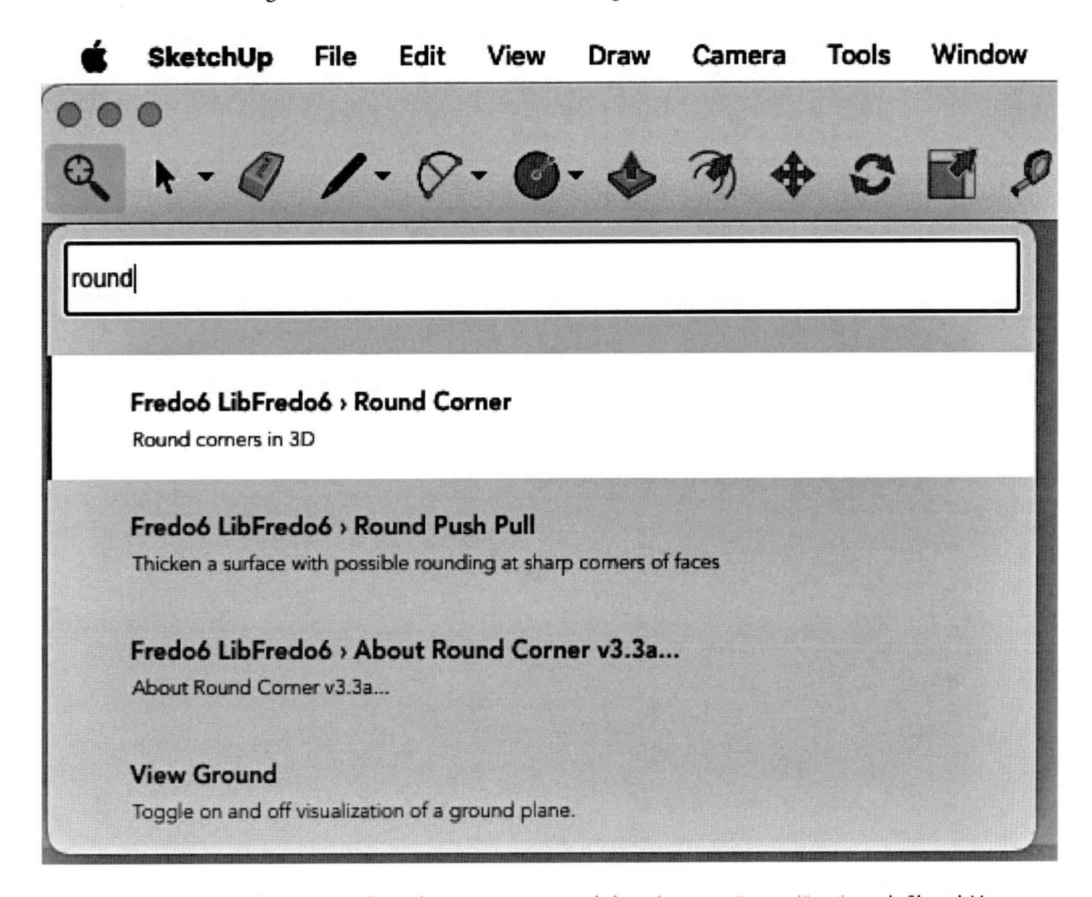

Figure 11.13 – All commands and extensions containing the term "round" in Search SketchUp

From here, you can just click on the item to launch the command. It is a great way to pull up a command and run it immediately. What do you do, however, when you want to find where an extension lives for future reference? This is where the **Help** menu **Search** command comes into play.

If you click into the **Search** field in the **Help** menu and type `round`, you will get a different list. This list will not list clickable commands as **Search SketchUp** does, but will list all of the commands, extensions, and stock, that appear anywhere in SketchUp, along with their location:

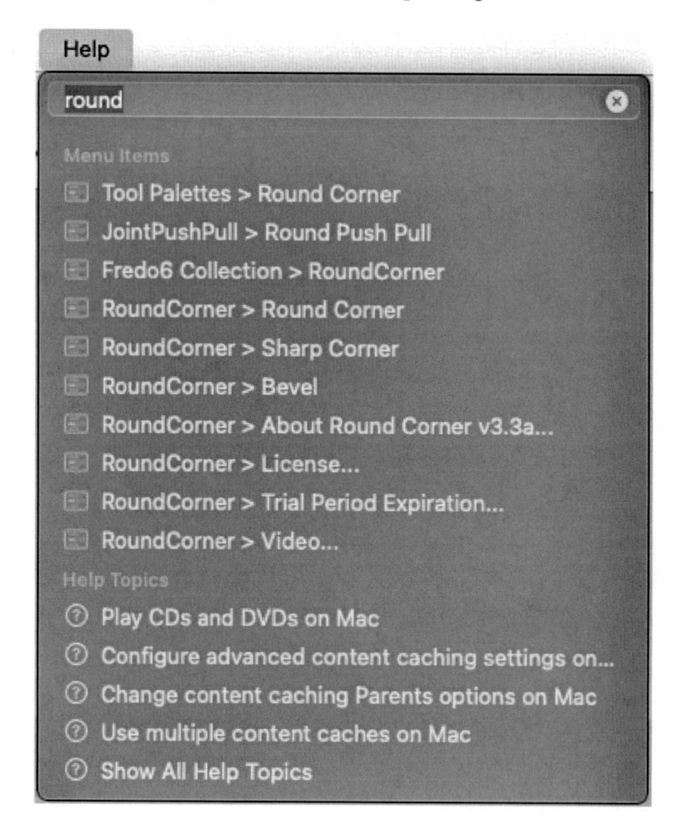

Figure 11.14 – Search results for the term "round"

As an added bonus, hovering your mouse over any of the results will actually pull up the location of the item. If the cursor hovers over the **Tool Palettes > Round Corner** entry, the **Tool Palettes** menu will pop out, and the submenu containing **Round Corner** will be expanded automatically:

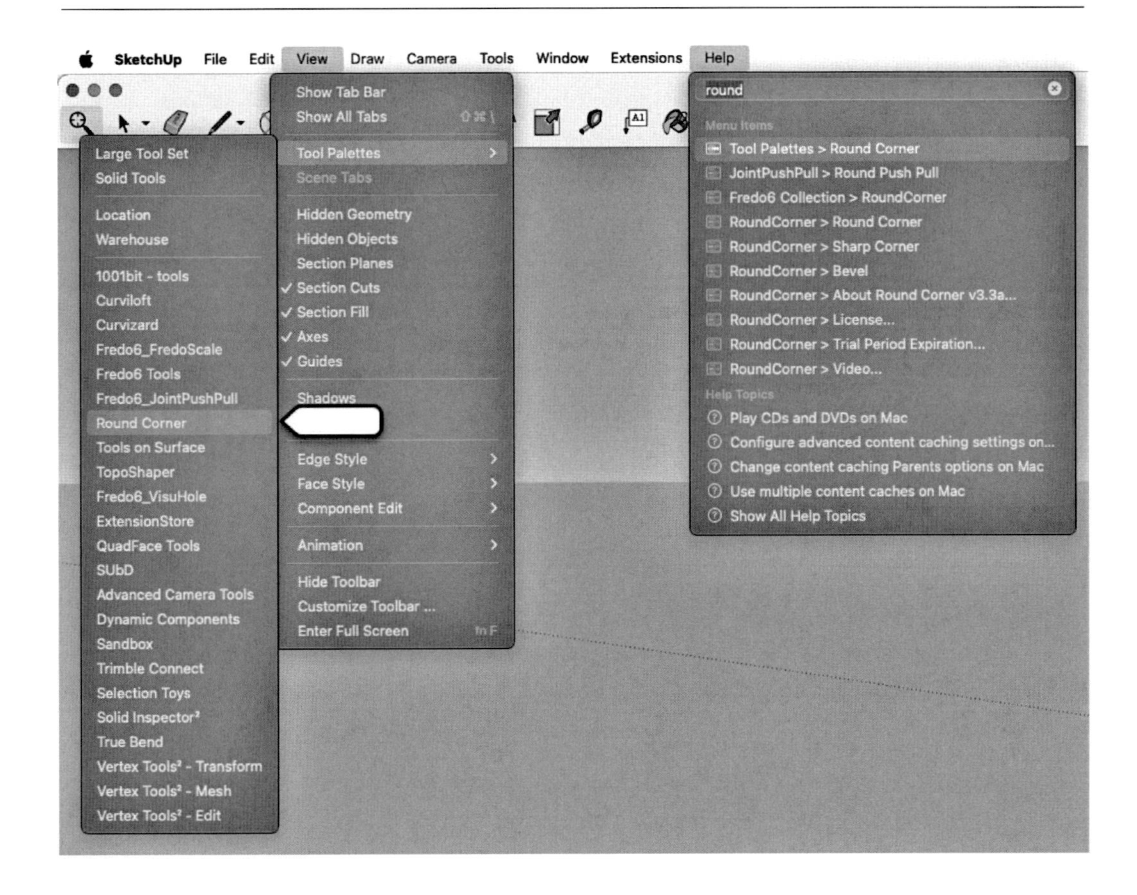

Figure 11.15 – Round Corner in the Tool Palettes submenu displayed from the search results

This is a great way to find exactly where a command lives in case you need to get to it again later.

Once you have your extensions installed and know where they are residing, it is time to assign shortcuts so you can get to them quickly and easily.

Assigning shortcuts to extensions

The process of assigning shortcuts to extensions is really just as easy as assigning to a native command, with one caveat. I figured that as I stumbled on this, it would be worth mentioning here before we wrap this chapter up. When it comes time to connect a shortcut key to an extension command, make sure you are assigning the proper command in the extension!

Reading this, it probably seems totally obvious, but there are many extensions out there that come with a whole slew of commands and options. Most extensions will actually give you access to a handful of commands that you can activate with a shortcut. In fact, there are very few extensions on my computer that only have a single command connected to them!

Simply put, when it comes time to assign a shortcut to a command, don't just pick the first command in the list that has the name of the extension. If you do, you may be pulling up the settings for your shortcut, rather than executing the needed command.

With that, we have covered the basics (and some intermediate tips) on working with extensions.

Summary

Just to reiterate, extensions are not a requirement for using or mastering SketchUp. They are, however, the key to leveling up your SketchUp ability and creating custom workflows. In this chapter, we discovered a few places for you to find extensions, went through how to install extensions step by step, and talked a little bit about how to best incorporate extensions into your workflows. A proper understanding of the extension marketplace will allow you to expand your modeling capabilities and help raise your SketchUp skills.

In the next chapter, *Chapter 12, Using 3D Warehouse and Extension Warehouse*, we will learn how to use SketchUp's integrated online repositories to find exactly what you need and how to share your work with the world!

12
Using 3D Warehouse and Extension Warehouse

A great functionality of SketchUp is the ability to use assets created by others right inside your model. The only downside here can be finding those assets! This is exactly why the SketchUp team maintains the warehouses. These online repositories will allow you to search for 3D models for use in your project or to find new extensions to help make a more efficient workflow.

In this chapter, we will look at the following:

- Understanding the two warehouses
- Logging into the warehouses
- Searching Extension Warehouse
- Finding things in 3D Warehouse
- Using 3D Warehouse models

There is a lot to cover in this chapter, so let's dive right in!

Technical requirements

For this chapter, you will need SketchUp Pro and access to the internet with a supported web browser (Google Chrome, Firefox, Safari, or Microsoft Edge).

Understanding the two warehouses

Before we can get into any specifics on how these sites work, let's spend a little bit of time understanding what purpose they offer and how to access them. In this chapter, we will specifically be talking about these two websites:

- Extension Warehouse (`https://extensions.sketchup.com`)
- 3D Warehouse (`https://3dwarehouse.sketchup.com`)

Both of these sites are owned and operated by Trimble and are available to everyone to view. While there are some similarities in the look and feel of these pages, they both serve very different functions. Let's dive into the purpose that each serves and take a look at how they are maintained, as well as specific uses that they may provide for you in your effort to level up your SketchUp game.

Extension Warehouse

We spent a whole chapter talking about extensions in *Chapter 11*, *What Are Extensions?*, so we probably do not need to get into the nitty-gritty about the purpose of extensions here. The important thing to know is why Extension Warehouse exists.

When the concept of extensions was introduced, people started creating their own extensions. After a very short amount of time, they started sharing them with each other. A little while after that, sites started popping up specifically for sharing these newly created extensions. While it was great to get to see a community being created and the sharing of things that people had made, there was a bit of a concern that there was not an "official" place to get these extensions.

This was the basic idea behind the creation of the Extension Warehouse. This could be a place where extensions that were on some level and "approved" by SketchUp staff could be made available to all SketchUp users. When SketchUp Pro 2013 was released, the first version of Extension Warehouse was released alongside it:

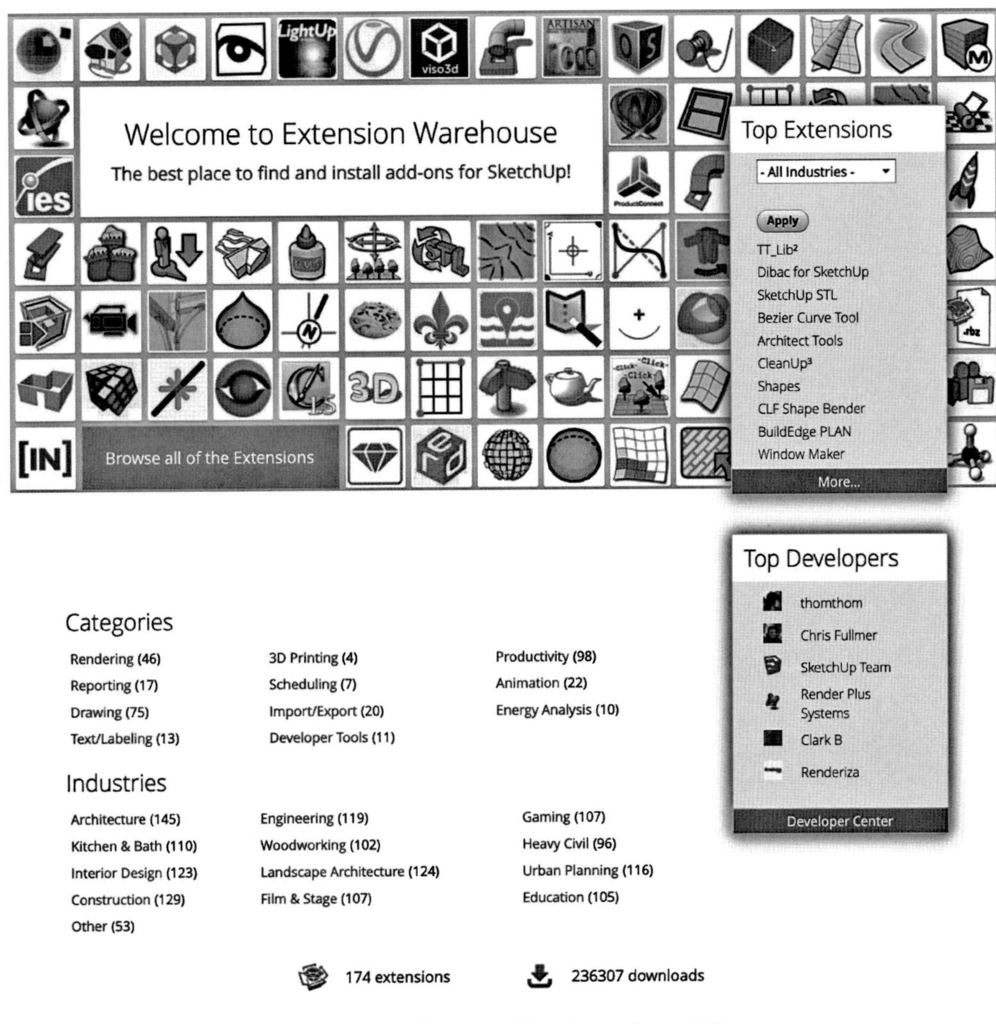

Figure 12.1 – Extension Warehouse from 2013

This was a great service for SketchUp users! Before Extension Warehouse, users had to know someone who knew someone to get an extension. Now, users did not need to find a creator of a specific extension on a forum to get their extension. They could just pull up and search a list of extensions on one page.

Extension Warehouse did something else for users; it provided peace of mind when downloading third-party software onto their computers. I have not heard many stories of malware introduced to anyone's computer by use of extensions, but I have heard (and experienced) issues with extensions that conflicted with each other. I have also experienced extensions that simply did not work once they were installed or did not work on my version of SketchUp Pro. Extension Warehouse was intended to address these issues. Every extension listed on Extension Warehouse was going to be reviewed by SketchUp team members to make sure that there were no issues like those I mentioned.

Now, this did not mean that every extension was exhaustively tested and verified to do exactly what the author said and never had any errors, but it was safe to download and install in SketchUp Pro. This offering was very well received by SketchUp users, and Extension Warehouse has grown over the past decade to include advanced search and listing capabilities and is now home to hundreds of extensions that people discover and download daily:

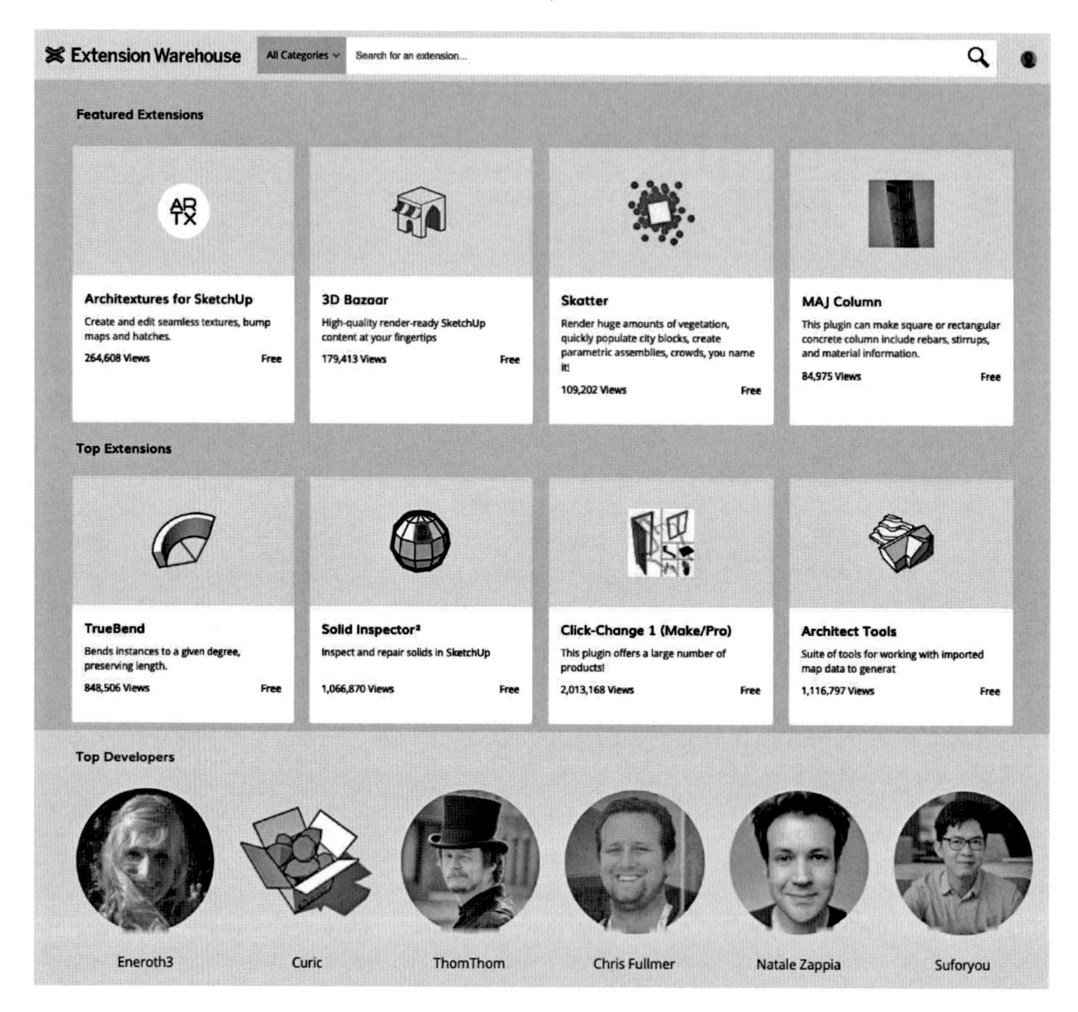

Figure 12.2 – Extension Warehouse as it exists today

Extension Warehouse is not the only place to get extensions for SketchUp, but it is still the only place to get extensions that have been reviewed by SketchUp staff and assured of installation without issue.

3D Warehouse

3D Warehouse is an online repository of SketchUp models created by users and product manufacturers all over the world. 3D Warehouse is an amazing place to discover models that can be placed into your SketchUp file or to share your creation with millions of other SketchUp users.

When first introduced (back when Google owned SketchUp), the intention was to use it as a place to store buildings created by users. It started as a fairly basic list of SketchUp files with thumbnails and a simple search function. Over the years, 3D Warehouse has grown in capability and content!

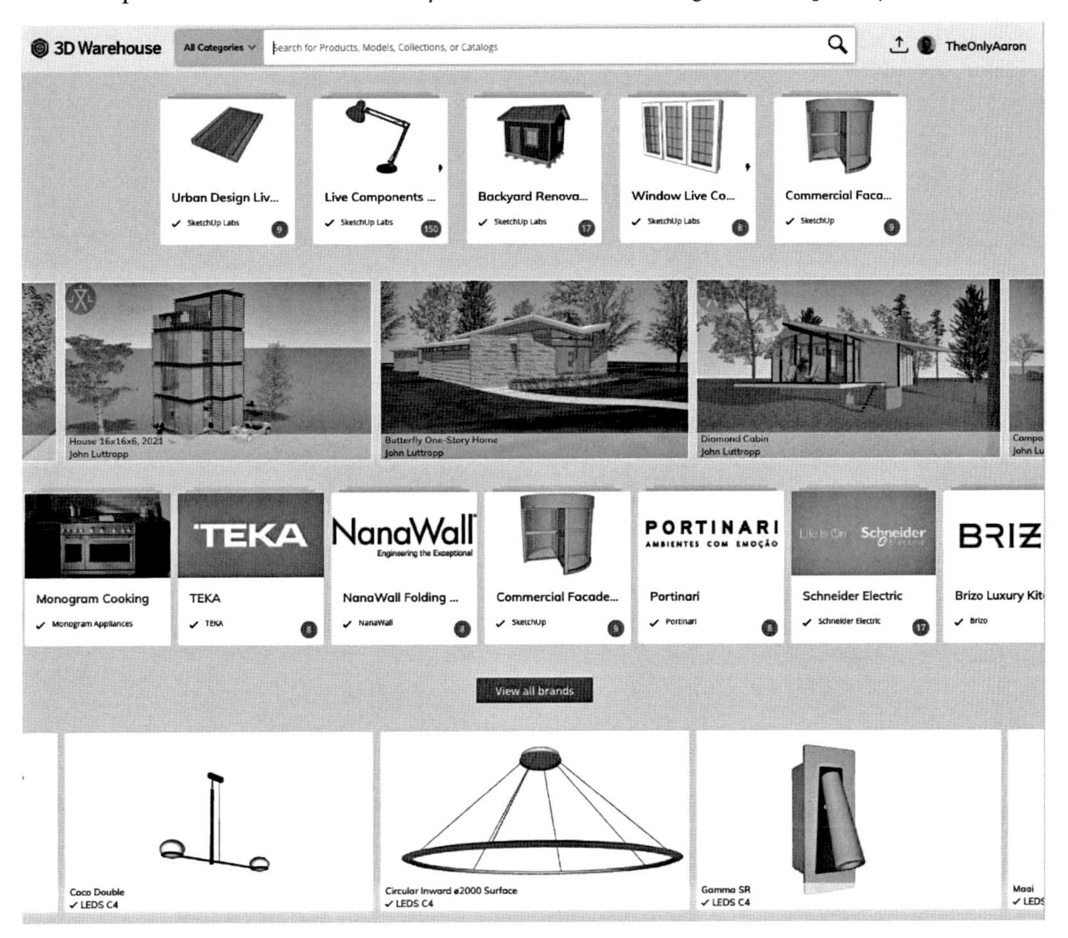

Figure 12.3 – Today's 3D Warehouse

The 3D Warehouse of today has a very detailed search capability and hosts millions of files. While a majority of the files hosted are from users, 3D Warehouse also contains a growing number of 3D models uploaded by product manufacturers. In some cases, these files are the exact files that are used in manufacturing!

Additionally, every single model on 3D Warehouse can be downloaded in the SketchUp file format for the last three versions of SketchUp, as well as in the Collada or GLB file format, regardless of what version the model was created in. Not only that but models can also be viewed in 3D from a mobile device directly from 3D Warehouse! This site has really grown and changed since its introduction all those years ago!

There is a lot that can be done with the warehouses, and we will be covering how and when to use each and the best way to navigate all that they have inside them. Before that though, we should talk about what you need to do to get access to their contents, and that means creating an account and logging in.

Logging into the warehouses

A few years back, there was a change to the warehouses – a user is now required to have an account and log in before downloading anything from either platform. You can still go to either warehouse, and browse without having to log in, but before you can download anything, you must log in with a valid account.

I know there are those out there that bemoan having to go through the process of logging into things or having to create accounts in order to get access to websites, but these people are missing the bigger picture. By logging into a warehouse, you are allowing it to know who you are and asking it to keep track of what you are doing. By having the warehouse remember which extensions you have downloaded, for example, you don't have to bookmark or remember them if you need to come back later and look at the listing page; Extension Warehouse already knows and remembers. Even better, Extension Warehouse will present you with a list of your extensions with a single click!

And 3D Warehouse will make your life even better, allowing you to upload and store private files and create and maintain your own folders with a model that you have found. This means you can peruse 3D Warehouse, and when you come across a file that you may want to use one day, you can simply mark it and "save it for later" without actually having to download and save anything!

We will cover the details of using this functionality in both warehouses in the *Searching Extension Warehouse* and *Finding things in 3D Warehouse* sections of this chapter. Right now, we need to cover the process of getting logged in.

Sign in options

When you first try to sign into either warehouse (or SketchUp for that matter), you have the option to sign in using your Trimble ID or using your existing Apple ID or Google credentials. If you have an Apple ID or Google account, you can use either one to log into SketchUp services. Logging in via these third-party services will give you exactly the same functionality and access as you would see when signing in with a Trimble ID, but without having to create an additional account.

What is a Trimble ID?

As this collection of software and online services that make up the SketchUp suite grows, a single sign-on solution is used to allow you to log in and take advantage of everything that is offered. This is what the Trimble ID is for. Once created, you can use it to log in to any of your SketchUp tools or services. A nice thing about the Trimble ID is that you may not need to do anything to get one. In fact, if you are running a recent version of SketchUp, you probably have one already!

The email and password that you use to log in to SketchUp is the same login information that you will use to log in to 3D Warehouse and Extension Warehouse.

Logging in to a warehouse

The process of logging in to the warehouse with your Trimble ID is a straightforward and simple process. With either the 3D Warehouse or Extension Warehouse open in a web browser, follow these steps:

1. Hover your cursor over the icon in the upper right, then choose **Sign In**:

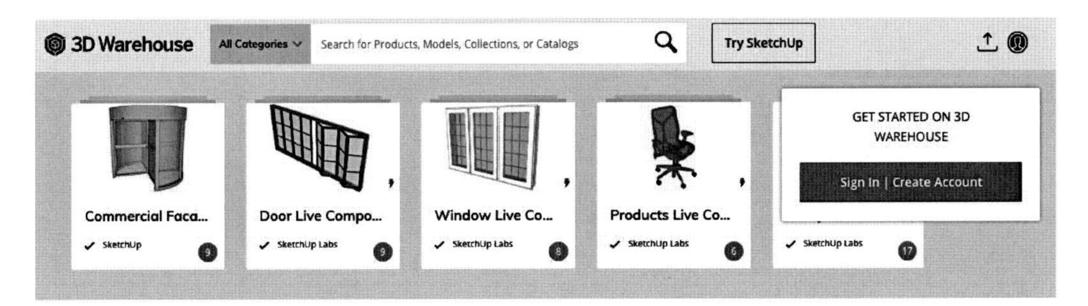

Figure 12.4 – The icon in the upper-right corner of the 3D Warehouse

   This will present you with an option to either log in with the email and password associated with your Trimble ID or log in using Apple or Google credentials. Logging in via Apple or Google credentials will take you off to a separate web page where you will log in. Once complete, you will be returned to the warehouse page. The rest of the steps presented here will walk through the process of logging in without using Apple or Google accounts.

2. Enter the email address associated with your Trimble ID into the **Username** field, then click the **Next** button.
3. Enter the password for your Trimble ID into the **Password** field, then click the **Sign In** button.
4. If you have two-factor authentication turned on for your account, you will be prompted for a verification code at this point. The code should show up as a text message to the phone number provided when you created your Trimble ID. If not, click the **Resend** link. Once you have the code, enter it into the **Verification Code** field and click the **Submit** button.

That is all there is to logging in. If you do not have a Trimble ID and want to get into a warehouse, you will need to create one.

Creating a Trimble ID

As mentioned earlier in this section, a Trimble ID is your key to all things SketchUp. Creating one should only take you a minute or two, and once done, you can use the same credentials for a very long time. Follow these steps for creating your Trimble ID:

1. Click the **Create a Trimble ID** link at the top of the **Sign In** page:

Figure 12.5 – Create a Trimble ID link

2. Enter your details for **First Name**, **Last Name**, and **Email**, then click the **Send Code** button.

   Remember that when it comes time to log into any service, your actual Trimble ID is this email address. Make sure this is an email that you have access to and will remember as your Trimble ID.

3. After a few seconds, you will receive an email to the address you entered. This email will include a 6-digit code. Enter that code into the field and click the **Submit** button.

   If, after a few minutes, you have not received the code, try clicking the **Resend** button. If you still do not receive the email, you may want to check your `Spam` folder. If you are still not receiving the email, then you may want to click the **Use different email** link to change to a different email.

4. Enter a password in the **New password** field and again in the **Confirm password** field, then click the **Submit** button.

   Note that the password does have a lot of requirements. Pick something that you can remember and will satisfy every requirement listed.

5. Set your account preferences including **Country**, **Time Zone**, and **Language**, then click the **Submit** button.

   If you want to keep your Trimble ID extra secure, you can choose to check the **Enable Multi-Factor Authentication** option. This will require anyone trying to use your Trimble ID to log in to verify themselves with a code sent to your phone.

With that, you are finished! You will be returned to the screen where you began account creation and can use your new Trimble ID (your email) and new password to log in!

In the world of software creation, we have this thing called the happy path. It is how we describe the process a user goes through when everything goes perfectly right and there are no problems at all. What we just covered is the happy path. There are, of course, problems that can occur when using your Trimble ID. I figured it was worth addressing a few of those in the next section.

Troubleshooting

I am not a member of Trimble technical support nor is my knowledge of the issues you may run into exhaustive. I do, however, spend a lot of time on our user forums (`https://forums.sketchup.com`) and have seen a few recurring issues that people have run into when trying to use their Trimble ID.

Wrong email address

When a user cannot log in and takes this issue to the forum, I would say that four out of five times, they are trying to log in with the wrong email address. I know that it can be difficult to keep track of our email addresses nowadays, as we all have three or four of them, but it is important that you remember which email address your Trimble ID was created with.

Multiple Trimble IDs

Consider this a variant of the previous issue, but there are people who have (intentionally or by mistake) created more than one Trimble ID and are logging in using the wrong one. This issue is usually described by a user trying to log into 3D Warehouse, then claiming that all of their models have been deleted. The issue is that the models were all uploaded using Trimble ID *A*, but the user is logged in with Trimble ID *B*.

No Trimble ID

This happens less often, but every now and then, a user will try to log in without ever creating a Trimble ID. If you have never gone through the steps listed in the *Creating a Trimble ID* section, then you do not have a Trimble ID to log in with.

Other issues

There are, of course, possibilities of other issues popping up when you use technology or accounts of any kind. If you are having an issue logging in with your Trimble ID, either swing by the forum (`https://forums.sketchup.com`) or contact SketchUp support (`https://help.sketchup.com/en/contact-support`) for assistance.

Once you get yourself properly logged into either warehouse, a world of new content will be available to you. Let's hop in and explore these sites, starting with the Extension Warehouse.

Searching Extension Warehouse

Extension Warehouse is a great place to find extensions, especially if you are just learning how to use them. Extension Warehouse has the ability to search through all the extensions in its database and allows you to view extensions by category, find extensions by a specific author, and see a list of all the extensions that you have downloaded in the past.

While it does have quite a few files available for download, it is a much simpler site than its sibling, 3D Warehouse. Since there are fewer items available for download, searching is simplified. While you can search for a specific term and look into a category for extensions, there is no need for too many filters to narrow your results.

Fewer search options do not mean that you do not have the tools to find what you need, however! Let's dive into Extension Warehouse and do some searching together right now. Since there are two ways to use Extension Warehouse, we will split this into two exercises. First, we will look at Extension Warehouse through a web browser.

> **Supported Web Browsers**
>
> There are lots of web browsers out there, and you will need to be running one that supports the warehouses in order to make sure everything is working as it should. The warehouses are best viewed in Google Chrome, Firefox, Safari, or Microsoft Edge. It should also be pointed out that you should browse the warehouses from the same computer that is running SketchUp Pro.

Extension Warehouse in a web browser

Since the original version of Extension Warehouse was only available through a browser, you can think of using it in a browser as Extension Warehouse in its purest form! Yes, it will run just as well when pulled up from inside SketchUp, but navigating from a web browser has always felt like a better way to explore Extension Warehouse to me.

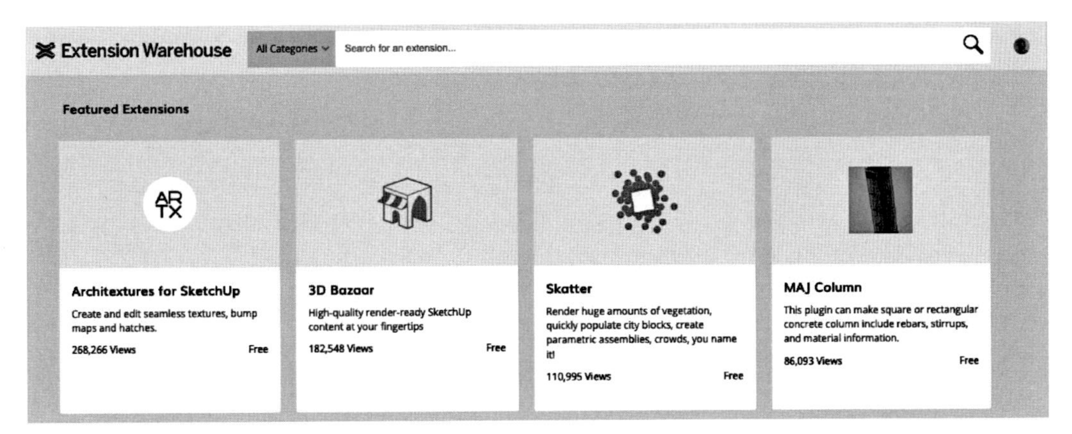

Figure 12.6 – Extension Warehouse's simple, clean front page as seen in Google Chrome

Before we dive in too deep, I do want to mention the main differences in using Extension Warehouse through a browser as opposed to using it directly from within SketchUp. The good news is that the functionality is the same, so there is no need to learn how to use the site in two different ways. The biggest difference is with navigation and access that you have to browser tools.

In a browser, moving forward and backward through tools is easier and sometimes quicker. Additionally, you can open links in new tabs through a browser. This can be nice when you are searching for extensions. In one tab, you can search for an extension. You can then open the results one at a time in their own tabs. This allows for a quick and easy side-by-side comparison of extensions. Finally, if you see an extension that you like the look of but are not ready to download, you can always bookmark the page in a browser.

There are, of course, downsides as well. In a browser, when you find the extension that you want, you will need to download the file through your browser, and then install it the next time you are in SketchUp. If you are accessing Extension Warehouse from within SketchUp, downloading and installing the extension is all done in one step.

For now, let's learn about Extension Warehouse through the browser interface. Let's start by finding an extension that we know we want. Let's find the listing page for the Bezier Curve Tool extension:

1. Start by opening Extension Warehouse in your browser. Go to the `https://extensions.sketchup.com` page (you can just type `extensions.sketchup.com` into your browser to get here).

2. Make sure you are signed in (follow the steps back in the *Logging into the warehouses* section if you need them). You will know that you are signed in because your logo will appear in the upper-right corner of the page.
3. In the search bar at the top, type `Bezier Curve Tool`, then press *Enter*.
4. You should see a list of a few extensions with one called exactly **Bezier Curve Tool** at the top. Your results should look similar to this:

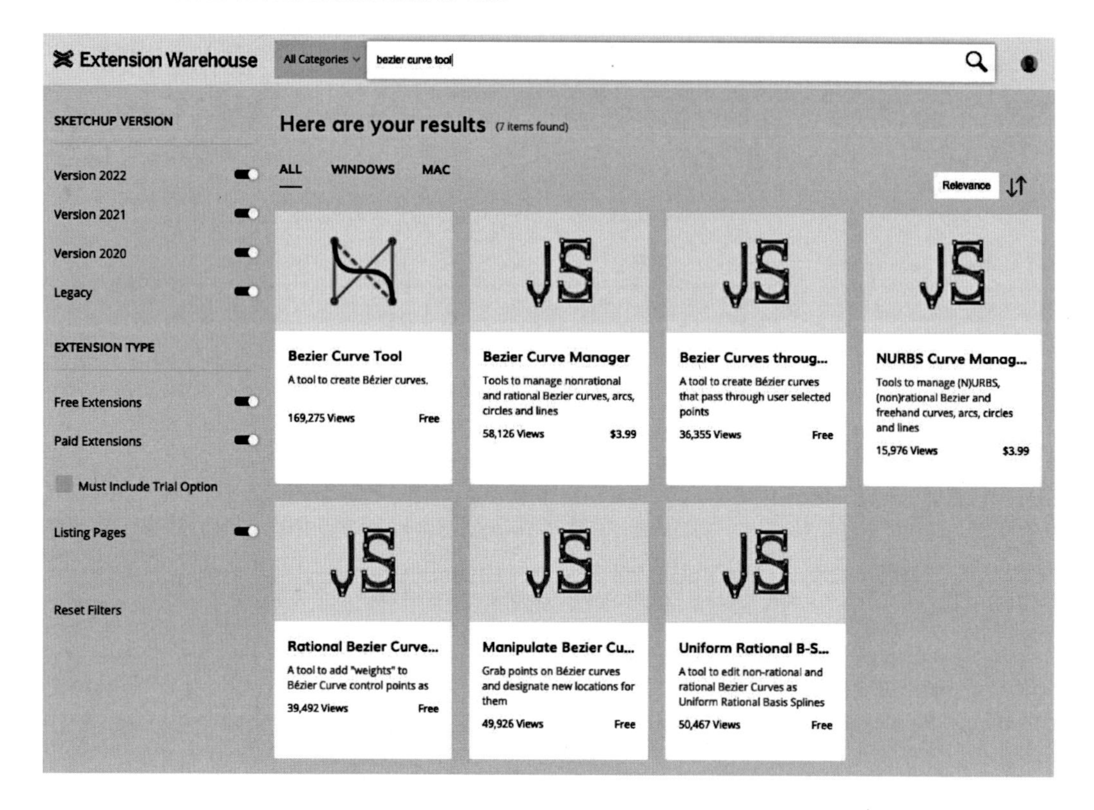

Figure 12.7 – Bezier Curve Tool search results

5. Let's take a look at the page for **Bezier Curve Tool** by clicking on it. You should see a page similar to this:

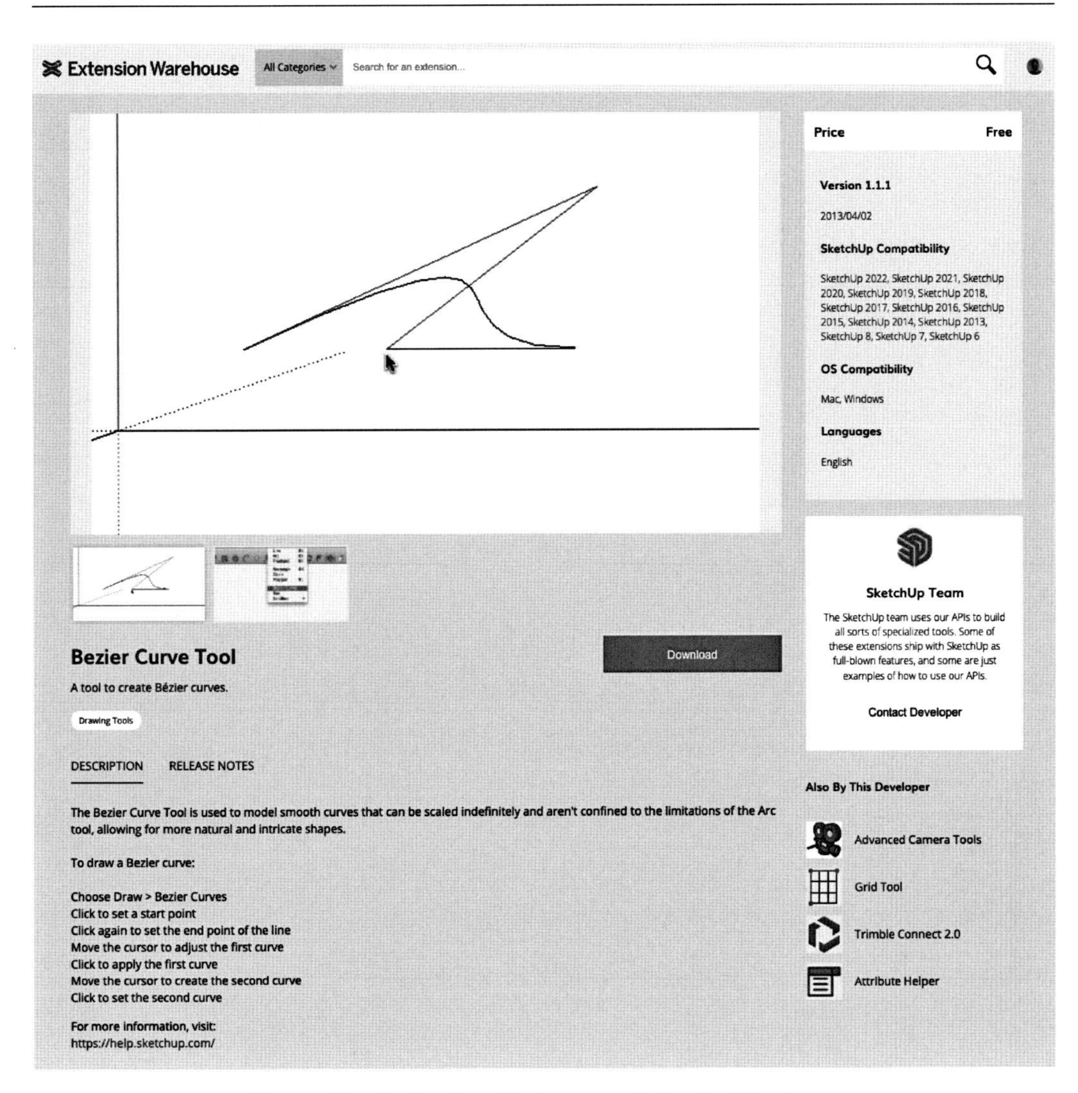

Figure 12.8 – Bezier Curve Tool page on Extension Warehouse

Notice that the page includes a wealth of information about this extension. It includes a few images or movies at the top of the page as well as a written description and release notes below. To the right are the specifics about the extension. Notice that this is a free extension, version 1.1.1, it was released on 2013/04/02 and supports every version of SketchUp back to version 6. This extension will run on Windows or macOS and is only available in English. Below that, we can see that this extension was published by the SketchUp team, and we get a list of other extensions by the author. Right in the middle of the page is a nice big **Download** button.

6. Let's download a copy of this extension by clicking on the **Download** button.

   When you download this extension, you will save a file called `su_bezier_1.1.1.rbz` to your computer. You may remember that this was the sample extension that we used in *Chapter 11, What Are Extensions?*.

Know Your Download Location

To some of you, this may be obvious, but fairly regularly, I see users asking for help finding an extension or model they have downloaded and now want to use. Often, browsers will use a single location to download files (on both my Windows laptop and my MacBook, I download into a `Download` folder). Regardless of where the files end up, you will need to know where they are before you can use them!

Simple enough, right? If you are given a link to an extension or a name from another SketchUp user, it is a simple process to go and find it, pull up its page, and download it. Now, let's go find an extension based on the functionality we know we want:

1. Let's get back to the main Extension Warehouse page by clicking on the *Extension Warehouse* logo in the upper left of the site.
2. Let's search for an extension that will allow us to bend geometry. Type `bending` into the search bar and press *Enter*. This will display something similar to these results:

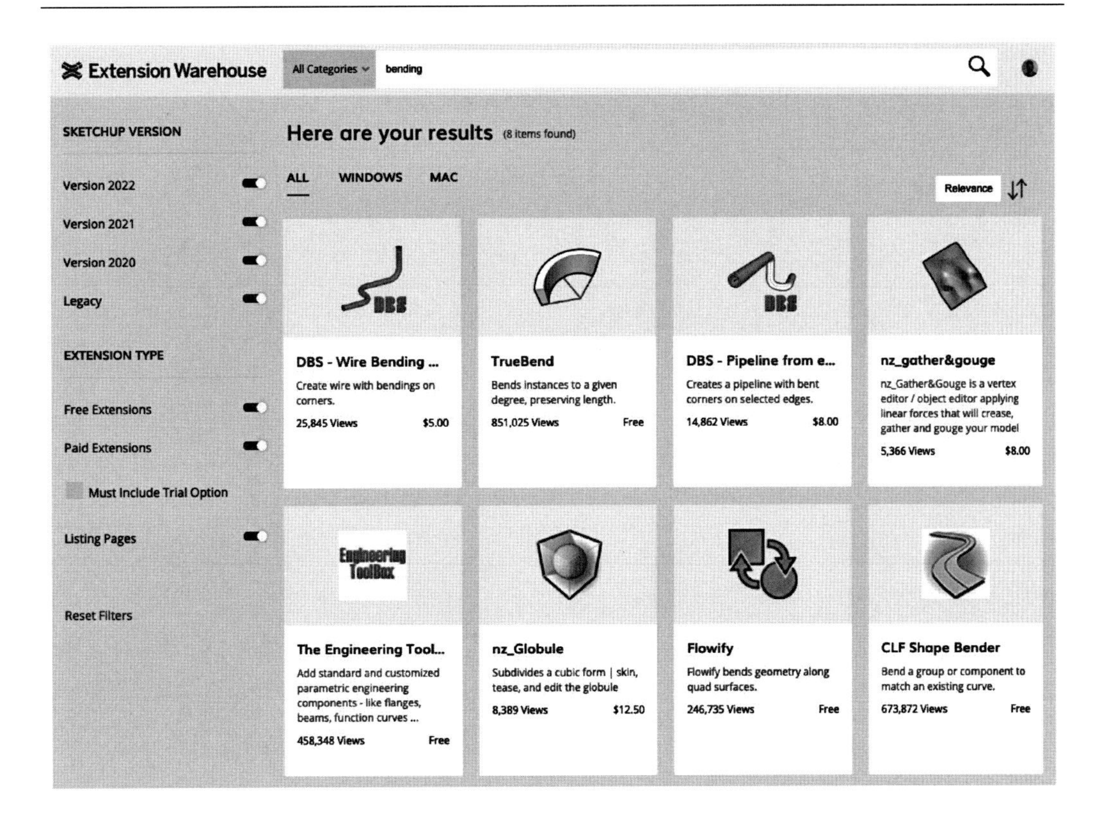

Figure 12.9 – Extensions related to the search term "bending"

Let's narrow down these results using the filter options on the left side of the screen.

3. At the top of the screen, you have the ability to toggle your operating system. Go ahead and click on **WINDOWS** or **MAC** as appropriate for your computer.
4. Since we are running the most recent version of SketchUp, let's toggle **Legacy** off under the **SKETCHUP VERSION** section of the page.
5. We are not looking for a paid extension at this point, so let's toggle **Paid Extensions** off under the **EXTENSION TYPE** section.

With that, we have narrowed our search result down to a single result:

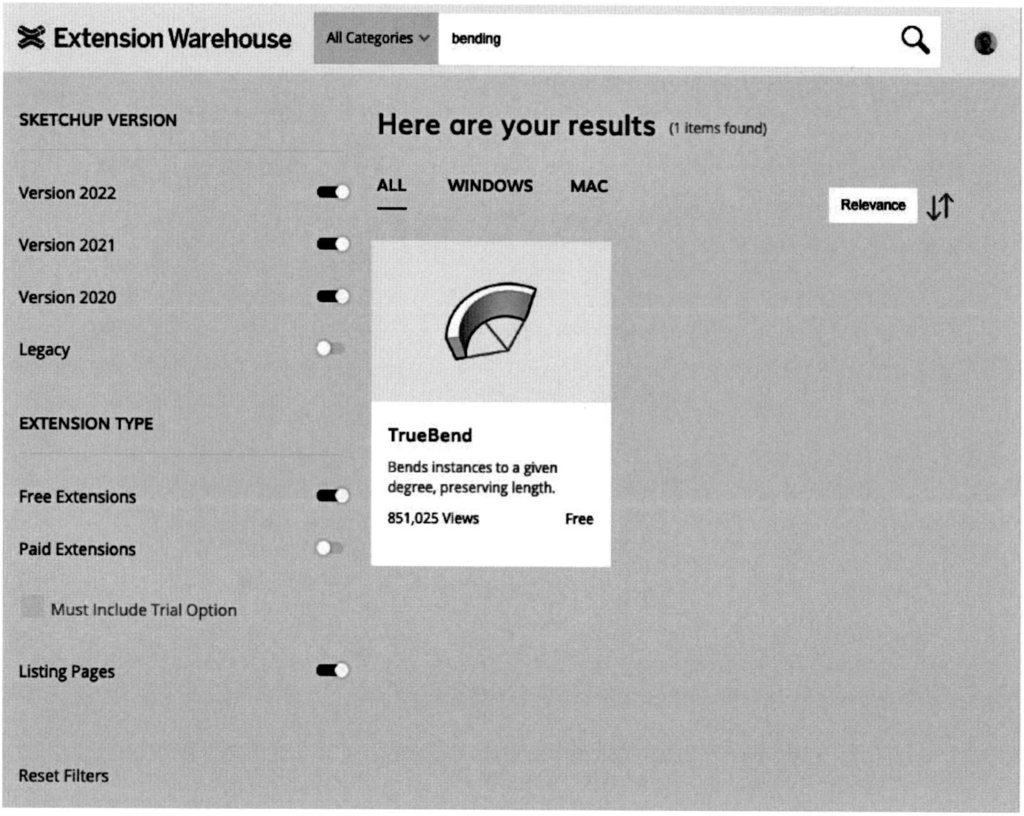

Figure 12.10 – A single result based on search criteria and filters

Though we did not need to use them, note that there are also options to include a trial version (if **Paid Extension** is toggled on) and an option to display listing pages. A listing page is a page for an extension that is not available for download from Extension Warehouse. These pages generally include the same information as hosted extensions but have a button that will take you to a third-party site for purchase or download.

Now that we found the extension that we want, let's download it. This is one of the biggest differences between accessing Extension Warehouse in a browser and from inside SketchUp. From the browser, we need to download the extension and save it to your computer so that it can be installed (the same way we installed an extension in *Chapter 11, What Are Extensions?*):

1. To download an extension, you must first bring up the page for the extension. From the search results page, click on the **TrueBend** extension card to show the **TrueBend** page:

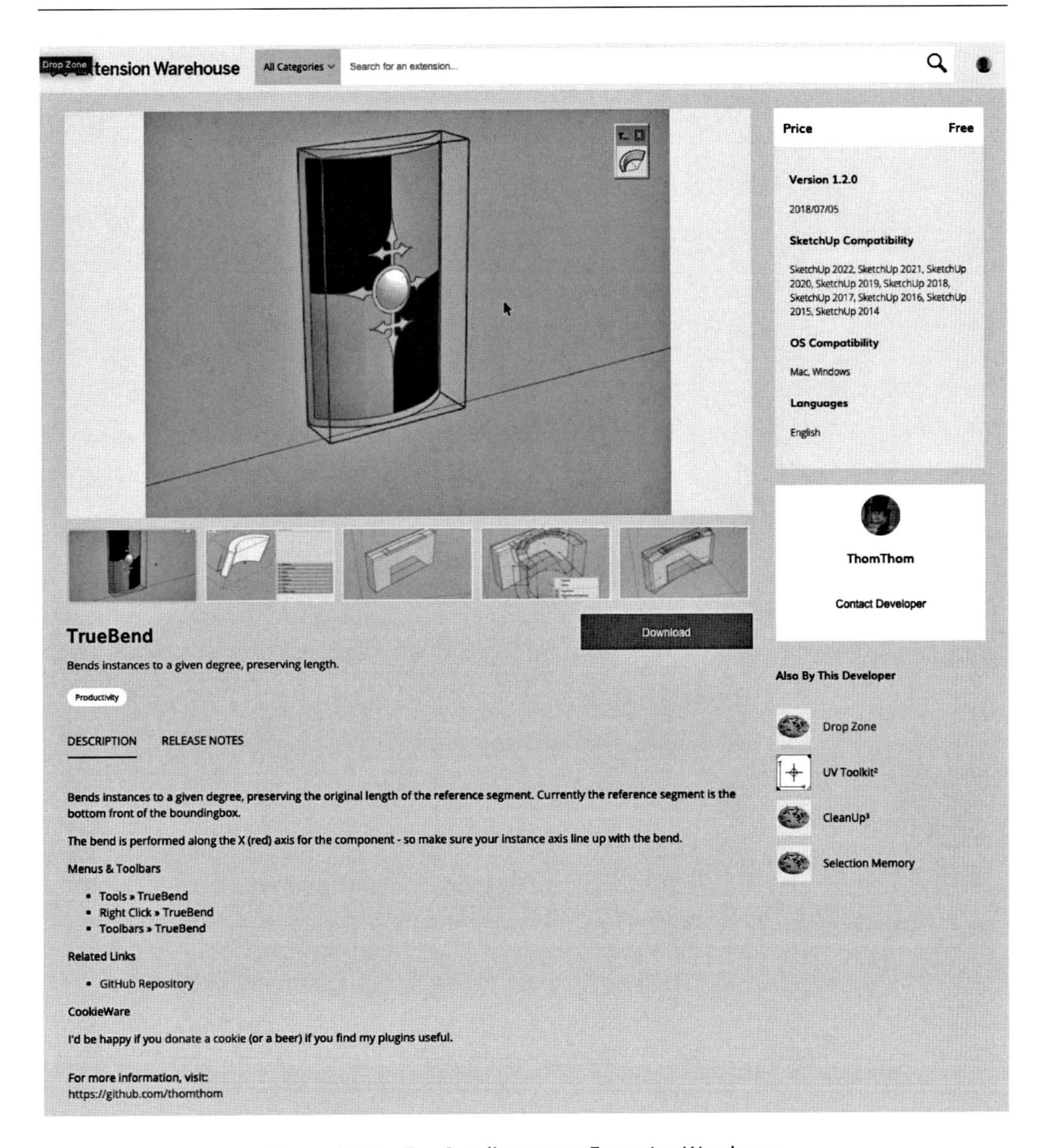

Figure 12.11 – TrueBend's page on Extension Warehouse

2. To download this extension, click the **Download** button.

3. The extension will download to your preset download location, and a message will be displayed letting you know whether the download was successful:

Figure 12.12 – Success message after downloading an extension

Note that this popup does contain a link to a SketchUp Help Center article that contains step-by-step instructions on installing extensions if you ever need help installing in the future.

TrueBend

While you have seen the page for TrueBend at this point, I wanted to mention what this extension is and the utility it brings to SketchUp. TrueBend is a simple way to deform a selected group or component by bending it symmetrically around its center. At this point, if you have followed the instructions, you can find a copy of this extension in your `Download` folder. I would recommend installing it and trying it out. Worst case, you do not see a need for it in your workflow and can uninstall it (a process that we will cover later in this chapter as a part of the *Extension Manager* section).

From here, you can open SketchUp and install this extension following the steps listed in *Chapter 11, What Are Extensions?*. This is not a bad workflow and your best option if you want to keep a copy of the extension file (either to archive or perhaps install on a second computer).

Another option at this point if you would rather not actually download the extension (perhaps you are browsing on your phone) is to bookmark this page through your browser. This will allow you to return to this page later without actually having to download anything, a nice option to have if you see an extension that you may want to come back and try out in the future.

Accessing the Extension Warehouse from within SketchUp is very similar, but with a few differences (and one major advantage). Let's switch over and find and install an extension from inside SketchUp, right now.

Extension Warehouse in SketchUp

When you click on the *Extension Warehouse* icon in SketchUp or choose **Extension Warehouse** from the **Extensions** menu, you will get access to the exact same site that you accessed through your browser. This can be nice, as you do not have to mess with switching to another program to look for extensions. Plus, though the **Extension Warehouse** window in SketchUp does not have all of the features of your browser, it does a good job of showing what you need to see while browsing for extensions:

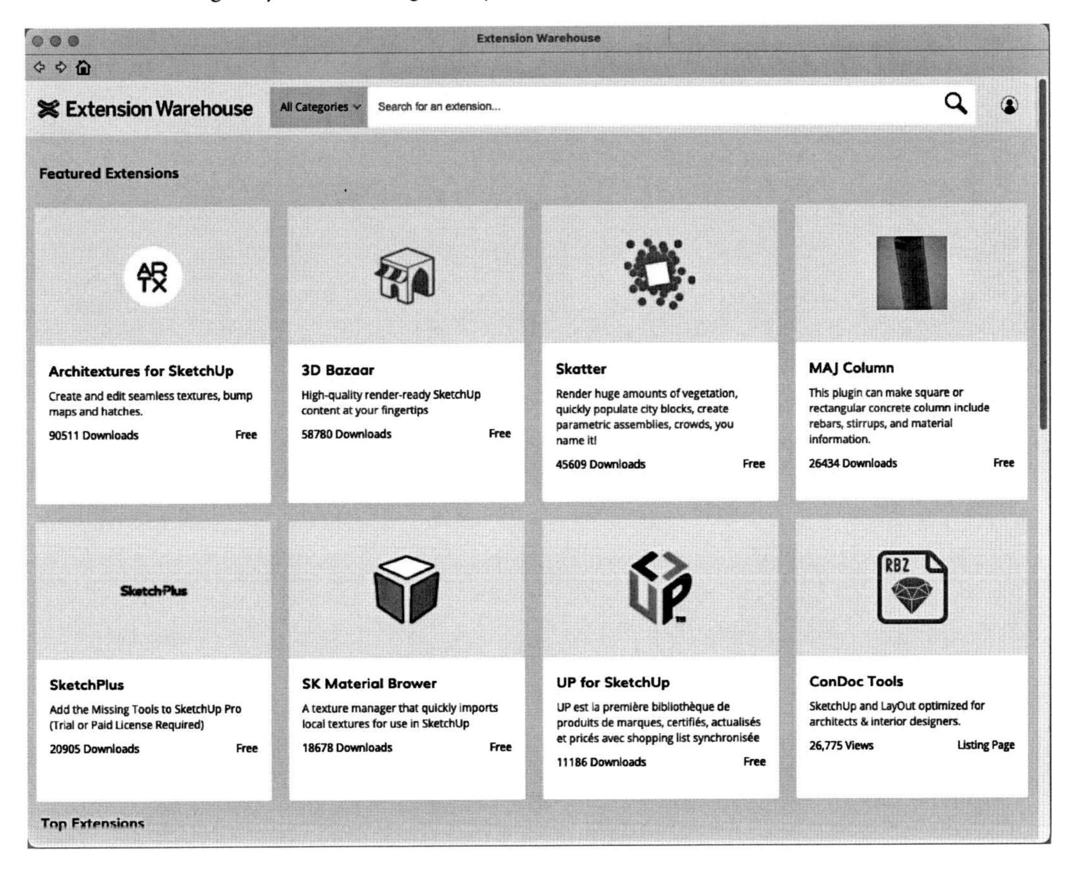

Figure 12.13 – Extension Warehouse as seen from inside SketchUp

Let's find and install another simple extension. We want to find an extension that will allow you to push a single line into a face, just like the native tool **Push/Pull** allows you to pull a single face up into a 3D mass. Let's open SketchUp and take a look:

1. Open SketchUp and click open the **Extension Warehouse** window (either with the icon or through the **Extension** menu):

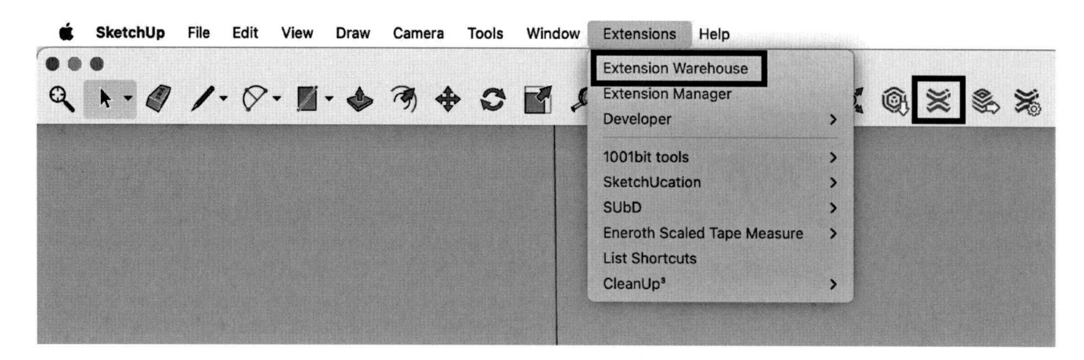

Figure 12.14 – The Extension Warehouse window can be accessed from the toolbar or menu

2. Since we are not exactly sure of the name of the icon we are looking for, we can search for a single term and see whether we can track it down. Enter the term `push` into the **Search** field and press *Enter*:

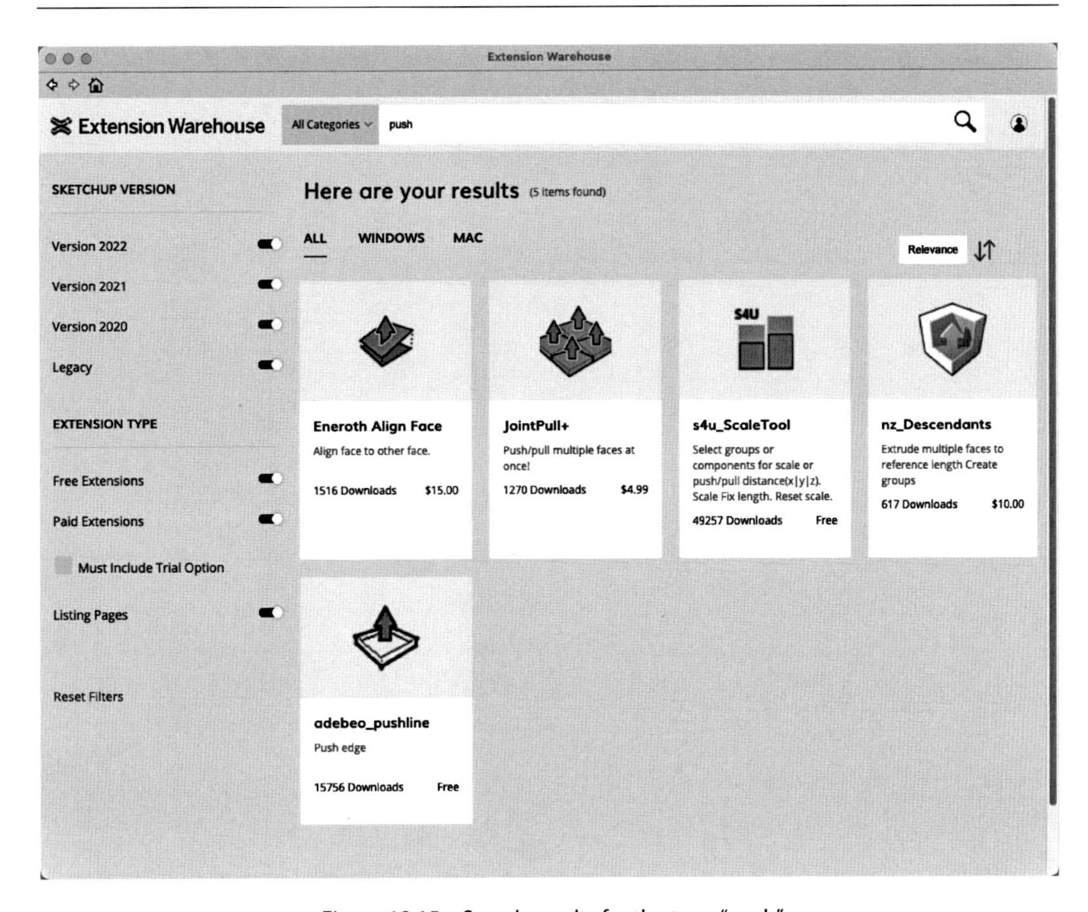

Figure 12.15 – Search results for the term "push"

This will return a few options that we can review and see whether we can find the extension that we are looking for.

3. At this point, we have the option to simply read the short descriptions on the extension cards or click on an extension to see more information. Go ahead and click on a few of the cards to check out the extensions. Note that you can use the arrows at the very top left of the window to navigate forward and back through pages, just like in a web browser.

4. After reviewing these options, it looks like **adebeo_pushline** is the extension that we are looking for. Click on that extension card to pull up the extension page:

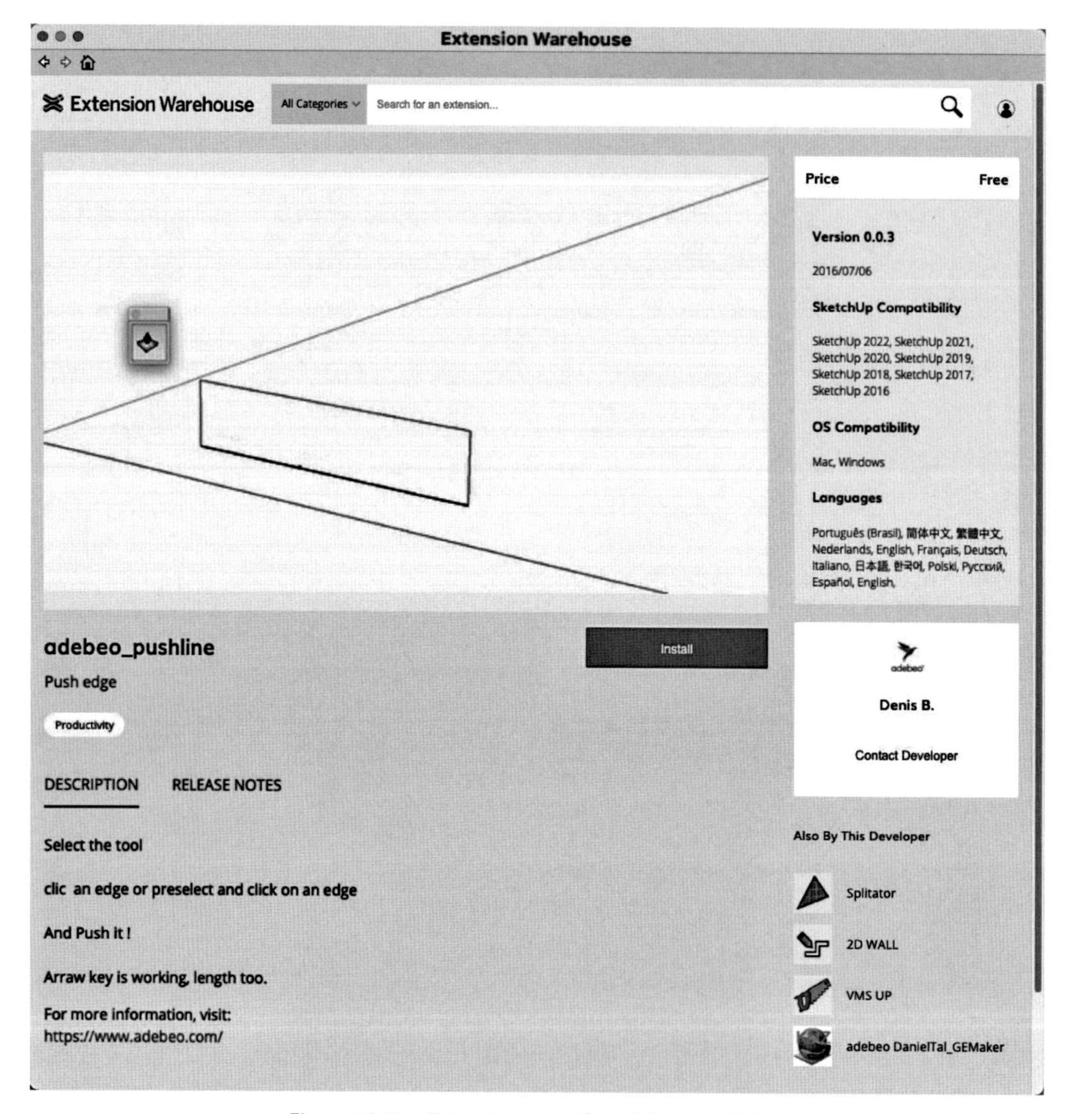

Figure 12.16 – Extension page for adebeo_pushline

5. Install this extension directly by clicking the **Install** button now.

That's it! The extension skipped the download process and was automatically installed directly into SketchUp and is ready for use. In this case, because it is a simpler extension, you should find it available to you right away in your **Extensions** menu. Some extensions will require you to restart SketchUp. This will be true, regardless of the installation method.

adebeo_pushline

Again, this extension is a simple but useful tool that may help you in one or more of your workflows. If you followed the steps listed, you will have it installed and can give it a try. If you do decide to use it, that's great! If you decide that you do not want it in SketchUp, follow the steps to uninstall an extension in the *Extension Manager* section of this chapter.

While there are advantages to both methods of browsing and installing extensions, I generally try to do so from within SketchUp. Being able to get an extension completely installed in a single click is a great option and saves me time! Regardless of your preferred method of getting extensions from Extension Warehouse, you will be able to take advantage of the **My Downloads** page, which we will explore next.

My Downloads

Whether you are logged in to Extension Warehouse through SketchUp or via a browser, you have the ability to view your **My Downloads** page. **My Downloads** is a list of every extension that you have installed or downloaded from Extension Warehouse while logged in with the current account (see information on Trimble ID in the *Logging into the warehouses* section of this chapter). This view (which can be sorted from newest to oldest or oldest to newest) allows you to click on the extensions you have previously used to return to their page on Extension Warehouse. If there is an extension that you want to remove from the list (say you tried an extension and did not like how it worked), you can click the **X** button on the right side to remove it from the page completely. This list will also give you information on paid extensions, including how much you paid, when you purchased, and, if applicable, when your subscription will expire.

This page is a great tool to use when you install new versions of SketchUp or migrate to a new computer. Rather than referencing what extensions are installed on an older version, you can scroll through **My Downloads** and redownload or install the extensions that you have used previously.

My Downloads is a great way to see the extensions you have used previously, but to manage the extension that you currently have installed, you will need to use **Extension Manager**.

Extension Manager

In addition to being the page that allows you to install extensions (as we saw in *Chapter 11, What Are Extensions?*), Extension Manager also shows a list of every extension that has been installed in SketchUp. Additionally, you can temporarily disable extensions or install updates from Extension Manager:

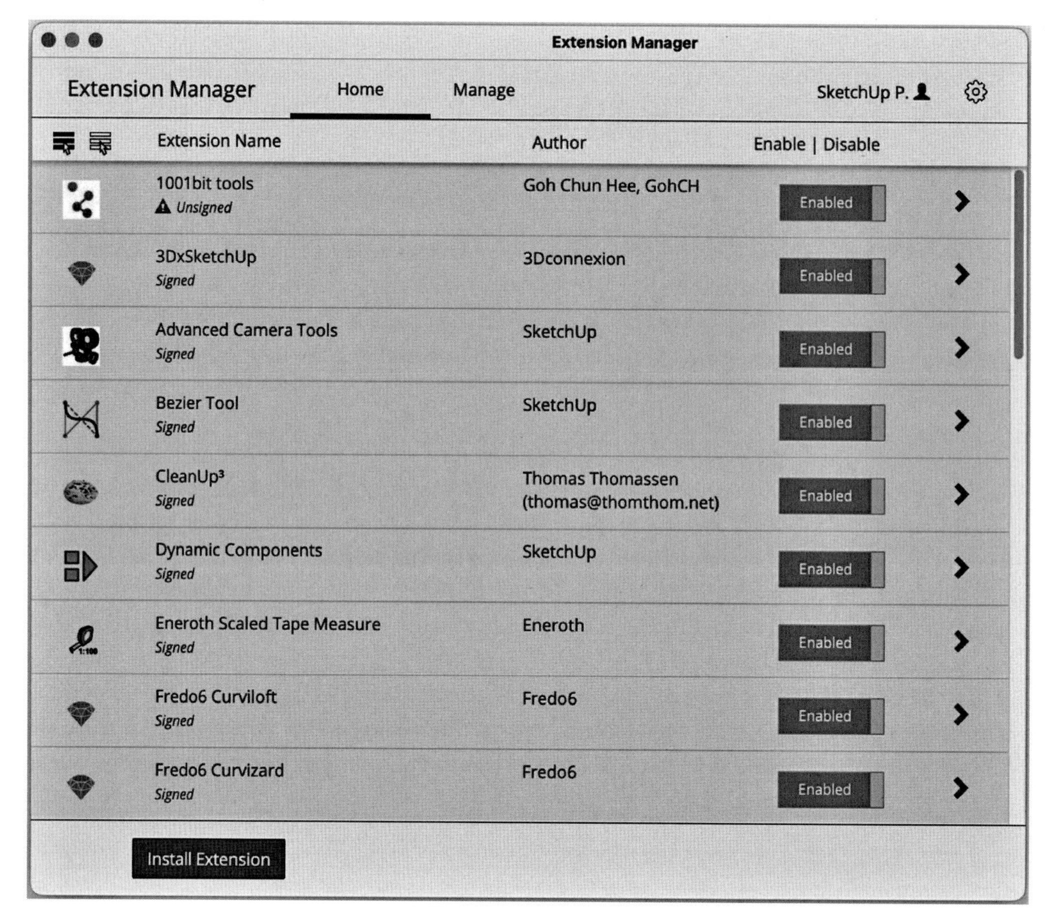

Figure 12.17 – Extensions Manager as seen on macOS

At the top of the **Extension Manager** window, you can switch between the **Home** and **Manage** tabs. The **Home** tab will display a list of all the extensions installed on this installation of SketchUp, including the icon, name, and author. To the far right is an arrow that will allow you to view additional details for an extension. Additional details will look similar to this:

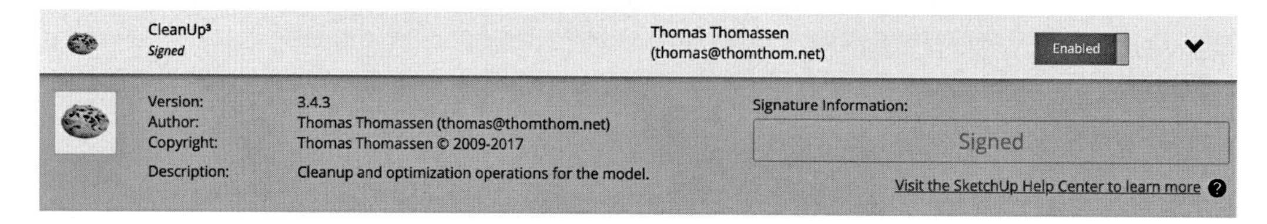

Figure 12.18 – Extension information for Thomthom's CleanUp³

This information includes things such as the version number and contact information for the author, as well as the extension signature status.

The only control on this tab is a toggle switch that allows you to enable or disable each individual extension. Toggling the switch to **Disabled** for an extension will turn off the extension until you enable it again using the same switch. Many extensions will disappear from SketchUp as soon as you click on this switch, while others may require you to restart SketchUp before they will disappear.

Why would you want to disable an extension? There are a few reasons. It is possible that one or more extensions conflict with each other. I mentioned back in *Chapter 11*, *What Are Extensions?*, that extensions reviewed before they are listed on Extension Warehouse will not conflict with each other, but it is possible that extensions downloaded from elsewhere may cause such issues.

Another reason that you may choose to disable extensions is to speed up SketchUp's initial load time. In general, SketchUp starts up quickly. As you load more and more extensions, that initial startup time can get longer and longer. The simple fact is that the more extensions that need to get loaded, the longer it will take to start SketchUp. If you struggle with long startup times, you may try to disable extensions that you only use occasionally. If this helps, you can leave them disabled and only enable them when you need to use them.

The second tab at the top of the **Extension Manager** window is the **Manage** tab. This tab will display the same list of extensions, but replace the **Enabled/Disabled** switch with buttons for **Update** and **Uninstall**:

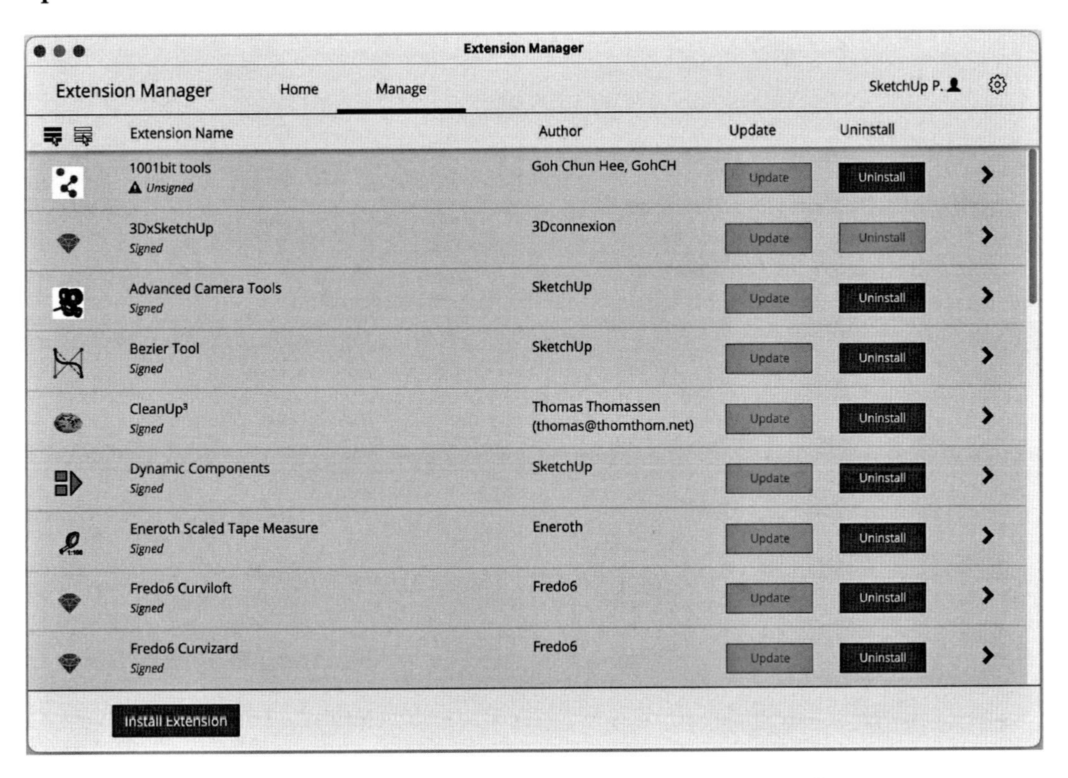

Figure 12.19 – The Manage tab in the Extension Manager window

When you install an extension through Extension Warehouse inside SketchUp, Extension Manager will keep track of the extension and let you know when a new version is available. In addition to displaying a popup when you first start SketchUp, the **Update** button will be enabled in the **Extension Manager** window. Clicking this button will automatically download and install the newest version of the extension. As with installing, some extensions may require you to restart SketchUp before they become available.

The **Uninstall** button is available for any extension that can be automatically uninstalled from within SketchUp. Clicking this button will display a confirmation popup before completely removing the extension from SketchUp:

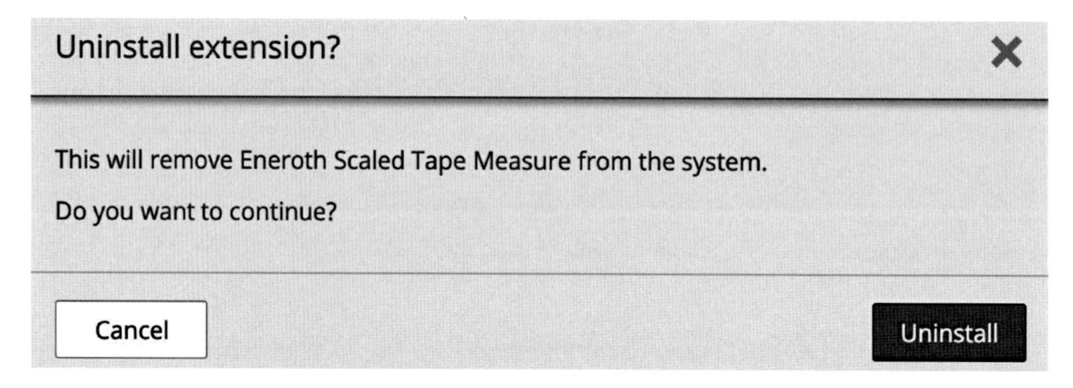

Figure 12.20 – Uninstall confirmation popup as seen in macOS

As with installations, some uninstalls will require you to restart SketchUp before the extension disappears from SketchUp completely. An uninstalled extension will not be listed in Extension Manager but can be seen as part of the list of extensions on your **My Downloads** page in Extension Warehouse.

Occasionally, an extension may be installed by a third-party software and the **Uninstall** button may be disabled. These extensions cannot be removed from within Extension Manager and may have to be uninstalled by the software that installed them or manually removed.

Extension Warehouse is a big, amazing world with so many new SketchUp capabilities available to you. Being able to find, install, and manage extensions is a new skill that is sure to help you in your quest to level up! Extension Warehouse is, however, only one of the online repositories you will want to get used to using. The other is, of course, 3D Warehouse!

Finding things in 3D Warehouse

3D Warehouse is the largest web repository of its kind. Boasting around 4 million files available for download, 3D Warehouse is a huge asset to designers, allowing them to simply download and place models into their files rather than having to model everything from scratch.

There is a downside to 3D Warehouse. 3D Warehouse's biggest asset is also its biggest problem. Since anyone can get an account and upload any SketchUp file, quality can vary greatly, resulting in some models that are great, and some models that are not so good. This means that to get the most out of 3D Warehouse, you will need to be savvy about how you search for and use the models you download.

I have, in the past, referred to 3D Warehouse as being like the Wild West. In the early years of 3D Warehouse, it was very easy for people to create whatever model they wanted, dump it there, and then leave. This ended up causing a lot of files that were of no value to end up sitting there, just taking up space. Fortunately, a lot more effort is being put into 3D Warehouse in recent years and it has made it a far more useful tool! From requiring sign-in to upload to expanding the search options, recent developments mean you will spend less time searching through uploads of slightly modified default scale figures and more time downloading and bookmarking useful models for your current or future project.

Just like Extension Warehouse, 3D Warehouse can be accessed through your web browser or from within SketchUp. There are a few differences between the two methods, so we will take a look at both right now.

Using 3D Warehouse in a web browser

Originally, 3D Warehouse was designed to be accessed as a web page. Navigating the pages and using the search tools feels very natural, and there are people out there who prefer to use 3D Warehouse from their favorite browser. The way that 3D Warehouse fills the browser window makes it much easier to browse:

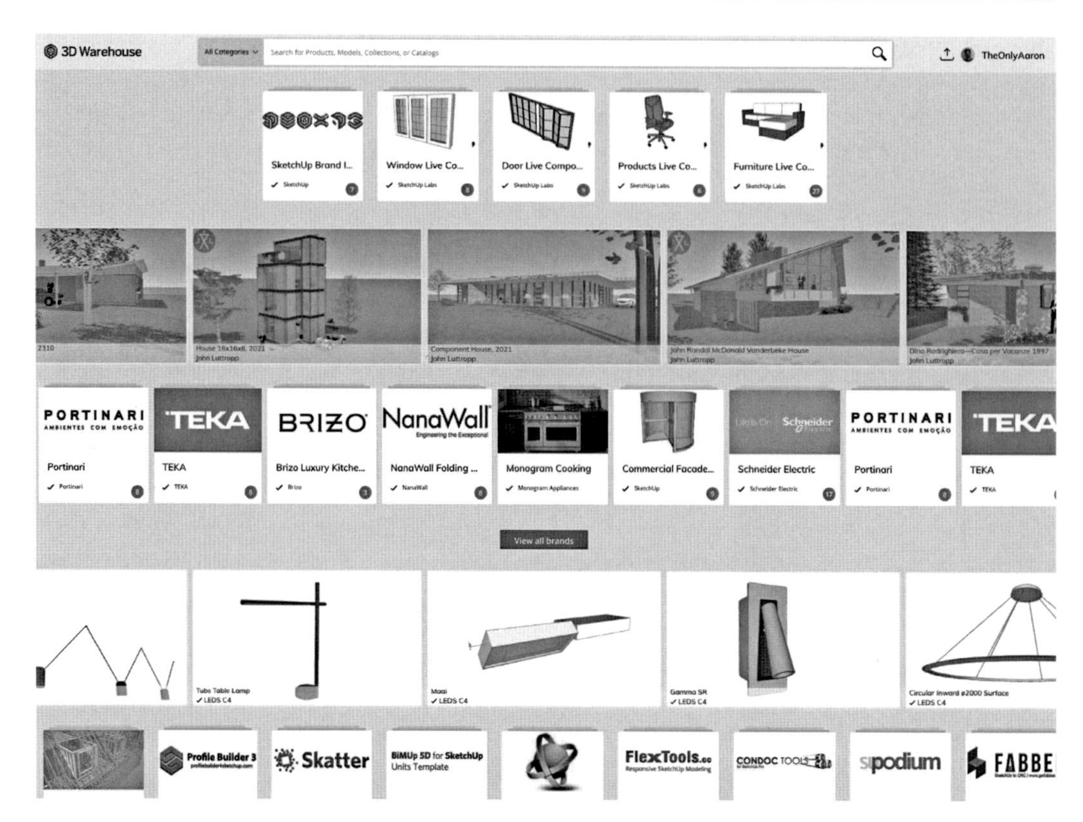

Figure 12.21 – 3D Warehouse as seen from Chrome

As with Extension Warehouse, the big advantage of using 3D Warehouse in a browser is the browser itself. From the browser, you can navigate quicker, open links in new tabs, or bookmark items that interest you for access later. An additional advantage is that using 3D Warehouse through a browser allows you to download files for use outside of your current model. This may not seem that great right now (isn't it better to download a file directly into your model?), but you will understand this advantage as you continue reading the chapter.

To get our feet wet in 3D Warehouse, let's head there together and see whether we can't find a nice couch (or sofa to some) to download and use in SketchUp:

1. Open your browser and go to `https://3dwarehouse.sketchup.com`.
2. In the search bar at the top, type `couch` and press *Enter*.

   This will return a list with a lot of results. In fact, this will return over 1,000 results:

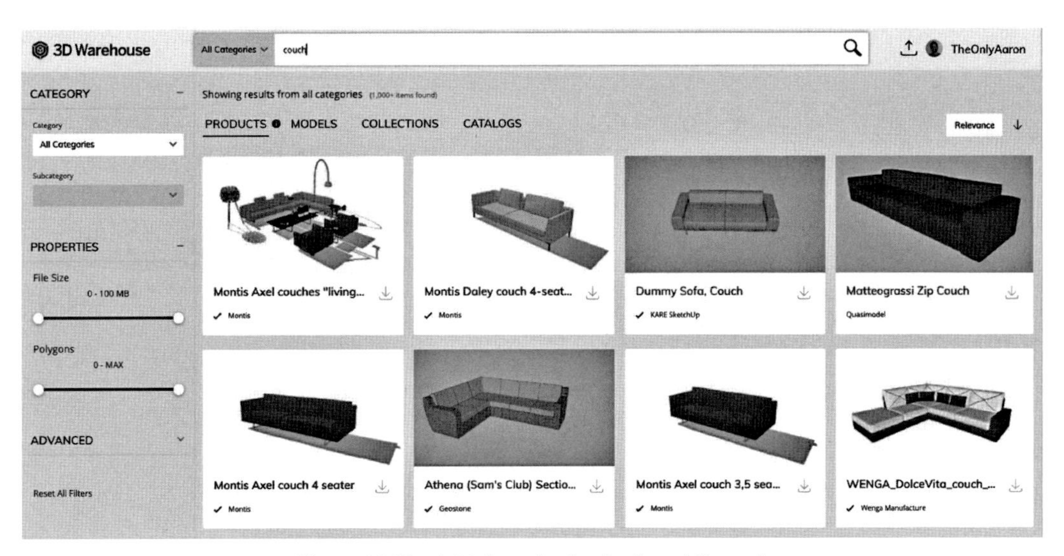

Figure 12.22 – Initial results for the "couch" search

Since 1,000+ results are too many to wade through, let's try to narrow down this list by using some of the search filters on the left side of the window.

Products, Models, Collections, and Catalogs

Toward the top of the page, you have the option to sort results by products, models, collections, and catalogs. Products are models uploaded by the companies that manufacture or sell the items that the SketchUp models represent. Models are files uploaded by users that do not necessarily have any connection to the creator of the product. Collections are folders created by users to organize groups for models. Catalogs are groups of product models created by companies.

First, let's start by limiting the search to furniture items only. I know, it seems odd, but when 3D Warehouse searches its database, any results with the searched term will come up as a result. This means someone's description of a living room that says **Living room with no couch** or a delivery truck named **Couch Brothers Deliveries** will both end up on the list. We can eliminate these sorts of results by forcing 3D Warehouse to only look in a selected category.

3. Click the **Category** dropdown and choose **Furniture**.

   This likely reduced the list, but the results are still over 1,000. Let's reduce it further by adding a subcategory.

4. Click the **Subcategory** dropdown and choose **Residential**.

   This reduced the results even more, but we need to filter even more. Let's refine our search by dialing in the requirements of the file we are trying to find. To do this, we can set ranges on two properties: **File Size** and **Polygons**. While both of these properties are useful, it probably makes sense to only set one of them when searching.

I Prefer Polygons

Filtering results by file size and polygons will effectively narrow the search to show you files of a certain size. I prefer to use polygons when searching because it is a reflection of the geometry in the model only. File size can be bloated by things such as materials and components in the file, while the polygon count represents just the faces that make up the model.

5. Slide the left handle below **Polygons** right to about **250**, and the right handle to the left to around **750**.

   This filter will have a big impact on the number of results returned. At this point, you may have around 100 total results.

Advanced Search Options

Below the property search options is a section titled **ADVANCED**. This section contains additional search options. These options include controls to search for files based on their creator, title, or date. These options tend to be more useful when you are looking for a specific model, rather than in the case of a general search.

6. Scroll through the results and click on a few models. Clicking on the card for a model will bring up a simple preview view of the model. This is one of the couches that was returned in my search:

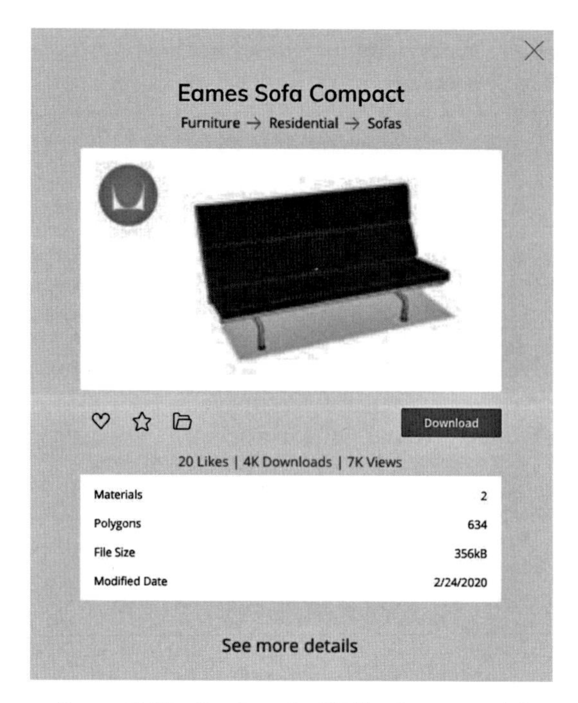

Figure 12.23 – Preview of a 3D Warehouse model

From this preview, you can see a larger thumbnail of the model and general information about the model. You also have the option to download the model or click **See more details** to bring up the full page for this model.

7. Let's take a better look at the model by clicking **See more details**. This will bring up the full-page view of this model:

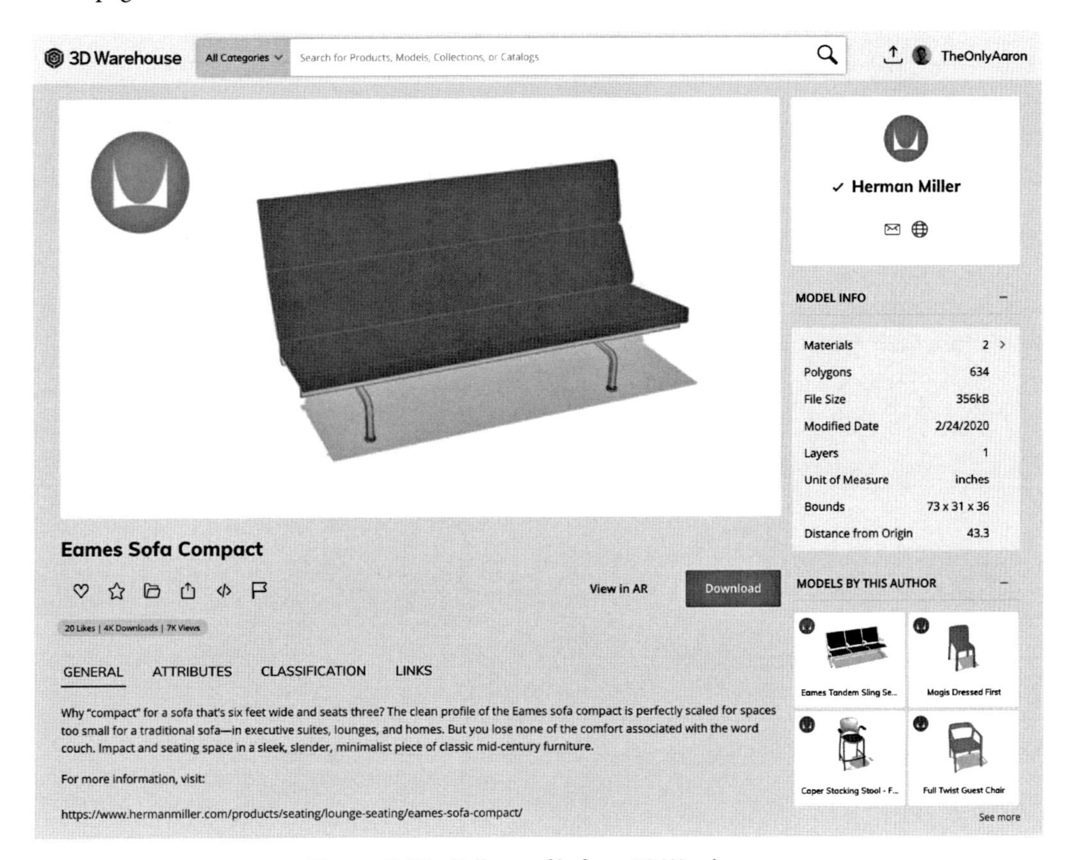

Figure 12.24 – Full page file from 3D Warehouse

The full page includes much more information about the model, the creator, and links to related models. Additionally, you can view models in **Augmented Reality** (**AR**) from this page. Let's go ahead and download a file for use later in this chapter.

8. Click the **Download** button on the page for the couch.

You will be prompted to select a version of the file to download. 3D Warehouse will offer to download to the previous two or three versions of SketchUp, as well as a GLB or Colada file. For this example, choose the version of SketchUp that matches your version of SketchUp Pro. With that, you have searched and downloaded a SketchUp model from 3D Warehouse. The next step from here is to import the downloaded model into SketchUp.

9. Open SketchUp, start a new file, and choose **Import...** from the **File** menu.
10. Navigate to your `Downloads` folder (or the location that you downloaded the file to in the previous steps) and select the couch file that you just downloaded.

    SketchUp will return you to your current model and connect the origin of the file being imported to your cursor.
11. Click in the model to place the couch.

As I mentioned earlier, I prefer to use 3D Warehouse from my browser. It can, however, be used from inside SketchUp, as well. While most (almost everything) of the interface is the same, the final steps of downloading and placing are much quicker. Let's quickly run through that process in the next section.

Using 3D Warehouse in SketchUp

To access 3D Warehouse from inside SketchUp, you can click the icon on the toolbar or select **3D Warehouse** from the **Window** menu:

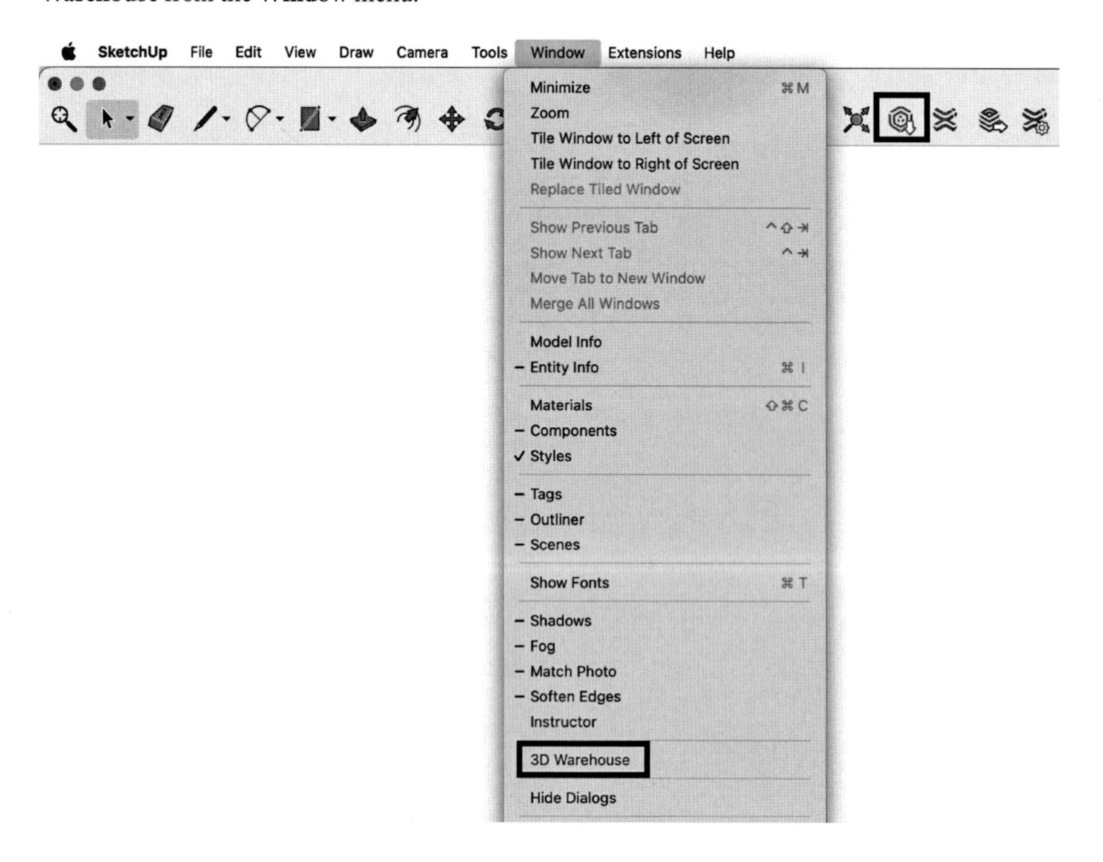

Figure 12.25 – Access 3D Warehouse from the Window menu or the toolbar

The **3D Warehouse** window inside SketchUp will give you the same access to search and view models as through the website. The big difference will appear when you are ready to download a model. When you click the **Download** button, rather than being prompted for the type of file to download, 3D Warehouse will ask whether you want to load the file directly into your current model:

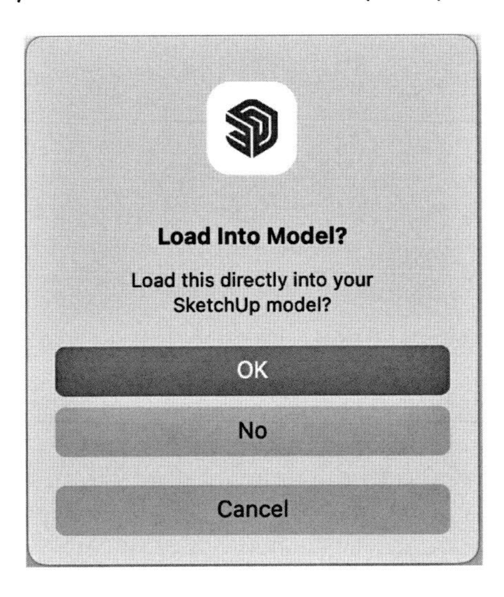

Figure 12.26 – Download options from 3D Warehouse inside SketchUp

Clicking the **No** button will prompt you to pick a location to download the file to. Clicking the **OK** button will return you to your model with the new file attached to your cursor. A single click will place the file into your current model.

Searching for and downloading models from 3D Warehouse is a great time saver, but only the beginning. A big difference between a novice and advanced user of 3D Warehouse is knowing how to examine the model and make sure it is what you want to use in your project. This is exactly what we will cover in the final section of this chapter.

Using 3D Warehouse models

Downloading a model from 3D Warehouse can save you a lot of time and energy. That can also add a lot of unwanted information and geometry to your model. This is the main reason that I always recommend downloading and reviewing a model before dropping it into your project. It is an extra step and can add a few minutes to your workflow, but it ends up saving you time when you compare it to having to deal with problems that can be introduced when placing bloated models into your work.

The process of reviewing and optimizing a SketchUp model could be its own chapter (or chapters), so we will have to use this section as a primer. I will run through the process of opening a model and list the top things to look for in your model. After that, we will look at a couple of options you have if you want to clean up a model. Of course, not every model is worth saving. Remember that there are millions of files available for download. So, if you start reviewing a model and find that it is going to be a lot of work to clean up, it might be the best plan to head back to 3D Warehouse and refresh your search!

Reviewing (and fixing) 3D Warehouse models

Remember back when I said that I preferred to search for models in my browser and download them? That's because I prefer to review my models before I drop them into an active project. To do this, I open an empty file and import the model to be reviewed. I do not open the model in SketchUp. This is an intentional step and an important part of this process. Importing it into an empty model will make the new model behave the same way that it will if I import it into my actual project.

In the next subsections, we discuss the things that I try to look at once I have imported a file from 3D Warehouse.

Excessive geometry

The biggest cause of lag in SketchUp is too much geometry on the screen. The more geometry there is in a model, the harder SketchUp must work to show it on the screen. This tends to be an issue when people start downloading entourage or set dressing from 3D Warehouse. Many of the models available for download look beautiful but may not have been optimized to the level of detail that you need for your specific model.

I remember downloading a model of a bowl of fruit for a kitchen model I was working on. The entire model (without the fruit bowl) was a few thousand faces (according to the **Model Info** window). The bowl of fruit I downloaded had over one million faces. This little detail that I was hoping to add to my model was over four times as much geometry as the entire rest of the model!

There is not a hard number for what is a "good" amount of detail for an imported model, and the proper amount of detail will change depending on the part that the downloaded file plays in your project, but there is a quick way to see just how much geometry is in a model.

The first thing I do when reviewing a model is to turn on **Hidden Geometry** from the **View** menu. If the model I am looking at turns black with all the dashed lines on its surface, I will consider another file for my project or see whether I can reduce the geometry in the model.

Manually reducing geometry can be a long road that may include redrawing sections of the model altogether. If this is a process that you find yourself needing to do on a regular basis, you may want to take a look at an extension that will systematically reduce polygons, such as Skimp or Transmutr (more on both of these extensions in *Chapter 13, Must-Have Extensions for Any Workflow*).

Unnecessary nesting

Nesting is when a group or component is inside another group or component. In some cases, this can happen repeatedly to the point that the actual geometry that makes up a model is a dozen or more levels deep. This does not necessarily cause a problem in a model (it does not cause significant file bloat or performance issues) but it can make models difficult to work with.

Checking for nesting is simple. After importing your file into an empty model, open the **Outliner** window and look at how many levels there are to the model. Again, there is not a specific number of levels a file should have, but if you see the base geometry for a piece of furniture is 15 levels deep, know that you may want to look for another model or spend some time removing a few levels of nesting before importing into your project.

Too many tags

Tags are essential for organizing any model. The problem comes when the tags used to create a model end up in a larger project. For example, when modeling a car, you may end up with tags for the wheels, the frame, the glass, the interior, and more. In a SketchUp model for a car, it would make sense to have a dozen or more tags. When it comes time to import that car into a model of a garage, those tags are likely excessive. In fact, in a garage model, there probably only needs to be one tag that controls the visibility of the car.

To check tags in your imported model, just open the **Tags** window. When I am doing this level of review, I will often delete all tags, leaving everything on the **Untagged** tag. When I import the file into my working model, I will create one new tag for the item.

Extra components

When files are imported into a model, they are imported as a component. Any components that are already in the file show up as additional components in the model. This may include important components such as the pieces that make up a piece of furniture (maybe the feel and cushions of a chair are components), but a file may also include components that were created or imported when the file was being created. Remember, components imported into a SketchUp model are saved as a part of the file, even if they are not in the modeling window.

To check for extra components, just pull up the **Components** window and look through the list. If all the components listed appear in the model you are looking at, you are probably good. If you see a bunch of components for extra parts that are not part of what you are hoping to import, you may need to clean up the file before importing it into your working model. The easiest way to do this is to use the **Purge Unused** command (just like we did back in *Chapter 2, Organizing Your 3D Model*).

Lots of materials

Like components, models can accrue extra materials as they are being developed. This, too, is pretty easy to identify. After importing a model, open the **Materials** window and look at the materials listed for **In Model**. If the materials listed can be seen in the model you are looking at, you could be done. If your model appears to have four materials, but the **Materials** window shows a few dozen materials, you should clean it up before adding to your project. This, too, can be done by running the **Purge Unused** command.

Developing the skills and experience to find exactly what you need out of 3D Warehouse will take time and practice. Once you have done so, you will find 3D Warehouse an indispensable resource. Quickly downloading and placing models can knock hours off a project and help you to level up!

Summary

Hopefully, this overview of the warehouses was a step toward adding to your workflows. In this chapter, we discovered what the warehouses were and where they came from. We saw what a Trimble ID was and how (and why) to sign in. Finally, we learned to search, download, and use the content found in both warehouses. While many people take advantage of both 3D Warehouse and Extension Warehouse, you may find that you rely on one more than the other, and that is just fine, for now. Eventually, developing a way of working in SketchUp that relies on both extensions and downloaded 3D Warehouse content will allow you to go far beyond your current abilities and help to take you to the next level.

While we will not be spending more time in 3D Warehouse in this book, our next chapter, *Chapter 13, Must-Have Extensions for Any Workflow*, will have you exploring some specific extensions and, most likely, using your new Extension Warehouse and Extension Manager skills to download and install some of them!

13
Must-Have Extensions for Any Workflow

In *Chapter 11, What Are Extensions?*, we talked about what extensions are, how they are made, and how to install them. Then, *Chapter 12, Using 3D Warehouse and Extension Warehouse*, showed us how to look for extensions in Extension Warehouse. Throughout these chapters, we looked at a couple of extensions, but they were just tools to help show you how the extension system works. In this chapter, we will look at a handful of specific extensions and how they function. We will be downloading these extensions and walking through how to use them in some step-by-step examples.

As we've already discussed, the world of extensions is a big one with hundreds of extensions created for all kinds of workflows. For the examples in this chapter, I have presented a few extensions that should be useful to anyone using SketchUp, regardless of the workflow. Additionally, the extension that we walk through is (at the time of writing) completely free to download and use.

The final section of this chapter includes a list of extensions that I believe will be useful for a lot of SketchUp users, but not everyone. Additionally, some of the extensions in the final section are not free to use. Many have free trials, so you can try them out before buying them, but I did not want to list step-by-step demonstrations of any extension unless I was sure that everyone could follow along.

In this chapter, we will cover the following topics:

- Selection Toys from Thomthom
- Bezier Curve Tool from the SketchUp team
- Curic Mirror from Curic
- adebeo_pushline from Denis B
- Exploring additional extensions

Technical requirements

For this chapter, you will need SketchUp Pro, a connection to the internet, and the `Taking SketchUp Pro to the Next Level - Chapter 13.skp` file, which is available at `https://3dwarehouse.sketchup.com/model/7ca74523-9566-45c2-b9f5-761f4630de52/Taking-SketchUp-Pro-to-the-Next-Level-Chapter-13`.

Selection Toys from Thomthom

Every once in a while, I will install an extension that becomes deeply entrenched in my SketchUp workflows. I am talking about the sort of commands that I use regularly. These are the kinds of commands that I end up feeling a little bit lost without. These are the kinds of commands that are designed to look and feel so much like SketchUp that I forget that they are even extensions!

This is the case when it comes to Selection Toys from Thomthom. There are times when I will try to use SketchUp on someone else's computer or work on a new installation and look for Selection Toys commands, thinking that they are a part of standard SketchUp!

Hopefully, you are intrigued at this point and want to learn more about this extension! Let's provide an overview of Selection Toys' capabilities, then try it out!

Selection Toys overview

Selection Toys adds an array of selection modifiers and filters to SketchUp. I know – way back at the beginning of this book, I said that every SketchUp model is made up of only edges and faces. If there are only two things, how can an extension with selection filters be useful?

Yes, there are only two "pieces" that make up SketchUp models, but those two pieces get added to other objects in SketchUp. When all is said and done, we have edges and faces in a model, but we can also put those edges and faces into groups and components. Additionally, we can add materials to those edges and faces. We can also organize the groups and components by assigning them to tags. Plus, we can weld edges into curves, or add things such as guides, sections, or images to the model!

In the end, if you were to take a working model and select everything on the screen, you may end up with the need to modify your selection with a filter – something such as, "*deselect all guides and dimension lines.*" This is what Selection Toys does. It does not take long to use Selection Toys before you realize the power a simple selection filter can have.

I use the term "simple filter," but the fact is, Selection Toys has a lot of options. The toolbar for version 2.4.0 has up to 45 different icons on it! I say "up to 45 buttons" because Selection Toys allows you to customize its toolbar so that you can see the icons for the commands you use, rather than everything all at once. To that end, it is this sort of function that makes Selection Toys so usable and functional. Everything from the icons to the way they are presented is well thought out and quick and easy to use.

Selection Toys has two groups of commands: **Select Only** filters and **Deselect** filters. You run Selection Toys commands by first selecting some portion of your model, then choosing to keep only certain items selected, or removing certain items from the selection.

Selected items can be filtered by the following:

- **Edges**: Toggle the lines in the selection on or off
- **Faces**: Toggle the surfaces that make up the selection on or off
- **Groups**: Toggle any groups in the selection on or off
- **Components**: Toggle any components in the selection on or off
- **Guide Lines**: Toggle any guidelines (created by the **Tape Measure** tool) in the selection on or off
- **Guide Points**: Toggle any guide points (zero length guides) in the selection on or off
- **Text**: Toggle any text (screen text or leader text) in the selection on or off
- **Images**: Toggle any imported images (not textures) in the selection on or off
- **Sections**: Toggle any sections in the selection on or off
- **3D Polylines**: Toggle any polylines (geometry created by certain extensions) in the selection on or off
- **Curves**: Toggle any curves (created by welding multiple edges together) in the selection on or off
- **Arcs**: Toggle any arcs (created by the arc tool or by breaking circles) in the selection on or off
- **Circles**: Toggle any circles in the selection on or off
- **Polygons**: Toggle any polygons (shapes created using the **Polygon** tool) in the selection on or off
- **n-Gons**: Toggle any n-Gons (these are polygons, including polygons that have been broken) in the selection on or off
- **Linear Dimensions**: Toggle dimension lines connected to edges in the selection on or off
- **Radial Dimensions**: Toggle dimension lines connected to arcs in the selection on or off
- **Front Default Material**: Toggle faces that have the default front material applied in the selection on or off
- **Back Default Material**: Toggle faces that have the default back material applied in the selection on or off
- **Hidden Edges**: Toggle edges marked as hidden (either by the **Eraser** tool or by using the **Soften/Smooth** window) in the selection on or off
- **Soft Edges**: Toggle edges marked as soft (either by the **Eraser** tool or by using the **Soften/Smooth** window) in the selection on or off

- **Smooth Edges**: Toggle edges marked as smooth (either by the **Eraser** tool or by using the **Soften/Smooth** window) in the selection on or off

You can imagine how helpful this can be when you are working on a model that is made up of all different kinds of items. Rather than having to spend time zooming and highlighting items using the *Shift* key and the **Select** command, you can do a quick window select, then use Selection Toys filters to modify the selection with one or two clicks. Plus, the filters can be run one after another! You can select everything, then deselect **Groups**, and deselect **Polygons**.

Selection Toys can also automatically find and select items that are similar to a single selected item. After selecting a single item, you can right-click the item, then choose **Select** from the context menu, and then select items that share the same materials or are in the same group! If that was not enough, there are also a bunch of selection options so that you can select copies of groups or components, or even convert group copies into copies of components!

So, overall, this is an incredibly helpful extension and I tend to use it, one way or another, almost every time I model. So, now, let's get Selection Toys installed and give it a try!

Installing Selection Toys

Since we have already covered the step-by-step process of installing extensions, I will not run through all of that again (this can be found in *Chapter 11*, *What Are Extensions?*, if you want to review it). Instead, I will let you grab the file and get it installed.

Selection Toys can be found in Extension Warehouse:

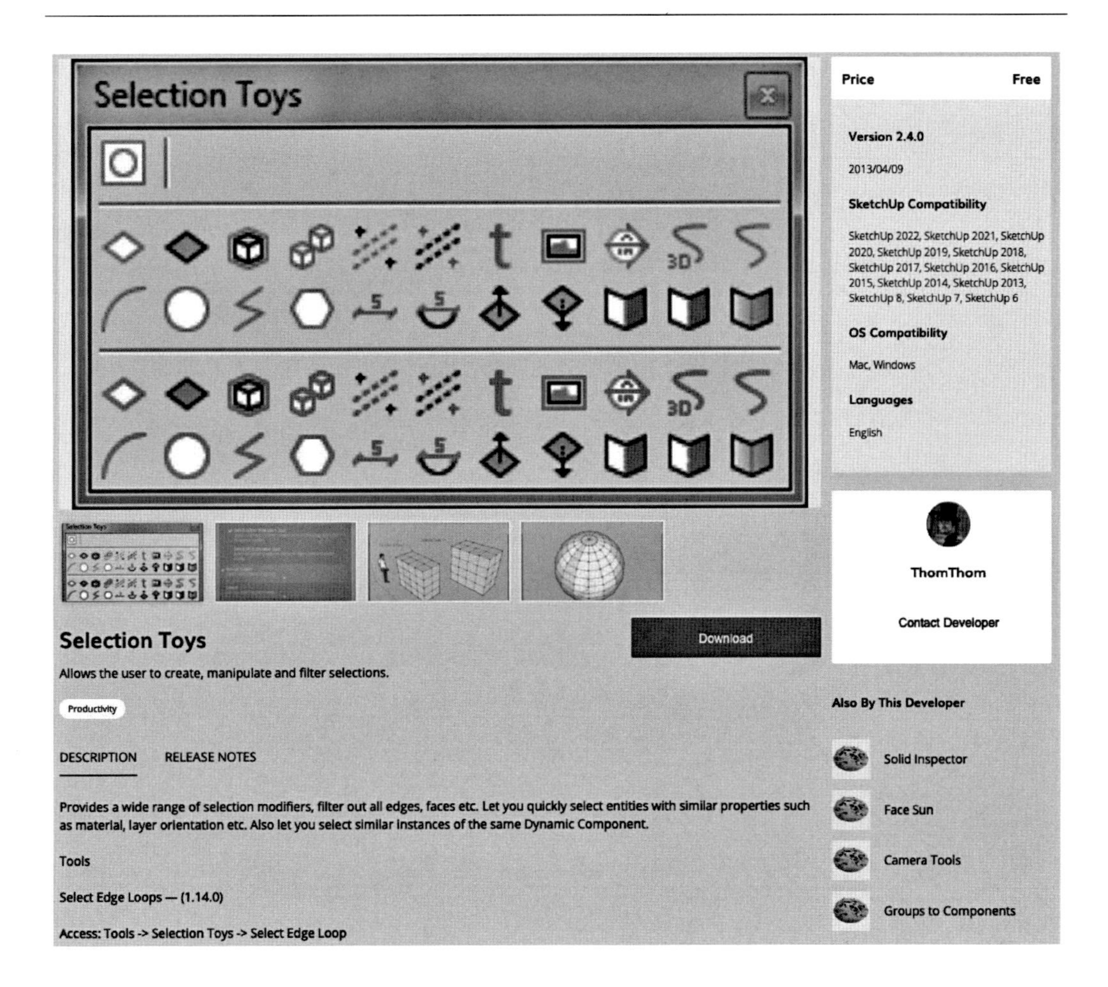

Figure 13.1 – Selection Toys version 2.4.0, as seen in Extension Warehouse

Download it using the following link or search for `Selection Toys` in Extension Warehouse: `https://extensions.sketchup.com/extension/c9266b2c-0b55-4d21-a0a4-72e23b8a0fb4/selection-toys`.

Once you have installed Selection Toys, we can try it out! We'll practice using it in the next section.

Using Selection Toys

Now that Selection Toys has been installed, let's try it out! The nice thing about this particular extension is that you can use it anytime you need to select or modify a selection, so you can use it on any model you are working on. Just to have something to play with though, let's open the `Taking SketchUp Pro to the Next Level - Chapter 13.skp` file mentioned in the *Technical requirements* section at the beginning of this chapter.

Let's hop in and start using Selection Toys:

1. Open the **View** menu and click either **Toolbar** or **Tool Palettes** and turn on the toolbar for **Selection Toys.**

   This will turn on a *big* toolbar. The default Selection Toys toolbar contains 45 buttons. Fortunately, this can be edited to show just the commands you want to see (more on that later in this section). Let's try a few of these buttons out.

2. Use the **Select** command to drag a window to select everything on the screen:

Figure 13.2 – The full Selection Toys toolbar

3. Click on the second button on the Selection Toys toolbar (**Select Only Edges**). Notice that everything except for edges is deselected.

   This command alone is worth installing Selection Toys for! Just think for a second how many clicks it would take to go through and pick out all of the edges in this model. The opposite command (**Deselect Edges**) is just as useful.

4. Use **Select** to highlight everything in the model again. This time, move to the middle of the bar and choose the **Deselect Edges** button (it is the first red icon on the toolbar):

Figure 13.3 – The Deselect Edges button

   Again, a super helpful command that will allow you to fine-tune your selection without spending a ton of time selecting items while holding down a modifier key.

At this point, go ahead and play around with the Selection Toys toolbar. Select everything and click on a filter or two to modify the selection.

> **The Select All Shortcut**
>
> Rather than dragging a select window to highlight everything, you can press *Ctrl* + *A* (on Windows) or *Command* + *A* (on macOS) to highlight everything in the model.

At this point, I would imagine that you get why I think that the functionality of this extension is something that any SketchUp user would value. Its toolbar, however, is a bit much. Let's take a look at how we can customize the Selection Toys toolbar:

1. From the **Tools** menu, click on **Selection Toys**, and then **UI Settings**. Scroll down to the section labeled **Toolbar**.

   Here, you will find a list of every single button on the toolbar and can toggle them on or off. This is a great way to turn that unwieldy toolbar into something that you can justify dedicating some screen space to! Plus, it is a great way to see and understand what each icon represents. I like the commands that allow me to turn **Edges**, **Faces**, **Groups**, **Components**, **Text**, and **Dimensions** on or off.

2. Turn off (uncheck) every command except for **Edges**, **Faces**, **Groups**, **Components**, **Linear Dimensions**, and **Text**.
3. Click the **Save** button.

   Since the UI for Selection Toys is loaded when you first start SketchUp, these changes will not be applied until you restart SketchUp.

4. Quit and restart SketchUp.

   Upon restarting, you should be greeted by a new, truncated version of the Selection Toys toolbar:

Figure 13.4 – Customized Selection Toys toolbar

What could be better than an extension toolbar that only contains the icons for the commands you want? How about running Selection Toys without any toolbar at all? Selection Toys offers something that few extensions do – the ability to run from either the toolbar or the context menu. Let's try doing a little more filtering without using the toolbar at all:

1. Select everything in the model one more time.
2. Right-click on any selected item on the screen.

   At the very bottom of the context menu, you will see options for **Instances**, **Group Copies**, **Select**, **Select Only**, and **Deselect**.

3. Click on **Select Only**, and then **Curves**.

   The selection filter turned off everything that was not a curve, even though we hid the **Curves** icon in the toolbar!

Personally, this is how I prefer to use Selection Toys (I never turn on the toolbar at all). Plus, not only can you run the selection filters, but there are a bunch of additional commands in there. The ability to select copies of a selected group or component, or use the command to convert selected copies of a group into components, can all be run from the context menu.

This section was intended to serve as an introduction to Selection Toys (I really cannot do an extension with over 60 commands justice in just a few pages). Hopefully, you have seen the value in this free extension and are willing to try it out. It is a great way to help you reach the next level with your SketchUp skills.

While Selection Toys is a great set of tools to help modify existing geometry, the next extension we are going to try out is all about creating geometry in new ways.

Bezier Curve Tool from the SketchUp Team

A Bezier curve is a parametric curve that's created by defining a set of control points to define a smooth curve. Unlike traditional SketchUp curves, which are always created as a portion of a full circle, a Bezier curve can have a non-symmetric curve or curve in more than one direction, based on the number and location of the control points.

Some curves can be quickly created with native tools, while others require a tool such as Bezier Curve Tool:

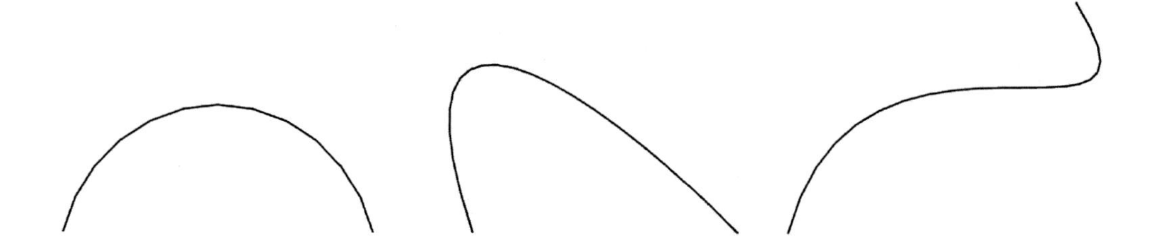

Figure 13.5 – The curve on the left is a standard arc, while the two on the right are Bezier curves

Having the ability to control the curve of a line like this opens up SketchUp models for creating new shapes and geometry that would be very difficult to achieve with just the standard **Line** or **Arc** command (or even **Freehand**).

Now that we have a basic understanding of the idea behind creating curves in this manner, let's dive into what this extension does.

Bezier Curve Tool overview

It is completely possible to create almost any geometry using the native commands in SketchUp. Possible, but not necessarily enjoyable. You can create lines, distort them, weld them, and create whatever curves you want. However, drawing them initially is much easier if you have the right tool. In the case of drawing arcs, Bezier Curve Tool is the right tool.

As an example, let's say that I have a simple kidney-shaped swimming pool that I want to model. Step one is drawing the footprint – a top-down outline of a shape that looks something like this:

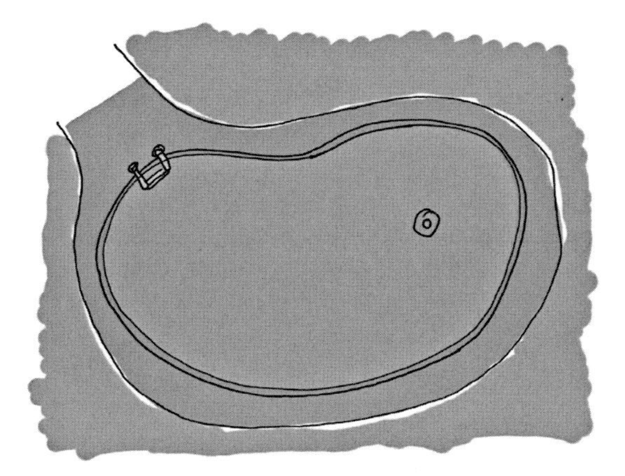

Figure 13.6 – The swimming pool that I want to create

Now, let's look at two different ways to create this shape. First, I can use a combination of **2-point Arcs** and edges drawn with the **Line** command to trace around this shape. In the end, I wind up with something like this (endpoints have been turned on so that you can see how many pieces were needed to create this shape):

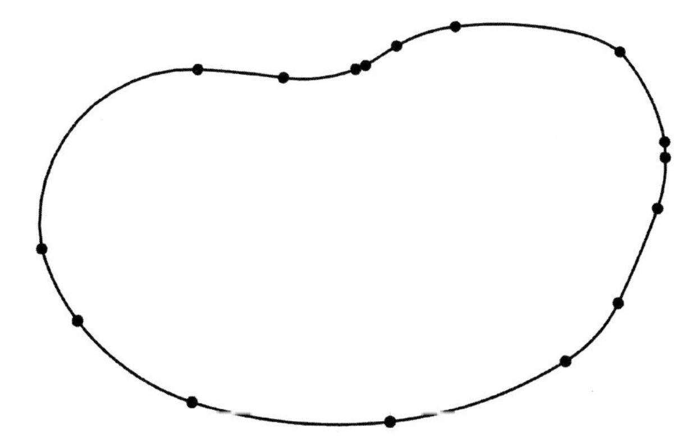

Figure 13.7 – Pool drawn with native SketchUp tools

Now, using Bezier Curve Tool, I can draw the same outline with just a few lines. Here is the same outline (with endpoints turned on again). You can see that far fewer steps were needed to create the footprint of the pool:

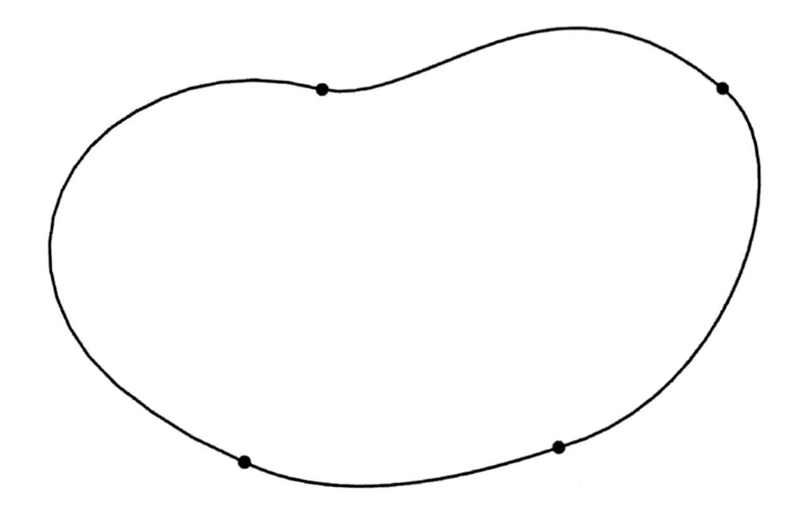

Figure 13.8 – Pool drawn using the Bezier Curve Tool extension

Bezier Curve Tool allows you to easily draw curving lines with much more control than you have with native tools, allowing you to create smoother geometry than can be achieved using only arcs.

Installing Bezier Curve Tool

Now that you understand the advantage of this extension, let's get it installed! As with Selection Toys, you should be able to install this extension by finding it in Extension Warehouse and clicking the **Install** button. Note that if you search Extension Warehouse for `Bezier Curve`, several extensions will come up. For this walkthrough, we will be using the extension called *Bezier Curve Tool* from the SketchUp team.

Here is a direct link to the page on Extension Warehouse: `https://extensions.sketchup.com/extension/8b58920d-0923-42f8-9c72-e09f2bba125e/bezier-curve-tool`.

Once Bezier Curve Tool has been installed, it can be found in the **Draw** menu. This particular extension does not have a toolbar or button that can be added to a custom toolbar, though you can assign a shortcut key to it.

Using Bezier Curve Tool

Let's try using Bezier Curve Tool to draw a couple of curves in the practice file. Open the `Taking SketchUp Pro to the Next Level - Chapter 13.skp` file and select the **Bezier Curve Tool** scene. This scene has three curves at the top that we will recreate. We will start by recreating the curve on the left:

1. Choose **Bezier Curves** from the **Draw** menu.

   This first curve will be drawn with four clicks. The first two will define the beginning and end of the line. The third and fourth clicks will pull the line out to define the arc. These clicks will be at these points:

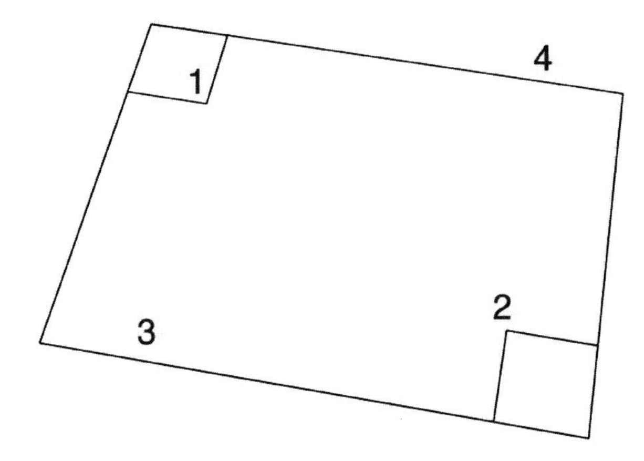

Figure 13.9 – Points used to recreate the first Bezier curve

2. Click point 1 (the inside corner of the top-left square) and then point 2 (the inside corner of the lower-right square).

   This will establish the start and end of the curve. Now, use two more clicks to set the curve of the line.

3. Click point 3 (the inferenced edge of the rectangle down the green axis from point 1) and then point 4 (the inferenced edge of the rectangle up the green axis from point 2).

   As you move the cursor to define points 3 and 4, notice how the curve changes and stretches to have the line follow. Upon selecting the fourth point, the curve will be created:

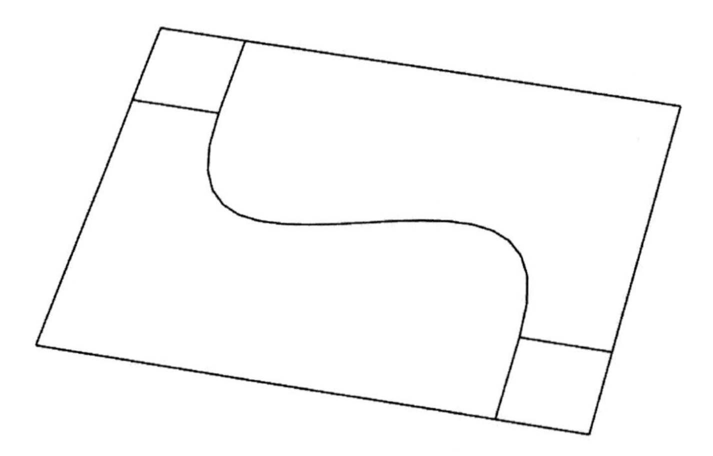

Figure 13.10 – Completed first curve

Let's do the same to create the next two curves. For the second curve, we will click four times, but at different locations:

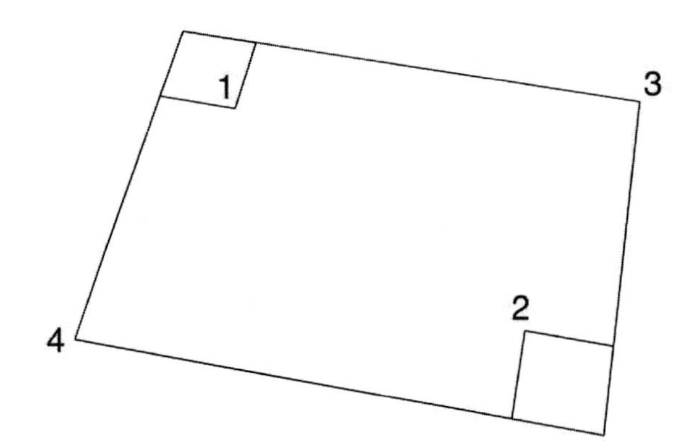

Figure 13.11 - Points used to recreate the second Bezier curve

4. Click points 1 and 2 the same way you did in the first example.
5. For point 3, click the top-right corner of the rectangle.
6. For point 4, click on the lower-left corner of the rectangle.

By moving the last two points, the resulting curve is significantly different:

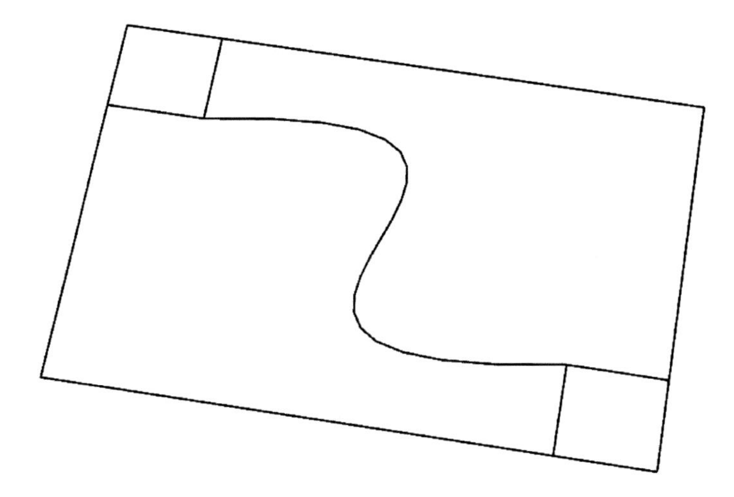

Figure 13.12 – Completed second curve

The past two curves have been created using two additional clicks. Bezier Curve Tool allows you to specify additional points, giving you even more control over the final curve. Let's tell the extension that we want more control points and create the final example.

7. Click **Bezier Curve** from the **Draw** menu and immediately type 5 and press *Enter*.

   Notice that the **Degree** field in the lower left changes from a 3 to a 5. Bumping this number will increase the number of control points that will be used to create the curve. Initially, we were defining three (the original line, plus two control points). This time, we will use five (the line, plus four control points). We will be clicking at these points:

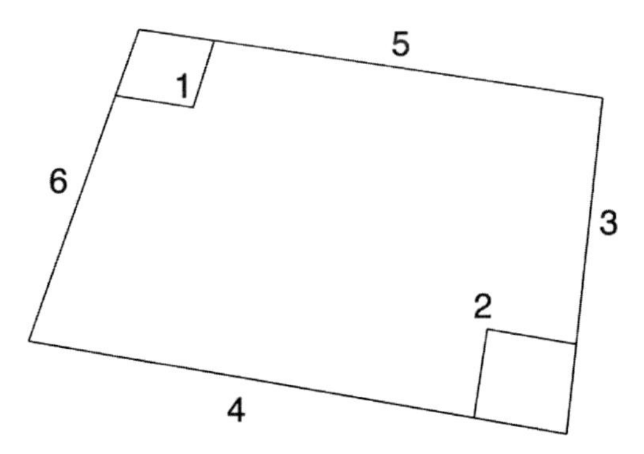

Figure 13.13 – The final curve will be created using these points

8. Start by clicking the same points for 1 and 2.
9. For point 3, inference the middle of the right-hand side of the rectangle.
10. For point 4, inference the middle of the bottom of the rectangle.
11. For point 5, inference the middle of the top of the rectangle.
12. For point 6, inference the middle of the left-hand side of the rectangle.

    Despite some crazy geometry while you were placing the control points, you should end up with a curve that looks like this:

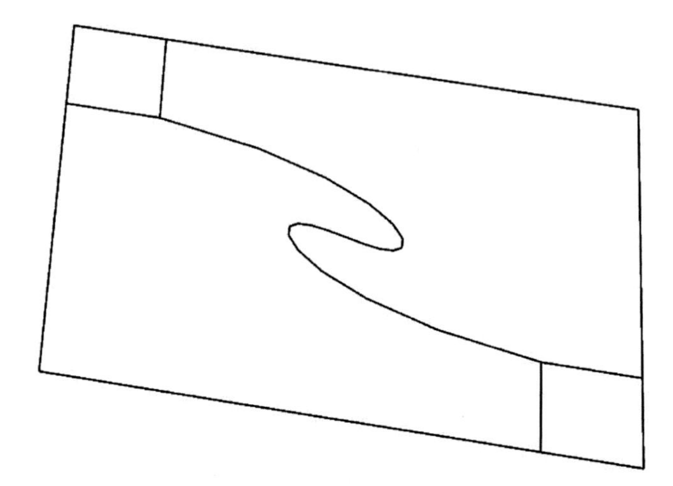

Figure 13.14 – Final curve

> **3D Bezier Curves**
>
> In the examples in this chapter, we drew curves that were constrained to a 2D surface. This was a "crawl before you walk" mentality since this section was only supposed to serve as an introduction to this extension. Do note that, using Bezier Curve Tool, you can draw curves that exist in 3D space the same way you can in 2D by choosing points that are at different heights. Remember, this was just an introduction. If you play with this extension, you will quickly see how much more can be done than what we were able to cover here!

Bezier Curve Tool allows you to create smooth arcing lines that can be pulled and stretched in any direction you like. This extension can become an essential piece of many workflows once you get the hang of using it and is a "must have" for many users.

Up next, let's take a look at an extension that is easy to use and learn and fills a gap in the standard SketchUp toolset.

Curic Mirror from Curic

Some users are quick to point out the lack of a mirror function in the native set of SketchUp tools. There *are* ways to mirror selected geometry, but not using a command that acts like the mirror commands seen in other drawing software. With the **Scale** tool, selected geometry can be scaled along any axis to -1 to effectively mirror it. You can use the **Flip Along** command found in the context menu to quickly flip selected geometry. While these commands do work, they require geometry to be squared up to a specific axis and can only mirror in a single direction.

This is where Curic Mirror comes in. Using Curic Mirror, you can mirror your selection along any geometry in the model. You also have the option to keep or replace the existing geometry, meaning that duplicating half a model is a single step!

Let's take a look at some specifics of how Curic Mirror can be used in your workflow.

Curic Mirror overview

Curic Mirror is the epitome of a simple extension. Once installed, a single button can be pressed to activate the tool (just like the native SketchUp commands). All you have to do to use the command is select the geometry you want to mirror (faces, edges, groups, components... really anything in the model) and click on the **Curic Mirror** button. The extension will allow you to move your cursor over any surface in your model. As you do, you will see a preview of what your selected geometry would look like if you selected that surface to mirror across. Clicking on a face will apply the mirror and you are done!

Much like a native SketchUp command, the base functionality is only the beginning! Just like native modification commands, Curic Mirror has a few modifier keys that can be used. If you press *Ctrl* (on Windows) or *Option* (on macOS), a little + will appear next to the cursor. Now, when you select your mirroring face, Curic Mirror will copy the selected geometry when it mirrors! This is a great tool to have when you are modeling symmetric geometry. Rather than modeling both halves of an item, you can model one half as a component, and use Curic Mirror to finish the model:

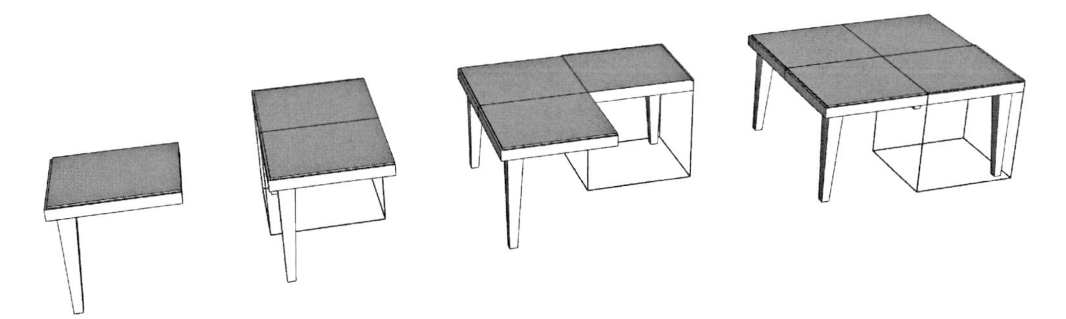

Figure 13.15 – A quarter of this table was modeled as a component and then mirrored three times

Additionally, Curic can mirror along the axes if you simply tap the arrow key associated with the axis color:

- *Right Arrow* to mirror along the red axis
- *Left Arrow* to mirror along the green axis
- *Up Arrow* to mirror along the blue axis

This is a great feature, especially if the model you are working on is aligned with the axis. It means that with a quick tap, you have mirrored your geometry. In the case that you have modeled geometry that had been rotated away from the axis, you can tap *Alt* (Windows) or *Command* (macOS) to toggle between using the world axis or the selected component or group axis to mirror along:

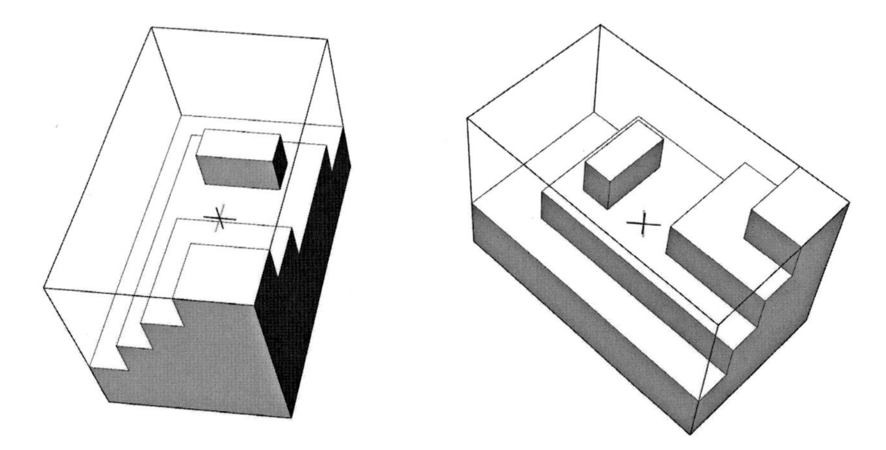

Figure 13.16 – The group on the left aligns with the world axis, while the group on the right has a rotated axis

Having the ability to mirror across either the work axis or the object axis helps address the issue of using SketchUp native tools to mirror. Unlike using **Flip Along**, which will only use the world axis, with Curic Mirror, you get to choose which axes you would like to use.

Unlike the previous extensions in this chapter, Curic Mirror is one that you can learn and start using to its fullest potential with just a few exercises. Before that though, we will need to get it installed!

Installing Curic Mirror

In Extension Warehouse, you can search for `curic mirror` to find the extension. At the time of writing this book, only a single extension shows up when you search for that term. Go ahead and get it installed by searching for it or by going to `https://extensions.sketchup.com/extension/72c3aa60-da8e-4a56-9f70-ab4bc706060c/curic-mirror`.

Using Curic Mirror

To see how this extension operates, let's jump to the next scene and mirror some geometry! Open the `Taking SketchUp Pro to the Next Level - Chapter 13.skp` file and click on the **Curic Mirror** scene.

This example contains four different sets of geometry. The two chunks of geometry on the left are made of loose geometry, while the two on the right are both groups. We will see how Curic Mirror copies along a face or the axes, as well as try out copying versus just mirroring selected geometry. Let's start by simply mirroring some geometry along a face:

1. With the **Select** tool, highlight the geometry in the lower right of the screen by either dragging a select window or triple-clicking to highlight all connected geometry.
2. Activate Curic Mirror by clicking the **Tools** menu, then **Curic**, then **Mirror**, and then **Mirror Tool**. Alternatively, you can click the Curic Mirror icon (this icon should be on your screen immediately after installing Curic Mirror).
3. Hover your cursor over the blue face.

   Hovering your mouse over a face will show a preview of what the mirror will look like. You can move your mouse around to different faces to get a preview of different mirroring options. Note that the mirroring face does not even have to be a part of the selected geometry (you can mirror across any visible face anywhere in the model):

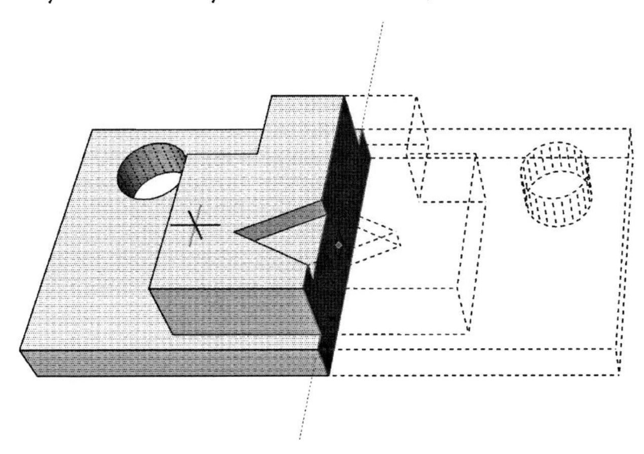

Figure 13.17 – Curic Mirror previewing the final mirror

4. Click on the blue face.

   Notice that the geometry has moved from its original position to where the preview lines were. This is how Curic Mirror will affect geometry, by default.

   Let's move to another section and this time, copy the selected geometry.

5. Use **Select** to select the geometry with the green face.
6. Activate **Curic Mirror**.
7. Hover over the green face.

   Before clicking on the green face to create mirrored geometry, tap the modifier key to toggle copy on (the modifier key is displayed in the status bar at the bottom of the screen). Tapping the correct modifier will display a small + next to your cursor.

8. Click on the green face.

   This creates a brand new geometry and places it next to the selected geometry. In this case, normal SketchUp rules apply, and the new geometry has been fused with the initial geometry:

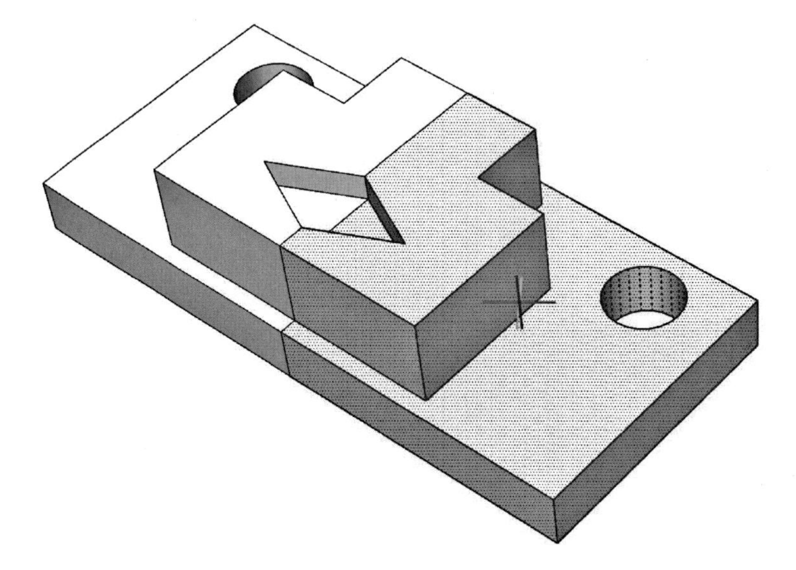

Figure 13.18 – Copied and mirrored geometry connected to the original geometry

Next, let's learn how to mirror geometry along the axes rather than a selected face.

9. Use **Select** to select the group with the yellow face.
10. Activate **Curic Mirror**.
11. Tap the *Right Arrow* key on your keyboard.

    Notice that the geometry immediately reverses. There are no confirmations or previews with this functionality, but since you are still in the **Curic Mirror** command, you are just a keystroke away from putting it back how it was.

12. Tap the *Right Arrow* key again to mirror the group back to where it was.

    Since you stay in Curic Mirror until you intentionally leave, it makes it very easy to apply multiple mirrors to selected items or geometry.

13. Tap the *Right Arrow* key, then the *Left Arrow* key, and then the *Up Arrow* key.

    Now, the group is mirrored along all three axes:

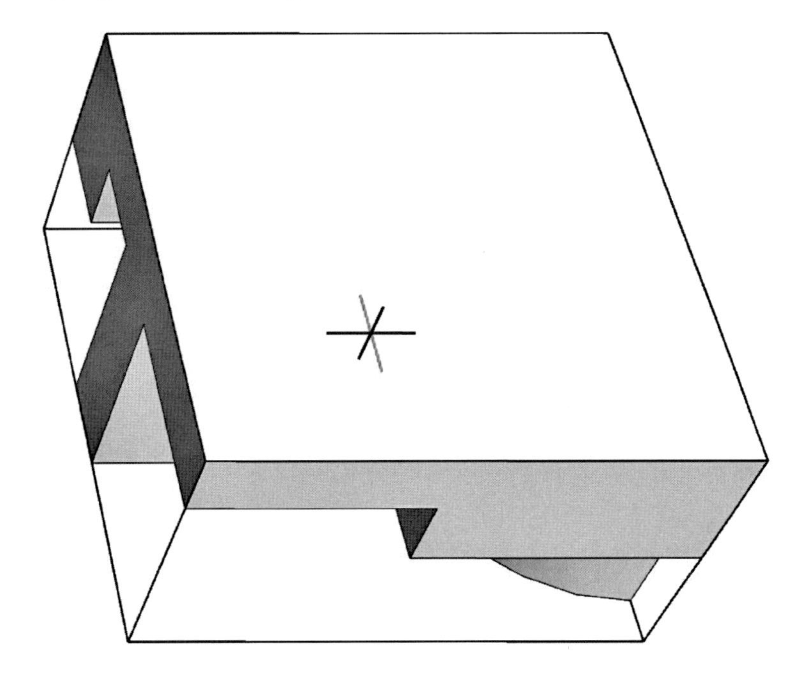

Figure 13.19 – Group mirrored along all three axes

Here, you can see how easy it is to try to mirror geometry and, if everything is not perfect, put the geometry back. This works great when the geometry is aligned with the global axis.

Finally, let's see what you can do when you are working with a group that has been rotated away from the global axis.

14. Use **Select** to select the group with the red face.
15. Activate **Curic Mirror**.
16. Tap the *Right Arrow* key on your keyboard.

    This causes the geometry to mirror along the global red axis. The result kind of looks like it rotated a little way around. While there are some cases where this might generate the desired result, it is more likely that you will want to mirror along the axes of the group or components instead.

17. Tap the *Right Arrow* key again to mirror the group back.

    Notice the little red/green/blue axis in the middle of the selected group? You can see that it is oriented to follow the world axis right now:

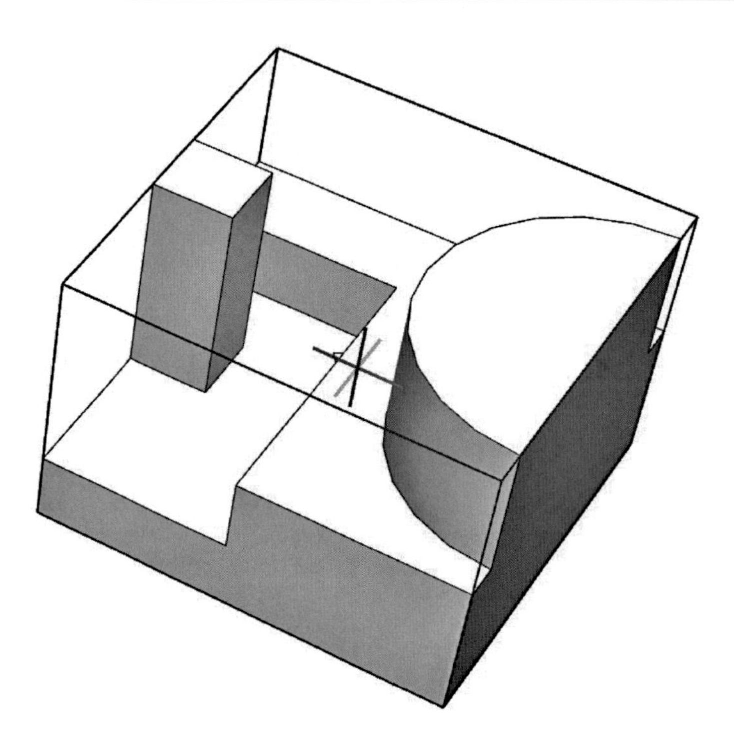

Figure 13.20 – Angled group mirrored along the world axis

18. Tap the Object Axes modifier key listed in the status bar (*Alt* on Windows or *Command* on macOS).

    Notice that the red/green/blue axes in the middle of the group changes to align with the geometry in the group! Now, if you tap any arrow key, the geometry will mirror along the axes that are inside the group rather than the world axis.

19. Tap the *Right Arrow* key on your keyboard.

    The group will mirror in the same manner as the group with the yellow face. This is a great option to have if you ever have to mirror geometry that does not align with the world axis.

With that, you have seen everything that this extension can do for you. Of course, knowing how a tool works and having mastery over the tool are two different things. Fortunately, Curic Mirror is a well-designed extension that will fit into many workflows. If you try using it a few times, I am sure you can pick up how to get the most out of it.

Now, let's move on to our final extension that is sure to fit into most workflows: adebeo_pushline!

adebeo_pushline from Denis B

The first three extensions we looked at in this chapter served as functions that could be seen as missing from SketchUp. While it is possible to run SketchUp without these (or any) extensions, they help make SketchUp feel more complete when they are added into the mix. The fourth extension that we are going to look at, adebeo_pushline, is one that I would place into a different category.

This extension adds functionality to SketchUp that I, personally, never thought I needed. Once I started using it, however, I immediately saw its value. adebeo_pushline allows you to interact with edges and arcs in the same way that you can use **Push/Pull** to interact with faces.

It seems like a great addition to SketchUp's basic tools, right? Let's dive a little deeper and take a look at how this extension functions.

adebeo_pushline overview

In general, if you have a line in SketchUp and you want to make it into a plane, there are two approaches. The first is to use the **Line** tool to add lines to the model and "draw" the sides of the face you want:

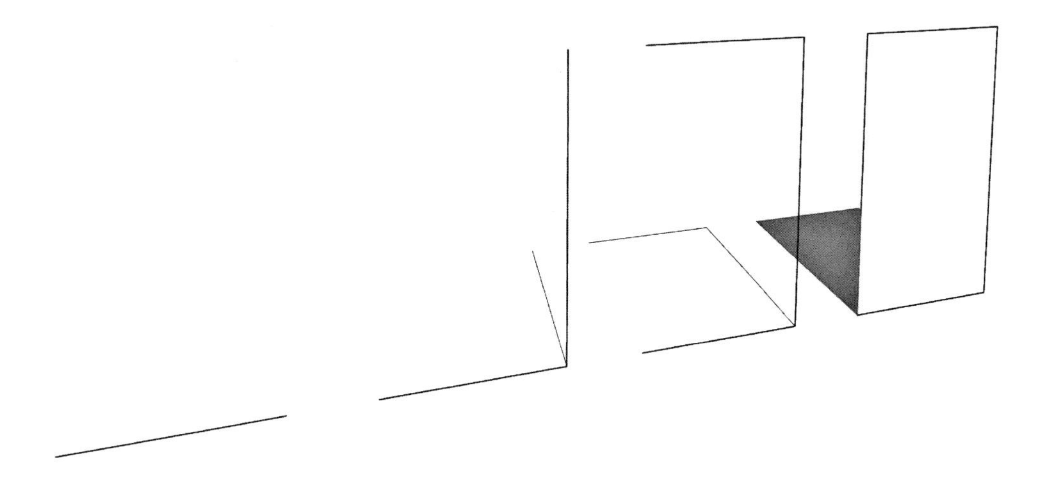

Figure 13.21 – Changing an edge into a face with three additional edges

This works great until you want to turn something such as a curve into a surface. With a curve, it is not as simple as drawing lines perpendicular to the initial edge. In a case like this, I would (before adebeo_pushline) add edges with the **Line** tool to make the arc into a face, then use **Push/Pull** to pull it up, and then use **Erase** to get rid of the extra geometry:

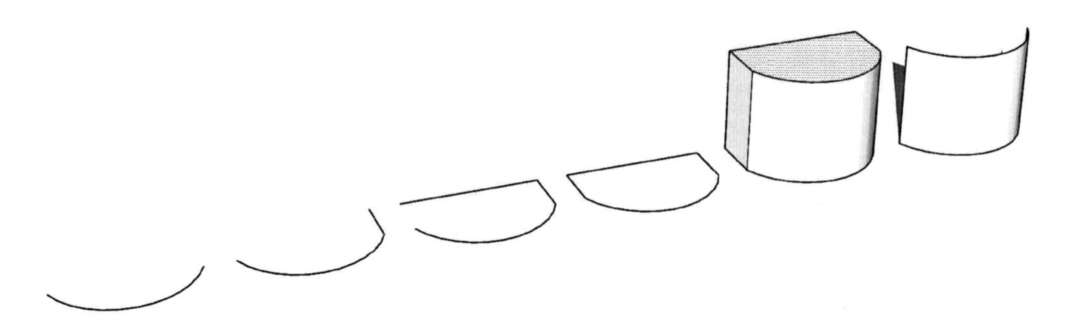

Figure 13.22 – Six steps to pull a 2D arc into a curved surface

Now, it is not hard to see these steps as fairly simple. If you had a good set of shortcuts and were focused, you are probably looking at 10 seconds or so to perform these steps. However, we are on a quest to take our SketchUp skills to the next level! We are not *OK* with spending 10 seconds performing six steps that could be done in just two clicks!

With adebeo_pushline, you can simply click on the geometry you want to extrude, then click again where you want it to extrude to! It literally works just like the native **Push/Pull** tool (with a few notable exceptions):

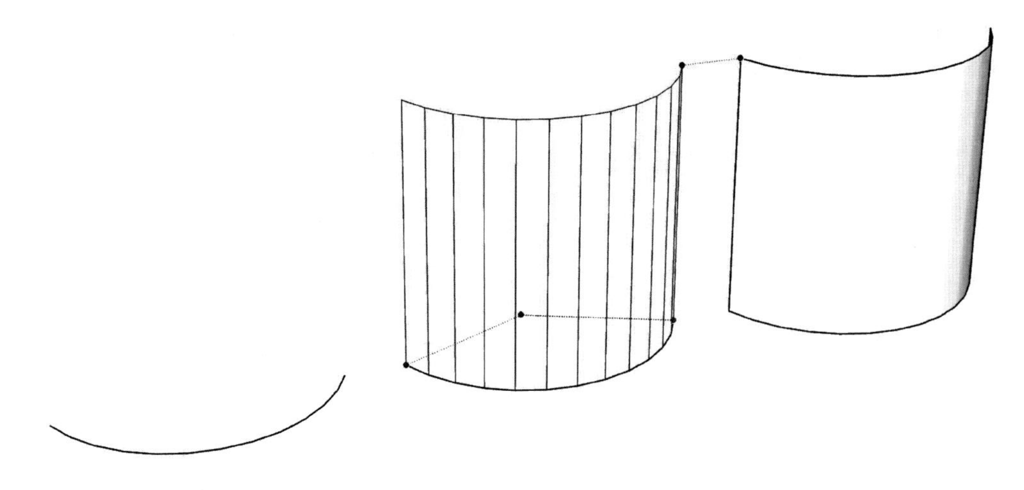

Figure 13.23 – Pulling a line up with two clicks using adebeo_pushline

With adebeo_pushline, you can specify an exact distance for the extrusion, use the arrow keys to lock to an axis, and use inferencing to help set the extrude length.

As for the exceptions I mentioned, two things are different from using the native **Push/Pull** tool:

- The first is that adebeo_pushline does not allow you to double-click on an edge to repeat the same extrusion length as the last time.
- The second difference from the native tool is that, with adebeo_pushline, you can preselect as many edges as you want and extrude them all at the same time!

This second difference more or less negates the first. Rather than having to double-click to repeat an extrusion on one face after another, you can just preselect all of the edges you want to extrude, then pull them all up into 3D at the same time:

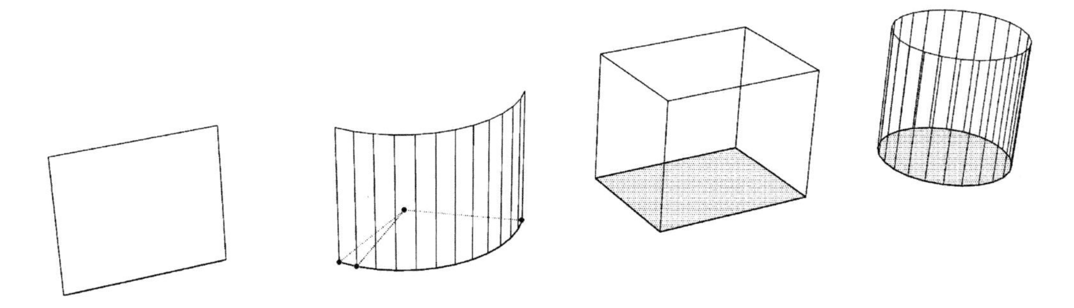

Figure 13.24 – Pulling multiple edges up at the same time

So, no, adebeo_pushline does not feel like it is completing the toolset offered by the native tools – it does feel like something extra. This does not, however, take away from the fact that it allows me to generate geometry quicker than I could without it, which is key to speeding up my drawing process and taking my SketchUp skills to the next level!

Installing adebeo_pushline

Like the previous extensions, adebeo_pushline can be directly installed in SketchUp from Extension Warehouse. To find it, you can search for `pushline` (at the time of writing, it is the only search result) or you can use the following link: `https://extensions.sketchup.com/extension/19789f54-7cc9-4ba1-a20a-3ccc642eae98/adebeo-pushline`.

Once installed, you can activate adebeo_pushline through the **Extensions** menu or by turning on a one-button toolbar.

Using adebeo_pushline

Let's practice using adebeo_pushline! Much like the previous extension, once you have practiced using it a few times, you will see just how simple it is to use. I would say that the only challenge I have with using adebeo_pushline is remembering to use it when I need to create faces from 2D geometry!

In the example `Taking SketchUp Pro to the Next Level - Chapter 13.skp` file, we can practice using the extension on a few different sets of geometry. Once we run through a few examples, you should start to see how you may add it to one or more of your workflows:

1. Click on the **adebeo_pushline** scene.
2. Start the extension by selecting the **Extensions** menu, then **Adebeo**, and then **PushLine**. Alternatively, you can click the icon on the **Adebeo** toolbar.
3. Click on the line on the left and move your cursor around.

   Notice that the line pulls up to follow your cursor, wherever it moves. This is great if your goal is to stretch a line into a surface from one point to another. In an example like this, you may want to simply pull it vertically into a face. If we want to pull the line, which is currently flat on the ground, up into a vertical face, we will want to constrain it to the vertical axis.
4. Press the *Up Arrow* key on your keyboard.

   Notice that the edges of the preview have turned blue, and as you move the cursor, the line only pulls straight up. Let's make a vertical face with this line, then pull out a few more.
5. Click to finish pulling the line up a little way above the ground so that it looks similar to the following:

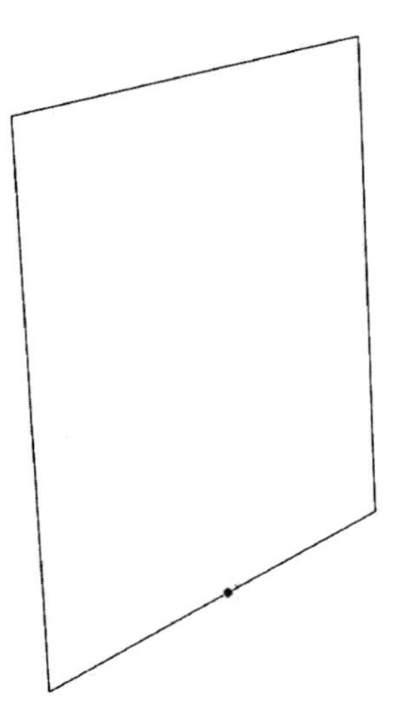

Figure 13.25 – Pulling a line vertically into a face

Inference locking can work in other axes as well. Let's try pulling the left edge of the new plane along the red axis.

6. Click on the left edge of the plane, then press the *Right Arrow* key, and move the edge to the left and click.
7. Click on the right edge of the original face, then press the *Left Arrow* key, and move the edge to the right and click.

You should end up with a series of faces that look like this:

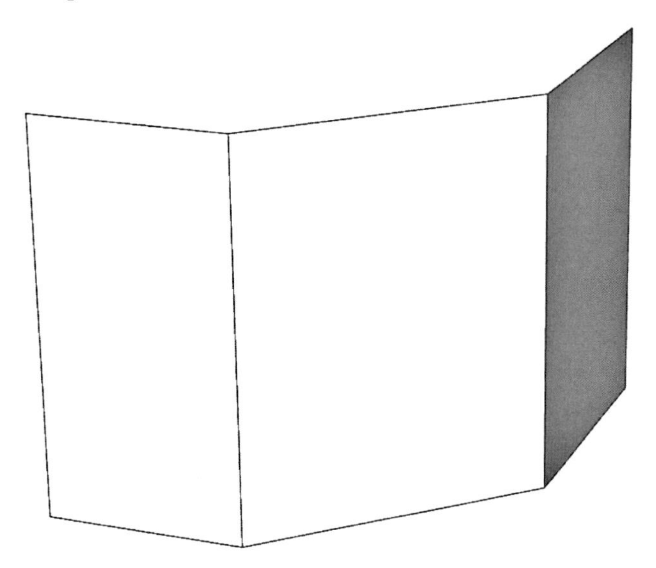

Figure 13.26 – Faces created from a single line with adebeo_pushline

So, that was pretty cool and a definite time saver, but let's take a look at how we can create more complex geometry.

8. Click on the leftmost edge of the curved line and pull it up (use the *Up Arrow* if you need to).

Notice that only a section of the curved line pulled up. This curve is made up of multiple arcs. Clicking on one curve to start pulling it up will require you to repeat the action for each section of the curve. If you wish to pull the entire curve up into a surface, you can weld the edges before you start.

9. Undo the previous step (either using *Replace with Ctrl/Command + Z* or the **Undo** command in the **Edit** menu).
10. Use **Select** to highlight all of the pieces of the curve.
11. Right-click and choose **Weld Edges** from the context menu.
12. Use adebeo_pushline to pull all of the connected edges up vertically:

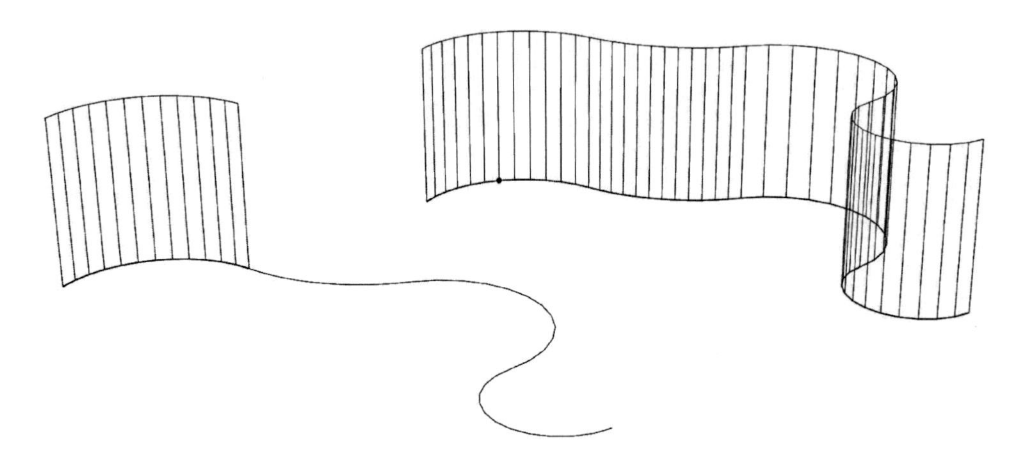

Figure 13.27– Pulling a single arc versus an entire curve

Welding edges is a great way to simplify pulling edges into surfaces, but it is not always the ideal solution. In *Figure 13.27* in the example on the right, let's say we want to pull out the top, left, and right edges that surround the "door" shape. If were to weld them into a single curve, we could pull them all out in a single step, but the result would be a weird, smoothed geometry that would not work when the faces meet at a right angle. Instead, adebeo_pushline allows us to preselect the geometry we want to pull out.

13. Use **Select** to select the left, right, and top edges that create the door shape from the box on the right.
14. Use adebeo_pushline to pull the edges out to the right (use the *Left Arrow* key if you want to lock to the axis). You should get something like this:

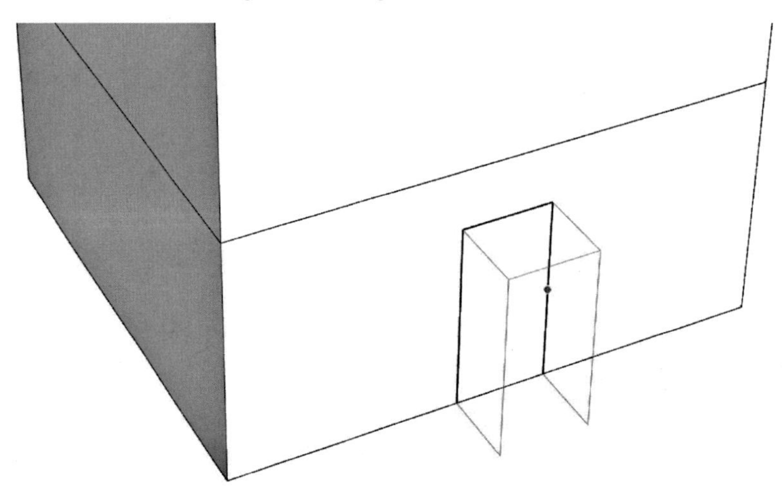

Figure 13.28 – Three preselected edges being pulled together

Preselecting geometry and welding are great tools to use alongside this extension, but there are some cases where you just have to perform multiple actions. For example, let's say that we want to pull that line above the door out to create a horizontal face that wraps around the building (maybe it will end up being an awning). Because this requires edges to be pulled in different directions, you cannot weld or preselect multiple edges to pull out together. In this case, you would have to pull out one side, and then the other:

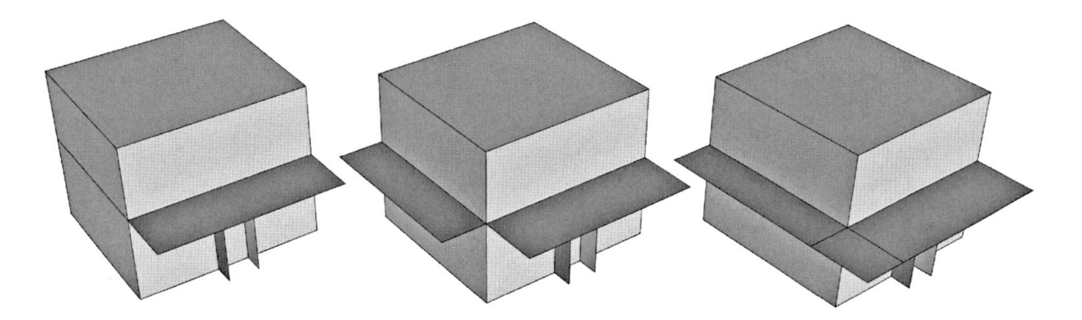

Figure 13.29 – To create this awning geometry, adebeo_pushline would have to be run three times

There may be cases where you can pull geometry out together and preselecting will be helpful, but because adebeo_pushline only allows you to move in one direction at a time, there will always be cases like this where multiple steps will be necessary.

2D Is Just the Start

I do want to point out that the geometry we are creating in this final example should be seen as the starting point. In general, it is not a good idea to leave 2D faces in a final model. For this reason, adebeo_pushline is a great extension to use in conceptual modeling or to try out ideas in a 3D "sketch." Of course, you can always use it to create preliminary geometry as we did with the door and awning, then come back afterward and use a command such as **Push/Pull** to add depth to the surfaces.

That was a pretty solid run-through of adebeo_pushline. Once you have used it a few times and are thinking about things such as how you should select geometry to pull into faces, you can see how it can be a real time saver.

Now that we have run through four of the (in my opinion) best free, general-purpose extensions that anyone could use, let's take a quick look at some of the other amazing extensions that are out there.

Exploring additional extensions

The previous sections of this chapter focused on extensions that most, if not all, SketchUp users would benefit from using. In this last section, I wanted to shine the spotlight on a few more extensions that I think are impressive. This is not an exhaustive list, as there are more extensions out there that do work great, but these are extensions that I have tried and can say are excellent.

There are a couple of things about these extensions that set them apart from the previously mentioned extensions. First, not every extension listed in this section will have a place in everyone's workflow. If any of the extensions here are of interest to you, I recommend checking them out and, where possible, trialing them. This brings me to the second difference; most of the extensions mentioned in this section do cost some money. Some can be purchased outright, while others require a subscription to function. All I can say is that if their capabilities add to your SketchUp modeling process, they are all well worth the money that they cost.

The goal here is to provide a brief overview of the extensions and the developer and license types and to let you know where you can get them. Wherever possible, I will share websites where you can learn more.

> **A Note About the Prices**
>
> While I would love to quote the prices of these extensions, I do not want to publish numbers that I cannot control and may change in the future. Prices often change over time and many developers will offer discounts or bundles of their software. Instead, I will let you know the license type and recommend that you follow the links listed to find the current price.

With that, let's take a look at some more extensions!

Solid Inspector from Thomthom

Solid Inspector (or Solid Inspector^2) is an extension from the developer Thomthom. This is a quick and easy extension that can be run on a group or component to identify whether it is a solid or not. If it is not a solid, Solid Inspector will tell you why it is not a solid. Additionally, it will, in some cases, offer to fix the problem and turn your group or components into a solid. If it cannot fix the problem (in some cases, the issues that prevent geometry from being solid are too complex for software to solve), it will highlight the issue so that it is easier for you to find and repair it manually.

This is a great extension for anyone doing their best to model solid geometry. While solid geometry is not a requirement in SketchUp, there are more than a few experienced modelers who will argue the benefits of keeping everything solid.

Here are some details about the extension:

- Developer: Thomthom
- License: Free
- Extension Warehouse page: `https://extensions.sketchup.com/extension/aad4e5d9-7115-4cac-9b75-750ed0902732/solid-inspector`
- Developer website: `https://evilsoftwareempire.com/`

If you do any 3D printing, then I would call this extension a must-have!

FredoScale from Fredo6

FredoScale is an extension from developer Fredo6 that will allow you to distort SketchUp geometry as if it was made of rubber. FredoScale is a suite of commands that will allow you to do things such as bend, twist, shear, or scale selected geometry. While you do have access to the **Scale** tool in the native set of tools, FredoScale gives you additional control and the ability to scale portions of selected geometry, rather than the entire selection, uniformly.

FredoScale is a great extension for anyone who needs to create geometry that is made up of distorted or bent shapes.

Here are some details about the extension:

- Developer: Fredo6
- License: Perpetual license (3 seats)
- SketchUcation PluginStore page: `https://sketchucation.com/plugin/1169-fredoscale`
- List of all Fredo6 extensions: `https://sketchucation.com/pluginstore?pauthor=fredo6`

If the **Scale** and **Follow Me** commands are not getting you the level of bending or sweeping geometry you need, you should probably check out FredoScale.

Artisan from mind.sight.studio

Speaking of smooth geometry, Artisan from mind.sight.studio is a toolset that allows you to create organic geometry from within SketchUp. Artisan transforms SketchUp into an organic modeling software with tools for subdividing and smoothing geometry. With this extension, you can create smooth geometry such as terrain, plants, soft furniture, or even character or animal models!

This is an amazing set of tools that will give you a whole new set of modeling abilities.

Here are some details about the extension:

- Developer: mind.sight.studio
- License: Permanent license (single seat)
- Artisan website: `https://artisan4sketchup.com/`
- Developer website: `https://mindsightstudios.com/`

If you have seen other modeling tools that brag about organic modeling or digital sculpting and wanted to see what they were like, give Artisan a try!

Skatter from Lindale

Skatter is an amazing tool from Lindale that allows you to add things such as vegetation or hair to your model. Skatter has some great tools for literally scattering items anywhere in your model you like. While the primary use of this extension is to spread vegetation around the model (Skatter comes with a library of grass, shrubbery, or even a forest full of trees), it can be used to "sprinkle" any components throughout your model. There are even tools that allow you to randomize the components as they are placed so that they don't look like a single model being repeated over and over.

Often, if you were to create a model full of vegetation, the goal would be to render it. Adding a full landscape of high-polygon, render-ready vegetation models can cause SketchUp to slow down or even lock up. Skatter provides the option to place special "render-only" components through the model that can be seen as low-polygon representations of the final rendered geometry. Skatter then sends the detailed geometry to your rendering engine in the place of the low-poly representation.

Here are some details about the extension:

- Developer: Lindale
- License: Permanent license (single seat) or subscription (single seat or floating seat)
- Website: `https://lindale.io/skatter`
- Developer website: `https://lindale.io/`

If you are someone who creates landscape models and does any level of rendering, you need to get a copy of Skatter. Quite frankly, it is a tool that anyone who is modeling any level of the landscape at a professional level must have in their toolbox.

Medeek Wall from Medeek

Medeek Wall is an extension from developer Medeek that is part of a larger set of extensions that will allow you to model and frame a structure inside SketchUp. Medeek Wall allows you to quickly generate framing for walls, including framing for openings (doors and windows), shear wall elements, columns, and blocking for things such as stairs. Medeek Wall even allows you to add electrical elements and generate material lists.

Here are some details about the extension:

- Developer: Medeek
- License: Permanent license or subscription
- Website: `http://design.medeek.com/resources/medeekwallplugin.pl`
- Developer website: `http://design.medeek.com/`

Medeek Wall is a powerful extension that will be of immediate value to anyone who is professionally designing or framing conventional structures. If this is you, I recommend trying out Medeek Wall and the rest of the Medeek extensions.

Architextures from Architextures

Architextures is an online repository of editable seamless textures. With the Architextures extension for SketchUp, you gain the ability to customize and apply them directly from within SketchUp. Before Architextures, I would have said that the only way to get really good-looking custom materials into your SketchUp model would be to take your own pictures and import them into your model. With Architextures, you can create perfect photo-realistic materials without ever taking a single picture!

Here are some details about the extension:

- Developer: Architextures
- License: Monthly or annual subscription
- Extension Warehouse page: `https://extensions.sketchup.com/extension/1a0e0f80-7186-48da-8dd4-f6337dac0873/architextures-for-sketch-up`
- Developer website: `https://architextures.org/`

Architextures is perfect for anyone creating a model with realistic-looking textures. If you do any level of interior design or even architecture modeling, Architextures is worth taking a look at.

Eneroth Auto Weld from Eneroth3

Eneroth Auto Weld from Eneroth3 allows you to select geometry and weld edges together where appropriate. Rather than having to precisely select the edges that need welding before running the native **Weld** command, Eneroth Auto Weld will find the edges that are appropriate to weld together and do it for you. Additionally, when installed, Eneroth Auto Weld will run automatically when **Follow Me** is run and will attempt to replace welded edge with true arcs or circles, where possible (as opposed to the generic curves created by the native weld).

Here are some details about the extension:

- Developer: Eneroth3
- License: Permanent license
- Extension Warehouse page: `https://extensions.sketchup.com/extension/16cd999d-050e-4910-b0a4-699f83decd75/eneroth-auto-weld`
- List of developer extensions: `https://extensions.sketchup.com/user/extensions/1438951607944306779129789/Eneroth3`

This is a simple yet powerful extension to have installed if you model curves or use **Follow Me** regularly. This is an extension that I wanted to dive into with a step-by-step walkthrough, but as it is a paid extension, I had to settle for telling you about it. If you find yourself regularly selecting and welding or smoothing geometry, Eneroth Auto Weld may well be worth the purchase price.

Transmutr from Lindale

Transmutr from Lindale is an importer of sorts. With Transmutr, rather than importing a file with 3D geometry and getting whatever the file gives you, you can finetune the imported geometry before it is dropped into SketchUp. This means that you can do things such as controlling the number of polygons in a model or converting materials as they come in with imported files. This can be a huge time and energy saver if you work with 3D geometry files regularly.

Here are some details about the extension:

- Developer: Lindale
- License: Permanent license (single seat) or subscription (single seat or floating seat)
- Website: `https://lindale.io/transmutr`
- Developer website: `https://lindale.io/`

Transmutr is an extension I would recommend anyone who imports files to check out. If you work with `.fbx`, `.obj`, `.3ds`, `.dae`, or `.stl` files and ever had issues with crazy geometry or materials making the imports cause problems in SketchUp, you should give Transmutr a try.

Vertex Tools from Thomthom

Vertex Tools from Thomthom turns SketchUp into a whole model modeling software. With Vertex Tools, you end up not editing edges and faces, but the vertices that define them! This method of modeling allows you to create and manipulate geometry in a way that can be subdivided (smoothed) systematically. It also allows you to stretch and shape geometry, which is limited when you are dealing with rigid faces. Vertex Tools is a set of commands that are intended to work alongside Thomthom's SubD extension but makes for a powerful editing suite all on its own. This process is not for everyone, and you may have workflows created around traditional SketchUp modeling.

Here are some details about the extension:

- Developer: Thomthom
- License: Permanent license
- Extension Warehouse page: `https://extensions.sketchup.com/extension/0d87cf96-04c1-4bbc-85d4-066491b69b60/vertex-tools-2`
- Developer website: `https://evilsoftwareempire.com/`

If you are familiar with other modeling methods (subdivision modeling or nurbs), then you may want to check out Vertex Tools.

Profile Builder 3 from mind.sight.studio

The final (I do wish there was room for more) extension on this list is Profile Builder 3 from mind.sight.studio. This is an impressive extension that allows you to create array geometry that will follow your cursor as you draw. Imagine the simplicity of pulling a rectangle into a wall. Now, imagine that wall has drywall and baseboard on the inside and sheathing and siding on the other. Imagine pulling a single line up an incline and having a fully detailed railing show up in that line's place! This is what Profile Builder 3 does. It creates a library of detailed entities, then uses them to create intricate models in just a few clicks.

I recommend Profile Builder 3 to anyone creating architectural models of any kind, but it can be useful to just about any SketchUp modeler.

Here are some details about the extension:

- Developer: mind.sight.studio
- License: Permanent license (single seat)
- Artisan website: `https://profilebuilder4sketchup.com/`
- Developer website: `https://mindsightstudios.com/`

Once again, I want to express that I have only had a chance to look at some of the available extensions, so I urge you to spend time looking into extensions that can help you take your SketchUp workflows to the next level.

Summary

This chapter was a little different from the others. I did not want to sound too much like an infomercial, but I wanted to get a little deeper into the world of extensions so that you might truly grasp how important extensions can be to you and your SketchUp workflow. If you followed along, you should already have four extensions installed (Selection Toys, Bezier Curve Tool, Curic Mirror, and adebeo_pushline) and ready to use. Additionally, you have seen my list of extensions that I think might be worth looking at for one or more of your workflows.

In the next chapter, we'll look at SketchUp's companion application, LayOut. There, we will provide an overview and get hands-on with LayOut.

14

Introduction to LayOut

LayOut is a two-dimensional layout software. LayOut's purpose is to allow you to take your 3D models from SketchUp and present them in 2D. This means giving you the ability to frame your 3D model on a page and add additional information via images or text. LayOut helps you present the ideas you have created with your SketchUp model to the rest of the world. 3D space is an amazing way to explore and plan, but when it comes time to convey information to others in a world where paper plans are the norm, a tool such as LayOut is the key to creating great-looking, easy-to-understand documents.

A copy of LayOut is installed alongside every single copy of SketchUp Pro. For this reason, it is always a little surprising to hear SketchUp users say that they have never used LayOut. This may make sense in some cases – that is, if a user is creating 3D assets for use in a video game, or is just using SketchUp to create geometry that is directly exported to a 3D printer. In most cases, however, professional workflows can require or be greatly enhanced by generating 2D documentation of models created in SketchUp, and with LayOut, generating those documents can be quick and easy, if you know what you are doing!

In this chapter, we will look at what LayOut is and the basics of how it works. Entire books have been written on creating LayOut workflows, so please see this as just a taste of what LayOut has to offer. The goal of this chapter is to give you an idea of what LayOut is and how it functions, rather than exhaustive training.

In this chapter, we will cover the following topics:

- Introducing LayOut
- Preparing a SketchUp model for LayOut
- Using LayOut with SketchUp models
- Adding detail to a LayOut document
- Generating output

Technical requirements

For this chapter, you will need LayOut and the `Taking SketchUp Pro to the Next Level - Chapter 14.skp` file, which is available at `https://3dwarehouse.sketchup.com/model/414ae4a0-c3a8-41dd-aee6-c3ca154e7b86/Taking-SketchUp-Pro-to-the-Next-Level-Chapter-14`.

Introducing LayOut

As mentioned in the introduction, many SketchUp users run SketchUp daily, have LayOut installed, and have at least seen the icon on their computer, but do not use it. Why is this? There are three reasons that I can think of:

- They need to use another program to generate their output. This makes sense for SketchUp designers that work in an office where output is standardized, and information passes back and forth throughout the firm and to other companies. Often, larger firms are required to have files generated from a specific program to satisfy the requirements of the companies they are working with or even the requirements of their government. In this case, it is understandable to have SketchUp as a piece of the grander workflow and create output in the software that is required.
- They use another program because it is the program they used before SketchUp. There are many SketchUp users that I have talked to who are not using LayOut because their firm has always created their drawings in AutoCAD, ArchiCAD, TurboCAD, and more. They will say that they already have templates created or they work with a drafter who prefers some other software. Frequently, these designers are already rocking the boat a little bit by choosing to create their models in SketchUp instead of another software, so I understand not wanting to attempt to push more change on their company.
- They don't know or understand LayOut. This is the response I have heard the most. Users opened it once, clicked a few buttons but did not understand what they were looking at, and closed it back down. Sometimes, these users are generating output by taking screenshots of the SketchUp screen for output or using the export command to create a JPEG file, then taking that into some other software to generate output. They have created a clunky workflow where they pass inconsistent imagery around when they have a tool that will allow them to create high-quality, consistent output already installed on their computer. The only reason that they are not using LayOut is that they do not know that it is a better solution.

I will go on record right now and say that every SketchUp Pro user should develop at least a passing knowledge of LayOut. Regardless of which of the three of these groups you fall into, you can benefit from working with LayOut. It may be that you do not generate entire plan sets using LayOut, but the ability to position and organize your 3D models into good-looking 2D imagery can be helpful to almost any designer.

Despite its inclusion with SketchUp, many users resist diving in and learning how to use LayOut. I have seen some people with beautiful models asking how to best use the dimension tool and the export option in SketchUp to generate a "plan." Others have complained that the dimension and text tools are too simple within SketchUp. The reason that it is not an easy process to generate a plan while inside SketchUp or why the text tools are so basic is that SketchUp was not intended to directly generate output. The text tools and dimension tools are there more as note-taking tools, giving you the ability to call something out for your reference. The image export options are there so you can get a quick snapshot of your model to share with a co-worker or show to a client to get quick feedback on a change you have made. When it comes to generating complete two-dimensional, detailed output that can be used to build or fabricate the subject of your model, you are looking for LayOut.

Take the following two figures as an example, both generated from the same model. *Figure 14.1* shows the first one, which has been taken directly from SketchUp. Dimensions were added, and then the whole thing was exported as a 2D image:

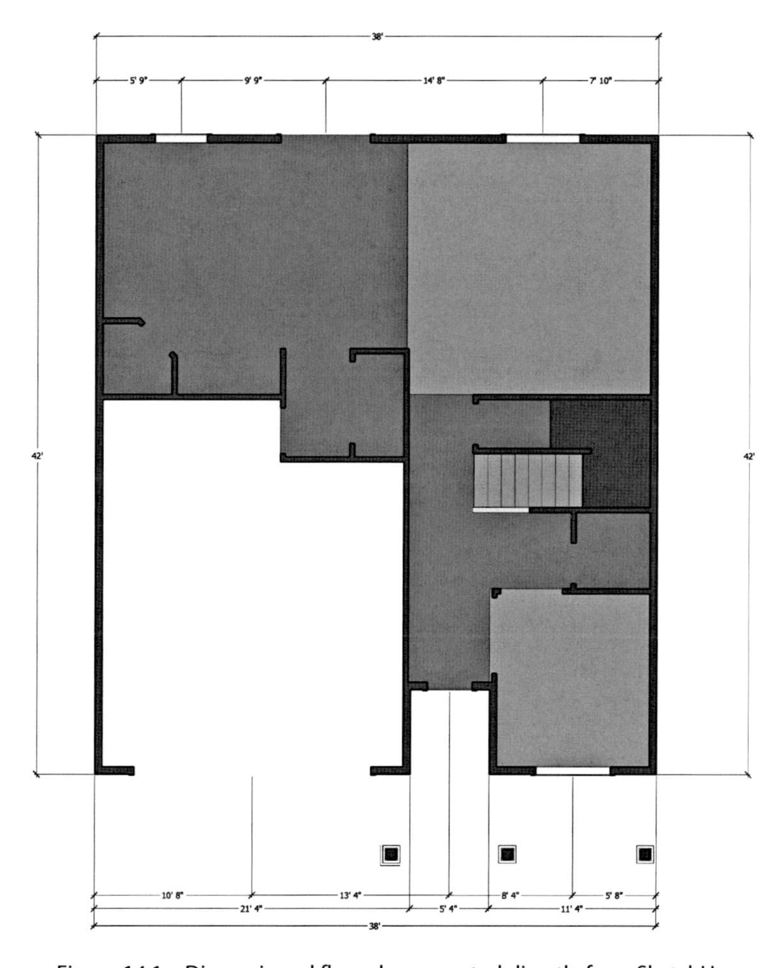

Figure 14.1 – Dimensioned floorplan exported directly from SketchUp

Not too bad, right? Well, consider that compared to a quick floorplan created in LayOut, like this:

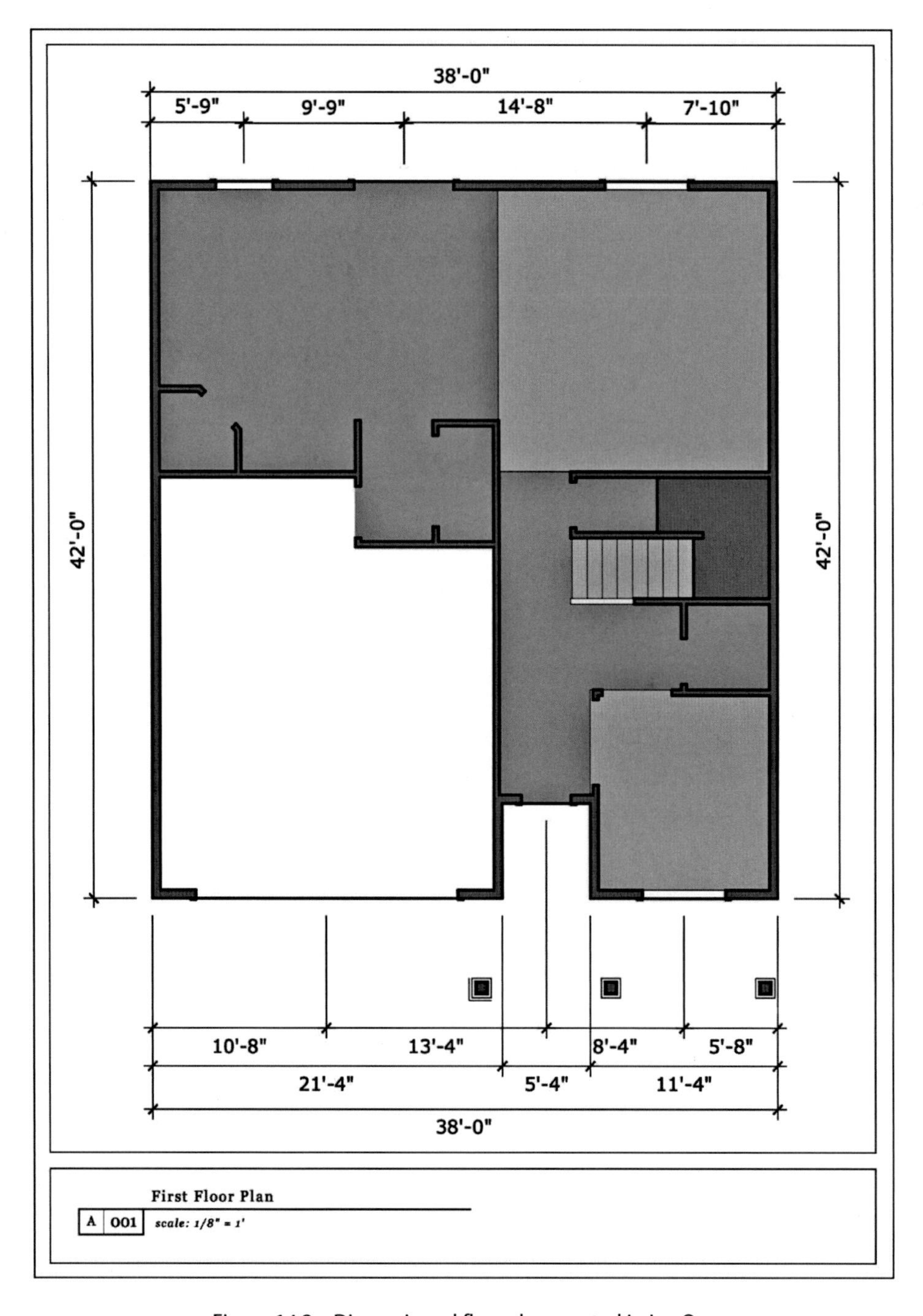

Figure 14.2 – Dimensioned floorplan created in LayOut

The biggest difference (besides the quality) was the amount of work that it took to generate. Once this model was imported into LayOut, the dimensions, text, and everything else were dropped onto the model image in half the time that it took to get the images drawn in the SketchUp model. This is not to say that the dimension tool in SketchUp is bad – it's just that the tool in LayOut was created and intended for adding detail to 2D drawings. This means you can generate better-looking output with less work if you use LayOut.

None of this is to say that using LayOut is simple. One of the things that I hear most often is "It's not like using SketchUp." This is true. It is not like using SketchUp because it is not SketchUp. LayOut is its own software with its own workflows and tools. Some of them feel like something in SketchUp while others are different. If you can start learning LayOut by letting go of the idea that it is just like SketchUp, this will be a much easier process!

With that, let's stop talking about using LayOut and start using it!

Preparing a SketchUp model for LayOut

While you could, in theory, start from scratch and just draw shapes right in LayOut, it was designed to work from an existing SketchUp model. While any SketchUp model can be inserted into a LayOut document, it will make generating documents much easier if you prepare your SketchUp model for LayOut. This does not have to be a complicated process and will look a little different for everyone, but we can run through a few things you can do in SketchUp that will save you time and energy once you get into LayOut.

Throughout this chapter, we will use the `Taking SketchUp Pro to the Next Level - Chapter 14.skp` SketchUp file to generate a LayOut document. Before we start looking at LayOut, let's take a look at this file, which has been prepared to generate output, and see the sort of work you may want to do when getting a SketchUp file ready to generate documentation using LayOut.

Following Along

While you can simply read the descriptions, you may get more out of this section by opening the file in SketchUp and exploring the details we review here. If you do choose to do so, try not to make any changes, or save the file, as doing so may make it impossible to follow the step-by-step instructions later in this chapter.

Scenes

If there is one thing that you will want to set up in your SketchUp model to use in LayOut, it is scenes. As a general rule of thumb, using LayOut will be easiest if you have a scene created for every view you plan to use in LayOut. If you plan on creating a plan that includes four elevations and two floorplans, then you should have at least six scenes in your SketchUp model. If you plan to add two sections, then that is another two scenes. Let's take a look at the scenes that exist in the `Taking SketchUp Pro to the Next Level - Chapter 14.skp` file:

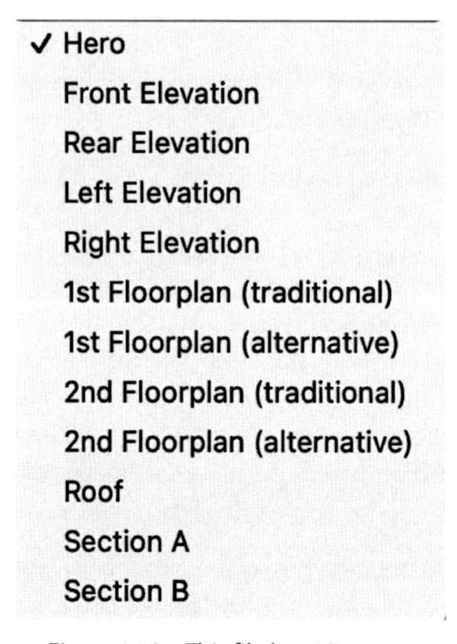

Figure 14.3 – This file has 12 scenes

This model has 12 scenes saved. There is a **Hero** scene, which includes an isometric view of the entire model (likely for use on a cover page). It includes a scene for each of the four elevations and two sections. It also includes scenes for plan views of the first floor, second floor, and roof. Notice that there are two versions of the 1st and 2nd floorplans (clicking on them will show two different ways that you can present floorplans using SketchUp and LayOut).

While it is possible to create views of any SketchUp model from within LayOut, it is easiest to do so from pre-existing scenes. Using scenes will make it easier to make changes in SketchUp and update your LayOut documents with the changes. Creating scenes in SketchUp also allows you to take advantage of all the skills you have developed for moving through a model in 3D space. While you can use LayOut to move a model around to get to a specific view, it is much easier to do in SketchUp.

Tags

While tag visibility will be set by scenes (assuming you are using them), LayOut gives you access to tag visibility as well. You may not use this to turn entire floors on or off once you get into LayOut, but it is a nice option to have if certain things are getting in your way while you're creating your drawings. Window casings, for example, may look good in your final drawings, but they may get in the way as you are adding dimensions. If they have their own tag, you can flip them off in LayOut, add your dimensions, then turn them back on before generating output.

Sections

Sections are a nice tool in SketchUp for working inside of a closed model. For LayOut, they are key to generating any drawing that is not outside your model. There are two scenes (**Section A** and **Section B**) that obviously use sections (click on them to see the view that is created). Sections are also used to generate floorplans. Click on any of the four floorplan scenes to see how horizontal section planes create a view down into the model.

As you explore the model, you will see that it has been cut using sections, but the section planes themselves are not visible! If you click **Section Planes** from the **View** menu, you can turn on all section planes and see where they lie. For each scene that was created using a section, one section plane was active, and the scene was saved with **Section Planes** turned off. This gives us a nice clean view inside the model without us having to see all of the section planes in the model.

Additionally, the sections in the floorplan scenes (all four of them) were used to generate a group that includes all of the geometry created from that section. The context menu's **Create Group from Slice** command will create a group of line work composed of visible edges in the section. This may seem redundant since the section cut is already visible in the scene, but it does serve a purpose once you get into LayOut.

When adding dimensions in LayOut, you can pull dimensions from any of the normal snap points that you have access to in SketchUp. Since a section is only a representation of what it would look like if the geometry was split at the location of the slice and not an actual break in the geometry, there is no geometry there to connect dimensions to. Generating the slice of geometry will create edges throughout the section so that you can add dimensions to any points seen on the screen.

Creating a group of edges from a section is another example of a step that takes a few seconds in SketchUp but can make detailing in LayOut much easier.

Parallel Projection or Perspective

Something else that is generally saved in your scenes is the camera type. Most views that are used for generating printed output are created using the camera set to **Parallel Projection**. If you click through the scenes in this model, you will see that most of them are using this camera type. There are certain view types where you may want to use a perspective view. In this model, the **Hero** scene, as well as the alternative floorplans, include cameras set to **Perspective**. These different camera types will give you options for when you want to show depth in your drawings and when you want to show a flat isometric view of your model. While it is possible to change camera types from inside LayOut, it is easier to see and preview from SketchUp.

Styles

Styles, once again, will be saved in your scenes. From LayOut, you can change to any style saved in the SketchUp file. LayOut does not, however, have the ability to create new styles or edit existing styles. For this reason, it is important to save any styles you plan to use in your output into your model before you head off to LayOut.

These are the top points that are worth thinking about as you finish your SketchUp model and prepare to generate documentation. It is worth spending a little time just poking around the model in SketchUp to see how it is organized before proceeding to the next section, where we will import this file into LayOut.

Using LayOut with SketchUp models

The primary function of LayOut is to import SketchUp files and display them in 2D. I say primary because it is possible to use LayOut to do other things as well. There are drawing tools, so you could use LayOut to create some basic 2D drawings from scratch. There are also tools to add supplementary information (text, symbols, and dimensions) to complement the 2D versions of your models. You can also use LayOut to import other files (text, spreadsheets, or files exported from CAD) to your layouts. All these other capabilities, however, serve to add to that primary process of importing SketchUp files and readying them for output on a printed or digital page.

To that end, the first thing we will look at is how to import and place a SketchUp model into a LayOut document:

1. Open LayOut.

   Much like SketchUp, the first thing you will see when you start LayOut is a welcome window:

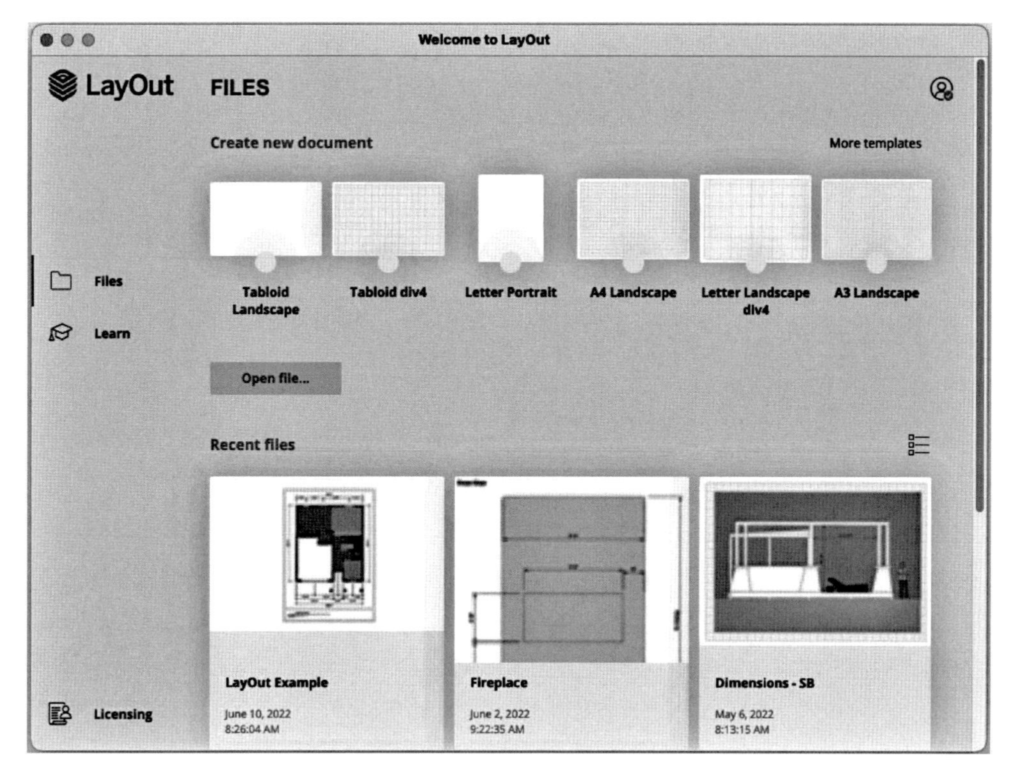

Figure 14.4 – LayOut welcome window, as seen on macOS

From this screen, you can choose a blank template to create a new file, open an existing file, or quickly open one of your most recently used files. For this walkthrough, we will create a brand-new document by choosing an existing template.

2. Select **More templates** from the top-right corner of the welcome window.

   This will display a page of existing LayOut templates for you to choose from. Notice that the top of the page has three tabs that you can click through:

   - **Paper** will display templates that are standard paper sizes (from A3 up through tabloid size sheets) with or without a grid displayed on them.
   - **Storyboard** includes a few templates of different aspect ratios that can be used for generating storyboards.
   - **Titleblock** contains the most options, with templates of different page sizes that include title blocks of different kinds (**Contemporary**, **Modern**, **Rounded**, **Simple**, **Simple Serif**, and **Traditional**). These title blocks include a combination of text that can be manually edited and special text that will fill itself in when a SketchUp model is added to the page called **auto-text**. This text is controlled at a higher level, allowing you to change all instances of the auto-text with a single command.

Picking Favorites

Notice that, just like SketchUp templates, each LayOut template has a white circle below it. Clicking this circle will add a heart symbol, indicating that this is the default for new files. Once selected, that template will be used anytime you start a new LayOut document.

3. In the **Rounded** section of **Titleblock**, choose the **Tabloid Landscape** template:

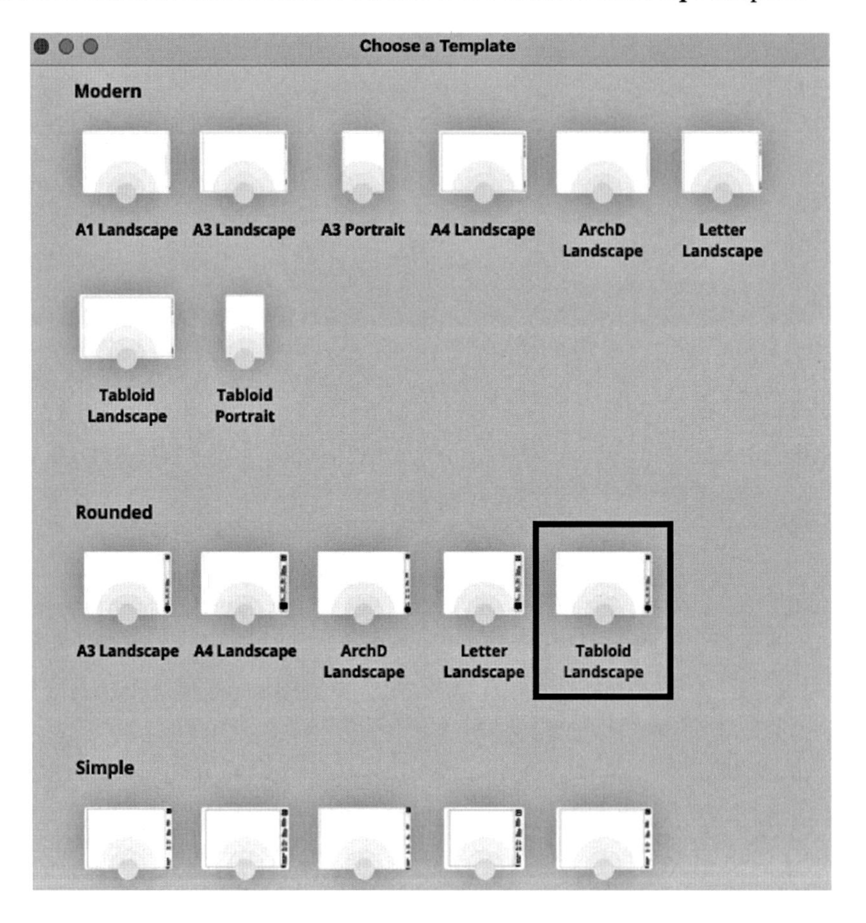

Figure 14.5 – The Tabloid Landscape sidebar is at the bottom of the template list

Exploring Templates

This example will explore LayOut using a specific template. The steps of this walkthrough will assume that you have selected the template listed in *Step 3*. However, I recommend that after you go through this walkthrough, you come back and explore the other stock templates. There are many options, and you may discover a template that you prefer over the one we are using in this example.

This will open LayOut, display the empty template, and give you access to the toolbars and panels that you will use to create your document (more on the UI elements in the *Adding detail to a LayOut document* section of this chapter).

Now that we have created a layout document, let's create a good-looking cover page with our SketchUp model on it.

4. In the **Pages** panel, click on **1: Cover Page**.

   This will make the cover page of the template active. Now, we just need to add our model to the page.

5. Select **Insert...** from the **File** menu.
6. Navigate to where you have saved the `Taking SketchUp Pro to the Next Level - Chapter 14.skp` file. Select the file and click on the **Open** button.

   This will drop a SketchUp model viewport into the middle of the page:

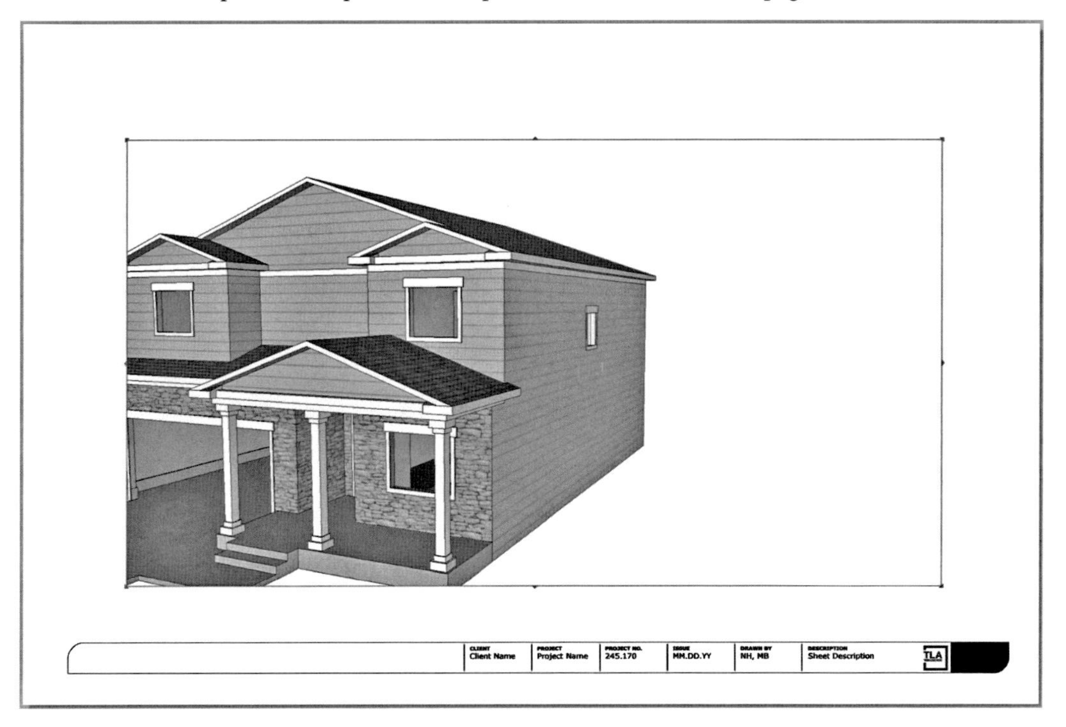

Figure 14.6 – Imported SketchUp model in LayOut

It is nice to see our model in LayOut, but it is currently not taking advantage of the size of the page. Let's stretch out the viewport so that it fills the empty page in the middle of the title block.

LayOut Terminology

A document created in LayOut is called a **LayOut file** and has the `.layout` extension. A single LayOut file can have multiple **pages** (a page is a representation of a real-world page of paper that your world in LayOut will be printed on). Each page can have multiple **viewports** on them (a viewport is a container that allows you to view a SketchUp model). Each viewport can be connected to a different SketchUp model (or different views of the same model).

7. Click on the **Select** arrow in the toolbar at the top of the LayOut window, then click on **Model Viewport**.
8. Click and drag the top-left corner of the model window up to the top-left corner of the page.
9. Click and drag the bottom-right corner of the model window down to the bottom-right corner of the page.

   There is not a "perfect" place to drag the corners, but we want the image of the model to fill the page. In the end, you should have something that looks like this:

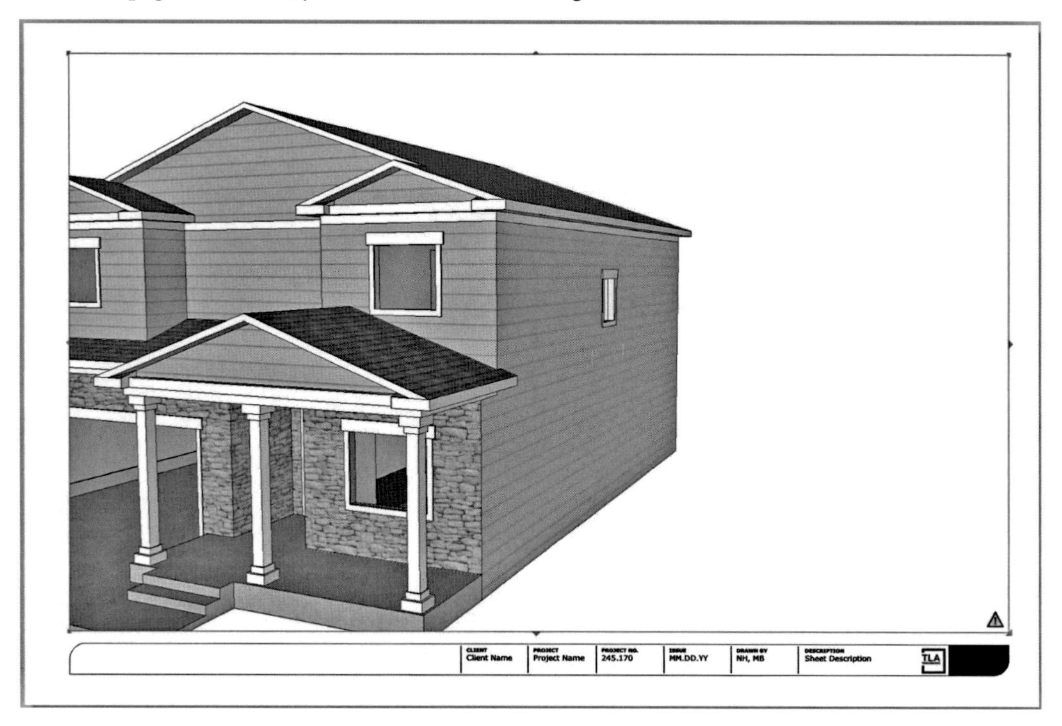

Figure 14.7 – Resized viewport makes for a bigger model

That Exclamation Point in a Yellow Triangle

Depending on your auto-render settings (more on this later in this section), you may get a little yellow triangle with an exclamation point in the bottom-right corner of your viewport as you edit your model. This icon shows up any time you do something that may make a change to how the content of a viewport should be shown. This is an indicator to you that the viewport needs to be re-rendered. If you are seeing this either before or after resizing the viewport, simply right-click on the viewport and choose **Render Model**.

The model is looking good, but it feels a little off center. Let's drag the model (not the viewport) over to the right a little bit so that the end of the driveway is not spilling out of view. To do this, we need to edit the 3D view of the model in the viewport.

10. Double-click on the model viewport.

    You will notice that the appearance of the mode changes (the edges look different and the axes show up). When you are in the 3D view, you can navigate the same way you can in SketchUp; the view you end up on will be the one that LayOut will use in the viewport. This can be done using the scroll wheel of a three-button mouse (scroll to zoom in and out, click to orbit, *Shift-click* to pan) or you can right-click on the model and choose a specific tool from the **Camera Tools** menu (**Orbit**, **Pan**, **Zoom**, **Zoom Window**, **Look Around**, or **Walk**).

11. Use your mouse or **Camera Tools** to change the view of the model.

    While you can get to any view of the model using these commands, we need to get to a single view (for tutorial purposes). Let's get back to the initial view we saw when we imported the model but zoom out so that the entire model fits in the viewport.

12. Right-click on the model and choose **Scenes** from the context menu. Then, click **Hero**.
13. Now, right-click on the model and choose **Zoom Extents**.
14. To exit the 3D view of the model, click anywhere outside of the model viewport.

With that, your LayOut model should look like this:

Figure 14.8 – Zoom Extents shows us the whole model

It should be noted that, while this first page contains only a single SketchUp model viewport, you can add multiple viewports linked to multiple SketchUp files if that is what you need. Whether you end up with a single SketchUp model viewport on a page or multiple, it is good to understand the link between these viewports and the files that populate them.

SketchUp file linking

Every SketchUp model viewport is linked to the `.skp` file that created it. If, at any point, you have changed the SketchUp file displayed in a viewport or if you want to change the file being referenced, you will need to modify the file references. These references can be found in the **Document Setup** window:

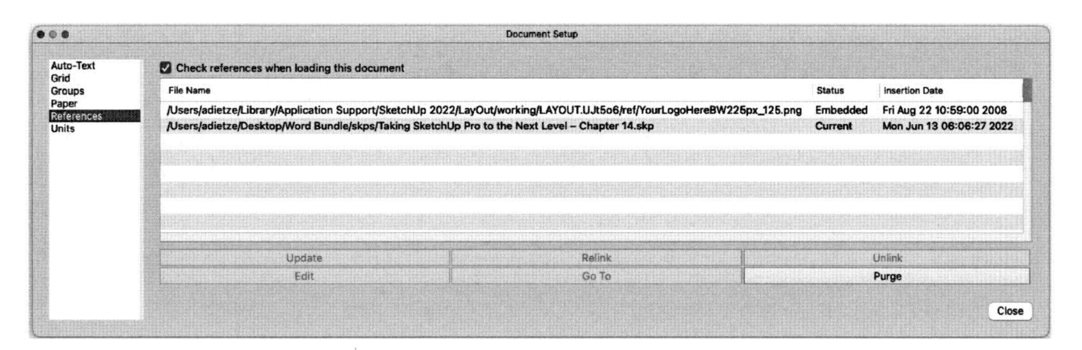

Figure 14.9 – The References tab of the Document Setup window

The **Document Setup** window can be accessed through the **File** menu (click **File** and then **Document Setup...**) and contains more than just references to external files. There are settings here for **Auto-Text**, **Grid**, **Groups**, **Paper**, and **Units** as well. While I will refer to some of these options later in this chapter, right now, I want to focus on how **References** works.

The **References** tab of the window will display a list of all external files (SketchUp files and other file formats) displayed in the current LayOut file. The list at the top will display a file path to the original file, as well as the status of the file and the date that it was inserted. The checkbox at the top of the window will cause LayOut to check referenced files every time this LayOut file is opened. If a file is ever missing or has changed from the version that was previously loaded, LayOut will color code the filename (blue for files that have been updated and red for missing files).

Here is a list of the six buttons below the list of references files:

- **Update** will become available if an updated file is selected from the list. Clicking this button will update the files in the model, displaying changes that have been made since the file was previously loaded.
- **Relink** will allow you to swap out the selected file for another. This is something that you may want to do if you have moved a linked file or want to change a file that is currently in your document for another.
- **Unlink** can be clicked to remove the active connection between a SketchUp model and its representation inside of LayOut. Normally, when you insert a SketchUp file into a LayOut document, LayOut maintains a connection between the files and allows you to keep the document up to date with the most recent information in the SketchUp files. If you choose to unlink the files, then the version of the SketchUp file in the LayOut document becomes detached from the original file and changes will not be shown.

- **Edit** will allow you to open the selected file in an external editor. For SketchUp files, this will launch SketchUp. For other files, LayOut will defer to your system's default editor for the selected file type.
- **Go To** will display the file at its location in Explorer (on Windows) or Finder (on macOS).
- **Purge** will remove any files in the list that are not currently being used in your document. This can happen if you are editing a document and you remove a viewport. This will get rid of the content visually on the screen, but the link to the file will still exist. Clicking the **Purge** button will remove any reference to the file.

Understanding the status of linked files is key to updating your LayOut document while changes are being made to the SketchUp model. If a change is made to a SketchUp model, you will need to select it from the list and click the **Update** button. This will let LayOut know that you want to import and use the more recent version of the file. Once that is done, you will need to select the viewport from your document and then right-click to choose the **Update Model Reference** option. This will tell LayOut to update the displayed model to the newer version.

The process of maintaining SketchUp models is a key part of using LayOut. While we cannot look at every single aspect of file maintenance in this one chapter, we can look at a few more options when it comes to working with a SketchUp model in LayOut.

The SketchUp Model panel

The main portion of any LayOut document is the SketchUp model. SketchUp models are displayed in model viewports, which have a lot of options as to what information is displayed in the document. All of this information is controlled using the **SketchUp Model** panel:

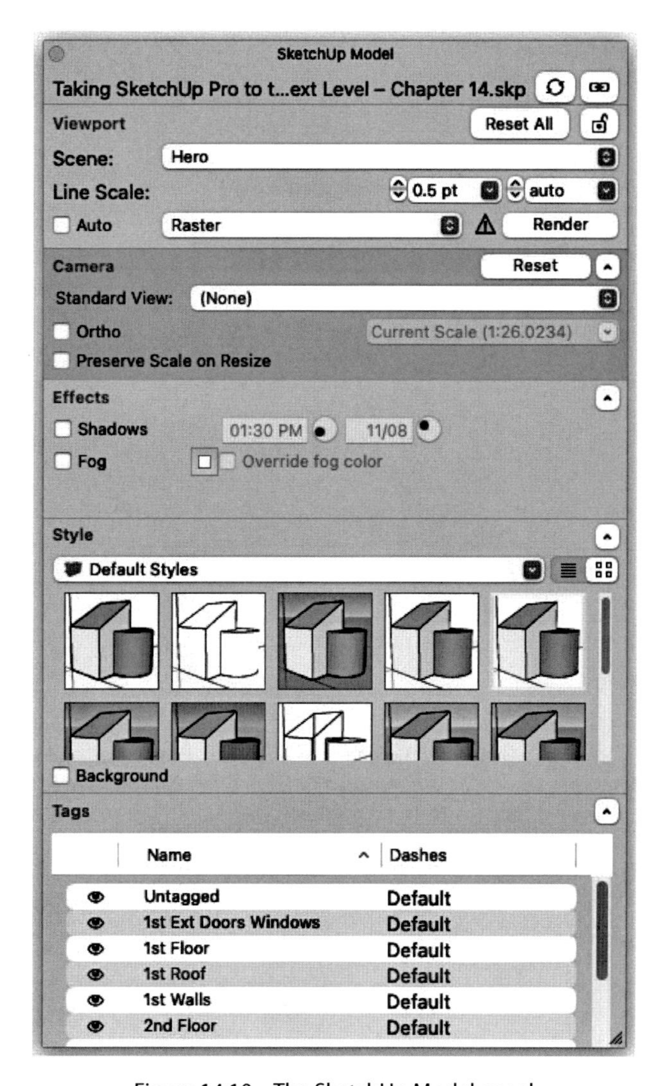

Figure 14.10 – The SketchUp Model panel

The **SketchUp Model** panel is a part of the **Default Tray** area on Windows and can be toggled on from the **Window** menu on macOS. This panel includes all of the information for a selected SketchUp model viewport. This panel is broken into five sections:

- The **Viewport** section is the top section of the panel and cannot be minimized. This section displays information such as the scene being displayed, line scaling information, and rendering options.
- The **Camera** section displays the current view. If the **Ortho** box is checked, it will allow you to set a scale for the viewport.

- The **Effects** section allows you to toggle the **Shadow** and **Fog** settings on and off.
- The **Style** section allows you to choose the style used to render the model in the viewport.
- The **Tags** section allows you to toggle specific tags on and off.

While it would be amazing to run through all of these sections in depth, for this overview, we are only going to take a look at the **Viewport** section and get an understanding of the rendering options. Make sure that the **SketchUp Model** panel is open and follow along:

1. Use the **Select** tool to select the SketchUp model viewport.

   The properties of the currently selected viewport will be displayed in the **SketchUp Model** panel.

2. Make sure that the **Auto** box is not checked.

   Auto will cause the viewport to redraw itself any time a change is made. In theory, this sounds like a good idea. However, since redrawing can be a time and resource-intensive process, it is better to manually tell LayOut when to re-render viewports. Any time I work with a SketchUp model viewport, I turn **Auto** off and manually render viewports by clicking the **Render** button to the right whenever I decide it is a good time to do so.

Name-Changing Render Button

The **Render** button will change its name based on the state of the selected viewport. If the selected viewport has changes that need to be re-rendered, then the button will be labeled **Render**. If the selected viewport is all up to date, then the button will read **Rendered**.

Notice that the rendering style (the bottom field in the **Viewport** section) is currently set to **Raster**. When set to **Raster**, edges can show up as kind of jagged or choppy:

Figure 14.11 – Raster rendering style

This is because everything in a raster rendering is represented by a single pixel that is a single color. This may not be the prettiest way to display something like a drawing of your 3D model, but it is very quick to display.

3. Click on the dropdown and change from **Raster** to **Vector**, and then click the **Render** button.

   Notice that the edges of the vector rendering are smooth, but we have lost the detail of the materials:

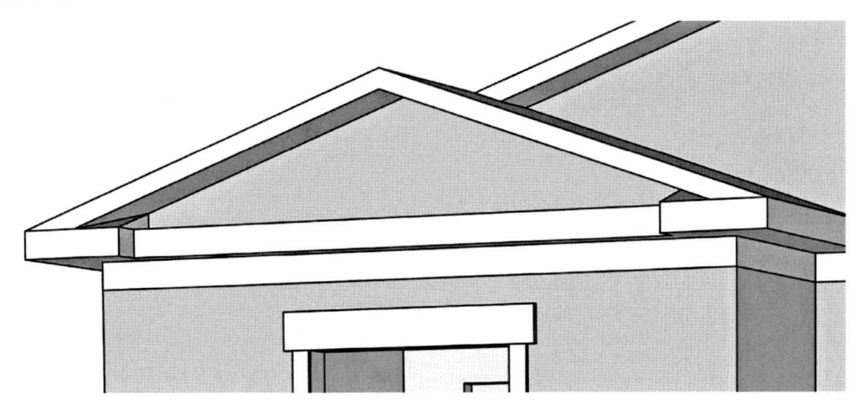

Figure 14.12 – Vector rendering style

Vector is a much better-looking rendering style for lines, but it does take longer to render. You also probably realized that we lost the materials from SketchUp and now everything is a single color. This rendering style can be a nice option if we want to skip over the details and just display line work.

4. Click on the dropdown again and change **Vector** to **Hybrid**, then click the **Render** button.

   This will display the lines from a **Vector** render over the top of the faces from the **Raster** render:

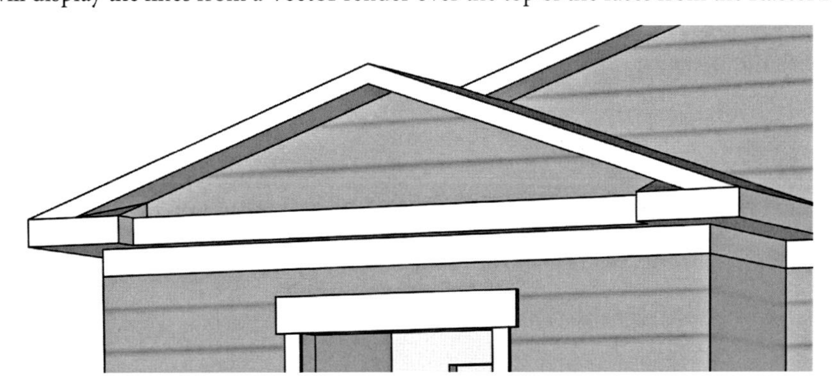

Figure 14.13 – Hybrid rendering style

This is the best of both worlds – smooth edges and detailed materials.

So, why give the option of rendering styles? Why not just always display **Hybrid**? By switching between styles, you have control over what is displayed in your output and how much time is spent rendering viewports! Many users will do all of their work while in the **Raster** rendering style because it is quicker. As you change what is displayed or update as changes are made, a Raster viewport will render quickly. Once the editing is done and it is time to generate output, the viewports can be switched to a **Vector** or **Hybrid** style. This can take a little bit of time to draw the edges but will make for a cleaner-looking output of the final documents.

It is important to know the different rendering options, but even more important is remembering to click that **Render** button as you make changes to your viewport.

Now that we have a basic understanding of how to work with a SketchUp model, let's run through the rest of what LayOut has to offer.

Adding detail to a LayOut document

Placing a 2D version of a SketchUp model onto a page is great, but the real benefit of what can be done with LayOut is in the detailed information you can quickly add to your drawings. These details may be added using the tools and panels in LayOut and include things such as dimensions, text, labels, callouts, or even 2D entourage. The majority of this chapter will cover the process of adding additional information to our LayOut document. Before that, though, let's take a quick run through the UI.

LayOut's UI is different from SketchUp (which makes sense as it serves a different purpose) but similar enough that you should not have too hard a time navigating the commands. Just like SketchUp, there is a toolbar that houses all of the commands and can be customized (unlike SketchUp, there is only one toolbar, and it is locked into place). Just like SketchUp, there are a series of panels that display information and allow you to organize your model. These show up in a customizable tab bar on Windows and as floating, stackable panels on macOS.

While we will not have the time in this chapter to run through examples of every command in LayOut, we can provide an overview so that you have an idea of where commands are located and what each one does.

Toolbar

As mentioned previously, there is a single toolbar in LayOut and it is at the top of the LayOut window, as shown here:

Figure 14.14 – The default toolbar as seen on macOS

This toolbar can be customized the same way that toolbars are customized in SketchUp (through the **View** menu) and contains commands that can also be accessed through the menus.

As we explore LayOut's UI further, you will see that the commands differ quite a bit from what SketchUp offers. The thing to remember with LayOut is that the commands are intended to allow you to position imagery and data on a page and add annotations. If you keep LayOut's core purpose in mind as you learn more about it, then the commands and UI will make a lot of sense.

Menus

The menus in LayOut house the commands in LayOut. Some of the menus are intuitive as they parallel SketchUp's menus, but a few might be confusing as they house commands that do not have an equivalent. Let's look at what resides in each menu:

- **File**: This menu contains standard file commands for opening, saving, and closing documents, as well as commands for saving templates and scrapbooks (collections of 2D drawing elements that can be dropped onto a document). Most importantly, the command to insert a file (as we did in the example in the *Using LayOut with SketchUp models* section) and **Document Setup...** (which brings up the settings for your document) can be found in the **File** menu.
- **Edit**: Standard edit commands are found here (**Undo**, **Redo**, **Cut**, **Copy**, and **Paste**), as well as options to duplicate, change layers of selected items, group, and explode. The **Edit** menu is also home to masking commands.
- **View**: Just like in SketchUp, this menu contains commands that allow you to look around your document. All the zooming commands are here, along with **Pan**. The **View** menu also contains commands for toggling the UI on and off, as well as a command that shows a grid on your document.
- **Text**: This menu is pretty self-explanatory. All of the commands that allow you to change how on-screen text looks are housed here. Additionally, the command to create auto-text sits in this menu.
- **Arrange**: Unlike SketchUp, which has you placing items in 3D, LayOut is focused on 2D layouts, which means that items will lay on top of one another. Commands to control the order of these items are found in this menu. Commands to align or space selected items can also be found in this menu.
- **Tools**: All of the drawing and annotation commands are housed in the tools menu, as well as the commands that allow you to edit or manipulate them. If you want to add something to your document that is not connected to an external file, it is added with commands from the text file.
- **Pages**: This menu contains all of the commands that are used to organize and navigate the pages that make up a LayOut document.

- **Window**: All of the panels used in LayOut can be toggled on or off through this menu. The **Colors** and **Fonts** windows can also be accessed here.
- **Help**: Standard help commands and information are available through this command.

> **Preferences**
>
> The **Preferences** window contains all of the LayOut settings that are not saved with the document. This is a very important window and is one of the few commands that changes based on your operating system. If you are on Windows, it is in the **Edit** menu, while macOS users will find it in the **LayOut** menu.

While many of the commands you will use are found in the menus (and in the toolbar), the most commonly used bits of the UI are its panels. These panels not only contain information about the elements that make up a LayOut document but also allow you to manipulate and modify the elements.

Panels

Very similar to SketchUp, LayOut has a stack of panels that display information about selected items and allow you to set item properties:

- **Shape Style**: This panel allows you to set the appearance of any items in LayOut. This includes any lines or shapes drawn with the tools. You can even use this panel to add borders to SketchUp model windows!
- **Scaled Drawing**: This panel allows you to create a new viewport in which you can draw at a specified scale. Normal items that are drawn in LayOut (using the tools) are created at a one-to-one scale. When you create a **Scaled Drawing** viewport, you can specify the scale of that viewport and draw it using real-life dimensions.
- **Pattern Fill**: If you create geometry using the tools and want to fill them with something other than a solid color (which can be achieved using the **Shape Style** panel), the **Pattern Fill** panel will allow you to pick a pre-made pattern to use as a fill instead.
- **SketchUp Model**: This panel allows you to set the properties and appearance of a selected SketchUp model viewport.
- **Dimension Style**: This panel gives you all the attributes of selected dimensions.
- **Pages**: This panel allows you to create, reorder, remove, and navigate the pages that make up your LayOut document.

- **Layers**: **Layers** are groups of items (anything from model viewports to dimensions to single lines drawn using the **Line** tool) that are placed in the order that they should be drawn. The **Layers** panel allows you to create, reorder, and remove layers in the document.
- **Scrapbooks**: **Scrapbooks** allow you to create reusable 2D drawing elements. The **Scrapbooks** panel allows you to navigate a library of pre-drawn elements that you can drag and drop onto any document.
- **Instructor**: This panel gives you information about the current tool. This is a great panel to have open as you learn LayOut as it actively changes the information displayed to what you need based on the activity you are currently performing.

This was a quick overview of the tools and panels. While it is great to know what commands are and where they are, learning how to use them is even better. Let's use the UI we just learned about to add another page to our document and add some annotations.

Using LayOut's tools and panels

Now, we will spend some time learning how to use the tools and panels to do some basic editing of our LayOut document. First, let's take a look at something simple: editing text.

Editing the project title

Let's start by changing the project title in the title block of our document:

1. Using the **Select** tool, double-click on **Project Name** on the right-hand side of the title block.
2. Notice that as soon as you double-click, the words change from **Project Name** to **<ProjectName>**. Any time you see words or numbers inside angle brackets, it is an indicator that auto-text is being used.
3. Click the **File** menu and then choose **Document Setup…**. Click on the **Auto-Text** option on the left.

   This will list all of the auto-text options in the document. Notice the **<ProjectName>** text that we have in the title block. To the right of the auto-text is the value that will be displayed. Currently, the text being displayed is **PROJECT NAME**.
4. Select the **<ProjectName>** auto-text from the list.

   Notice that the information about the auto-text populates the bottom portion of the window. From here, you can replace the existing text.

5. Replace the existing text with `Example House`, then click the **Close** button.

   When we return to the LayOut document, the **<ProjectName>** auto-text will still be highlighted.

6. Click outside of the selected text to close it.

   Notice that the text that replaces the auto-text is **Example House**. This is great because we can use this auto-text in place of using a final project name. Then, as the project matures, if the owner names change, or the address changes, we can update the project name in one place and all instances will be updated.

Next, let's let see what it takes to add more pages to our document.

Adding pages to a LayOut document

To add more pages, we will start by copying the SketchUp model viewport from the cover page onto our first inside page:

1. Select the SketchUp model viewport and then click the **Edit** menu and choose **Copy**.
2. In the **Pages** panel, click on **2: Inside Page**.
3. Click the **Edit** menu and choose **Paste**.

   This will place an exact copy of the model from the cover page on our first inside page. This is a good start but not what we want. First off, we will have to resize the viewport just slightly.

4. Use the **Select** tool to drag the corners of the viewport so that they fill the page.

Now, let's change the view of the model on this page to a plan view of the first floor.

Changing the model viewport

We have a new page with a new model viewport, but we need to change the view of the model that we are seeing. Follow these steps:

1. Select the SketchUp model viewport.
2. In the **SketchUp Model** panel, change **Scene** to **1st Floorplan (traditional)**, then click the **Render** button.

   As shown in the following screenshot, this does not look right:

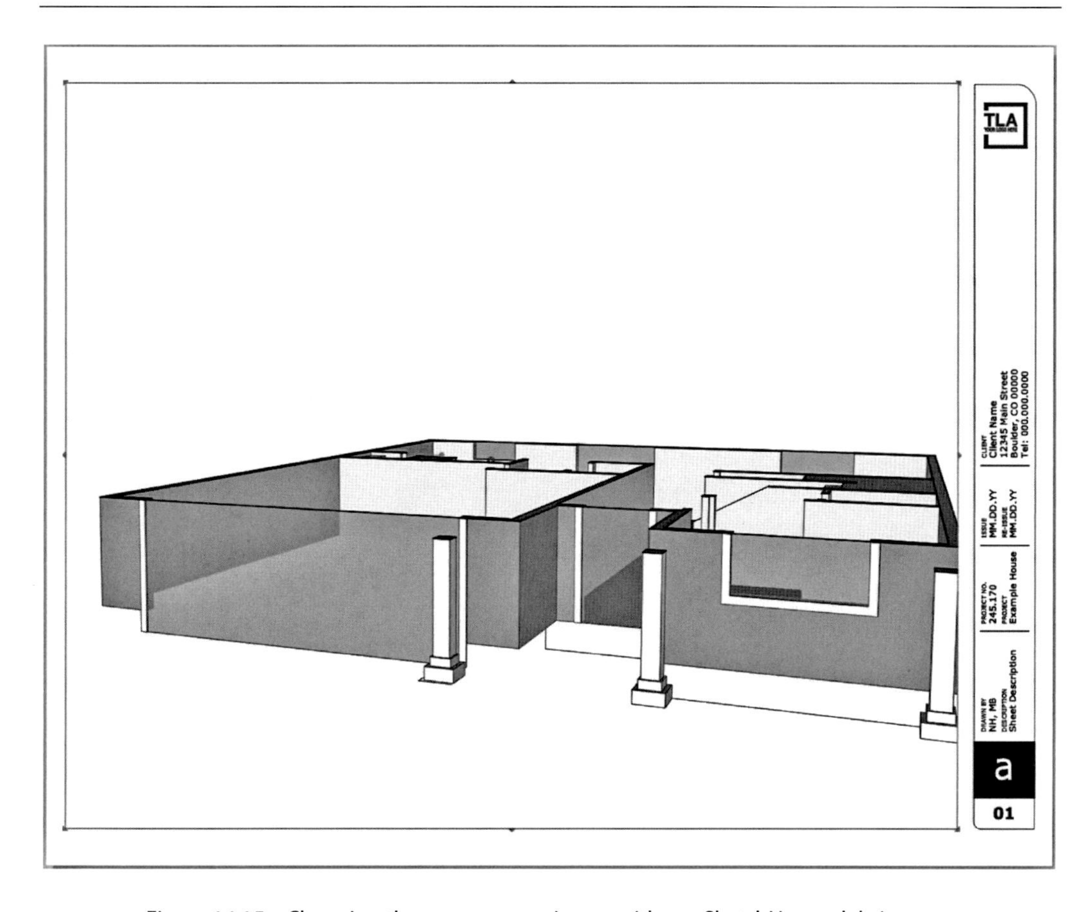

Figure 14.15 – Changing the scene causes issues with our SketchUp model viewport

The problem we are seeing is because we changed our camera setup in the copied SketchUp model viewport. When we double-clicked into the viewport and changed the view of the model, we told the camera not to be where the scene said it should be. Right now, the camera is in that same user-defined location, despite having changed scenes. Fortunately, the fix is very easy.

3. Click the **Reset All** button in the **SketchUp Model** panel, then click **Render**.

   This puts the camera back where the scene says it should be, and re-rendering the viewport gives us the view from above that we would expect in a floorplan.

   Now that we have the view we want, let's set a scale for this drawing.

4. With the viewport selected, expand the **Camera** section of the **SketchUp Model** panel.
5. Click the **Scale** dropdown (to the right of the **Ortho:** checkbox) and choose **1/8" = 1-0' (1:96)**.
6. Make sure that both the **Ortho:** and **Preserve Scale on Resize** boxes are checked, then click the **Render** button.

You should have a page that looks like this:

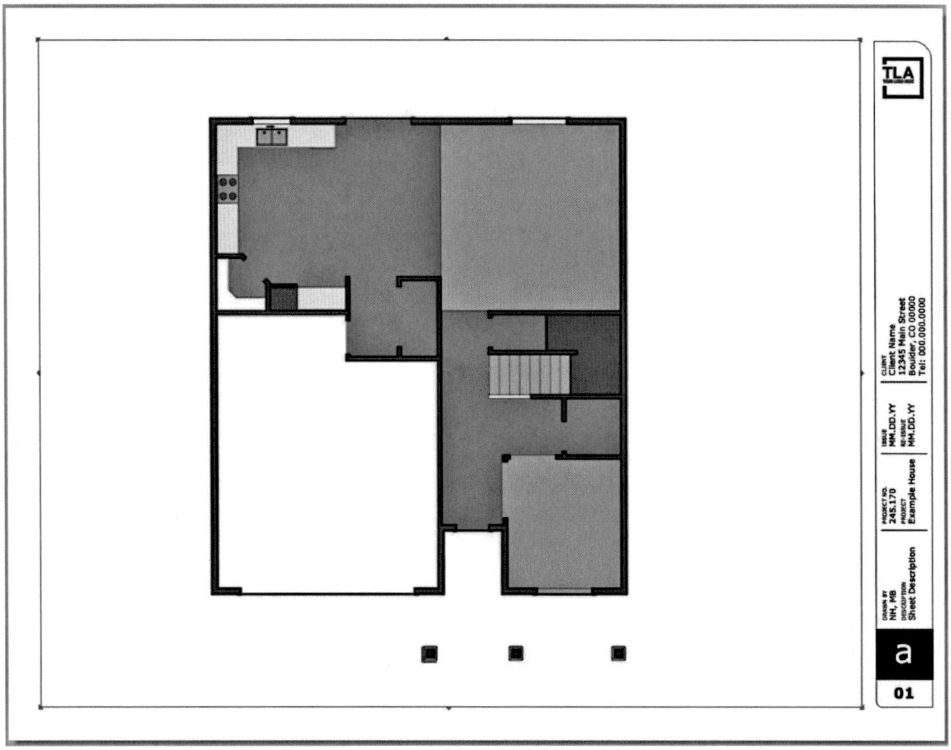

Figure 14.16 – Scaled floorplan

The drawing is smaller on the page, but if we had to manually measure anything from a printed version, we could be assured that it would be the proper scaled measurement.

Let's keep going!

Editing the page description

Now, let's change the page description on the title block, add a few dimensions, and add a title and scale callout to the drawing. First, we will look at editing the **Sheet Description** text:

1. Use **Select** to click on the words **Sheet Description**.

   Before we change anything, notice that the text highlights with a light blue box. Look at the **Layers** panel while this text is selected. Notice that the **Unique Elements** layer has a little light blue box next to it. This is an indicator that the selected element is on this layer.

2. Replace the selected text with `First Floor`, then click outside of the text box.

Next, let's add some dimensions to this drawing.

Adding dimensions

I will walk you through creating the first string of dimensions across the top of the page, then let you get through the rest of the dimensions on your own (I know you can do it). The first thing we are going to do before we draw a single dimension is create a layer to put them on (this will make them easier to edit or remove if needed). Follow these steps:

1. In the **Layers** panel, click the + button.
2. Type `Annotations`, then press the *Enter* key.
3. Drag this new layer to the top of the list of layers.

   We now have a layer where we can place all of our annotations (dimensions, text, callouts, and so on). Make sure that this layer is active (the active layer has the little pencil icon next to its name) as we add annotations to the drawing.
4. Start the **Dimensions** tool either by choosing the commands from the **Tools** menu or clicking on the button on the toolbar.

   To simplify this first example, I will refer to the points on the back wall according to the following diagram:

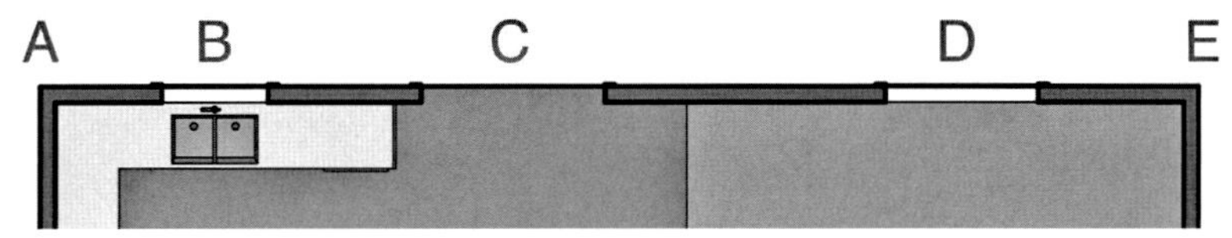

Figure 14.17 – Points we will be dimensioning

5. Click on point A, then again on point E (inferencing should give you a small green dot at either point).

   The first two clicks will pick the points to dimension between. The third and final clicks will allow you to place the dimension line.
6. Click again a little way above the back wall (make sure there is enough space to place another dimension line between this dimension and the wall) so that you get something like this:

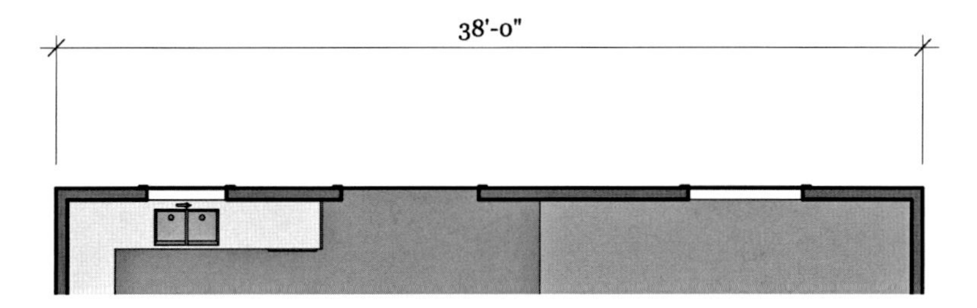

Figure 14.18 – Overall dimension on the back wall

This looks good so far! Let's add another row of dimensions along the back wall from the ends of the wall to the center of the door and windows.

7. Click point A, and then point B (referencing the midpoint of the window opening).
8. Click again to place the dimension line halfway up from the wall to the existing dimension.

This will give you the start for the second string of dimensions:

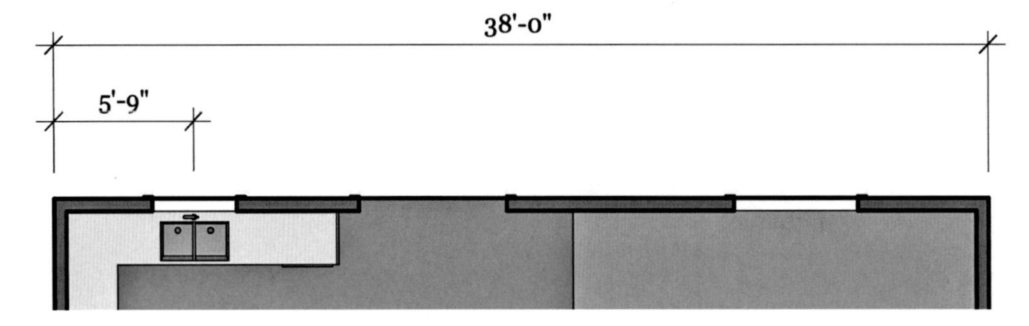

Figure 14.19 – Dimensions to the center of the first window

Now, let's add another dimension (from the center of the first window to the center of the backdoor). To keep this string of dimensions aligned, we are going to draw this dimension backward.

9. Click on point C, then on point D, then on the end of the previous dimension.

Drawing the dimension backward (from the "next point back to the previous point") makes it much easier to pick the third point and have the new dimension align perfectly with the previous dimension.

10. Use this method to draw a dimension from the center of the door to the next window and from the window to the end of the wall.

When you finish, you should have a string of dimensions that looks like this:

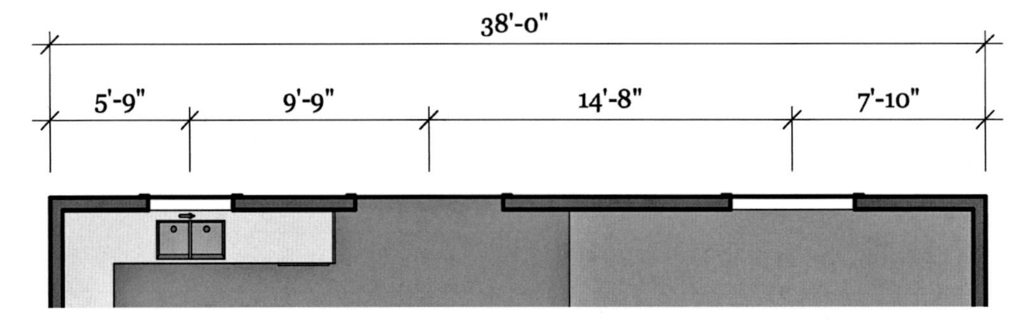

Figure 14.20 – Final dimensions along the back wall

11. Use the **Dimensions** tool to add dimensions around the outside of the house.

    When you finish, you should have a set of dimensions that looks like this:

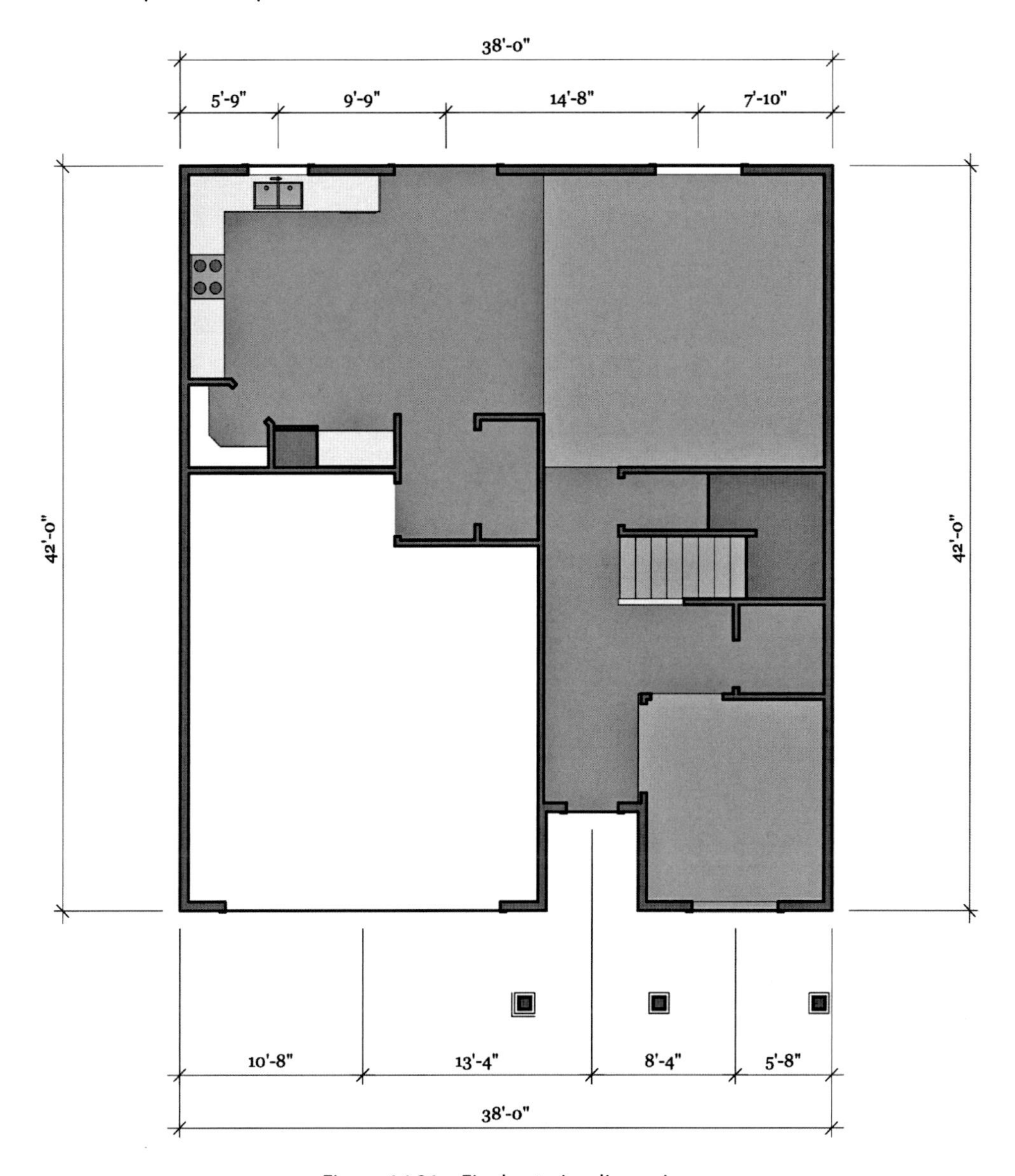

Figure 14.21 – Final exterior dimensions

Our floor plan is looking great! Let's add one more touch so that we can call this floor good!

Adding door and window callouts

Let's add a few callouts for the doors and windows on this floor. We will add labels for each door type in a circle and windows in hexagons. We will start with the front door:

1. Select the **Circle** tool from the main toolbar.

   Before we draw the circle, let's make sure that it is going to be drawn the way we want. We will set the properties of the circle using the **Shape Style** panel.

2. Open the **Shape Style** panel.
3. Make sure that the **Fill** and **Stroke** buttons are turned on and that **Pattern** is turned off. The **Fill** color should be white, and **Stroke** should be set to black. If this is not the case, click on the color swatch and choose the proper color from the colors window.
4. Now, draw a circle, just like you would draw one in SketchUp, just above the front door.

   Assuming you followed every step, you should have something that looks like this:

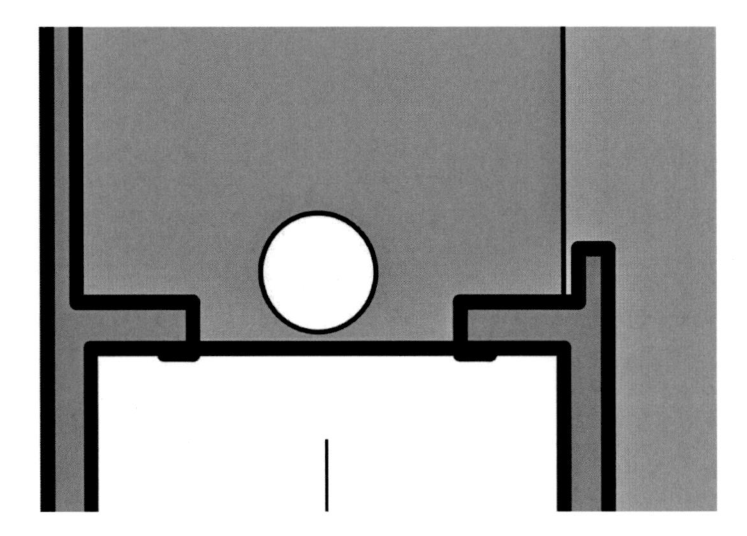

Figure 14.22 – A simple circle above the door

   Next, we need to add some text to this callout.

5. Choose the **Text** tool and double-click on the center of the circle. Then, type `A` and press the *Enter* key.

   Double-clicking when you are using the **Text** tool sets the point where you want to start typing. You can also use the tool to drag out a text window that you will fill with text. In this case, we set the start point and then typed a letter starting at that point. Based on your current settings, this text may be too big or too small for the circle we created (most likely too big). Let's change the font size and get it centered with the circle.

6. Use the **Select** tool to highlight the text we just created.
7. Click on the **Window** menu and choose **Text Style** if you are using **Layout** on a Windows computer, or **Show Font** if you are working on a Mac.
8. Use the **Font** window to increase or decrease the size of the font so that it will fit into the circle, then close the window.

   Now, we just need to get the text aligned with our circle. We could use **Select** to highlight the text and then drag it so that it mostly lines up with the circle, but there is a quicker and more accurate way.

9. Use the **Select** tool and the *Shift* key to multi-select both the circle and the text.
10. In the **Arrange** menu, click on **Align**, and then **Vertically**.
11. Again, in the **Arrange** menu, click **Align**, and then **Horizontally**.

    This will center the circle and the text vertically and horizontally. The last step here is to move it so that both items together are centered over the door they are calling out. Let's make sure that they stay connected by putting them into a group.

12. With the text and circle still selected, right-click on them and choose **Make Group**.
13. Now, use the **Select** tool to drag the callout so that it is above the center of the door. Note that you can use the left and right arrow keys on your keyboard to fine-tune the placement.

    This works perfectly for this door, but we are going to need a callout on every door on the first floor! We will do this by copying the current group and placing a copy at each door.

14. Use the **Select** command with the copy modifier (*Control* on Windows or *Option* on macOS) to select the callout group and start to drag it away from the front door to the next doorway.

    There are eight total doors on this floor. Use the **Select** tool with the copy modifier to place the callout group at each of these locations:

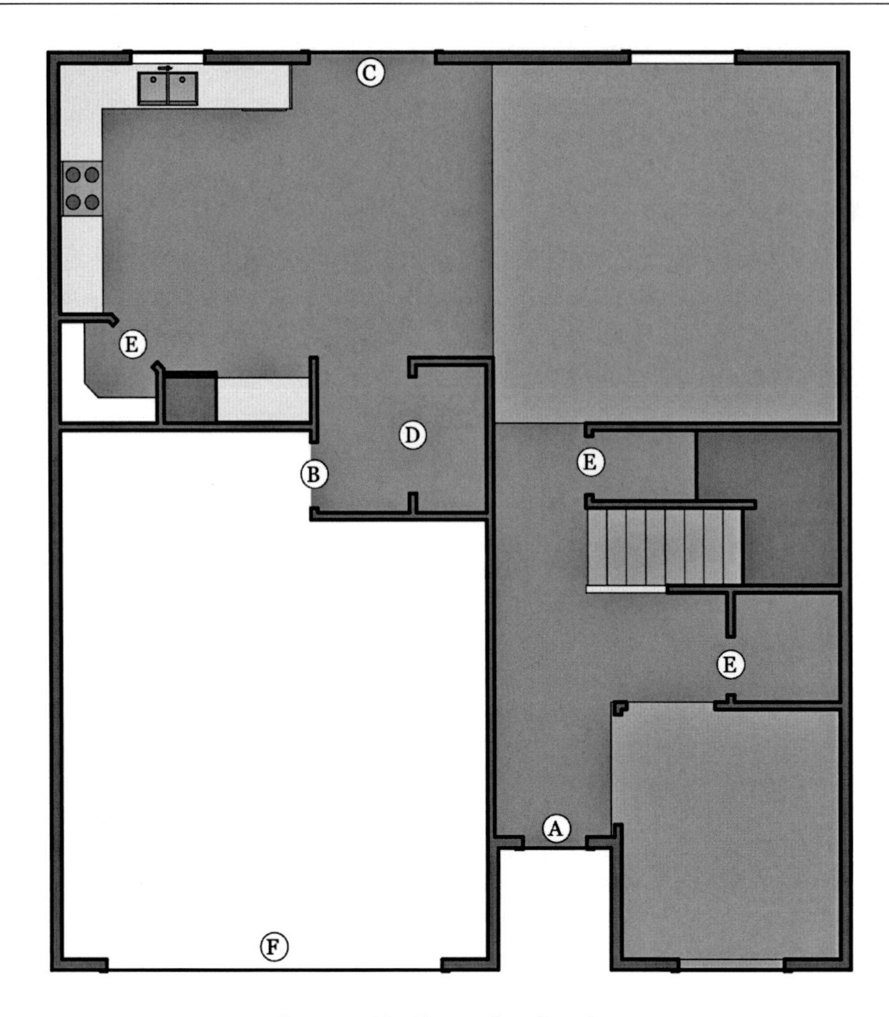

Figure 14.23 – Door callout locations

Once you have placed all the callouts, you need to change the text in the circles based on the door. Use *Figure 14.23* to place unique callouts for each door size.

15. Use **Select** to double-click on a callout group. Double-click again to enter the text box. Then, type the letter for that door.
16. Click outside of the group two times (one to close the text box and a second time to close the group) to finish.

There are only three windows on this floor, so we will need a second callout for windows. For this one, let's use a hexagon symbol with a number callout.

17. Use the **Polygon** tool and the **Text** tool to create and place a new group next to each of the windows, as shown here:

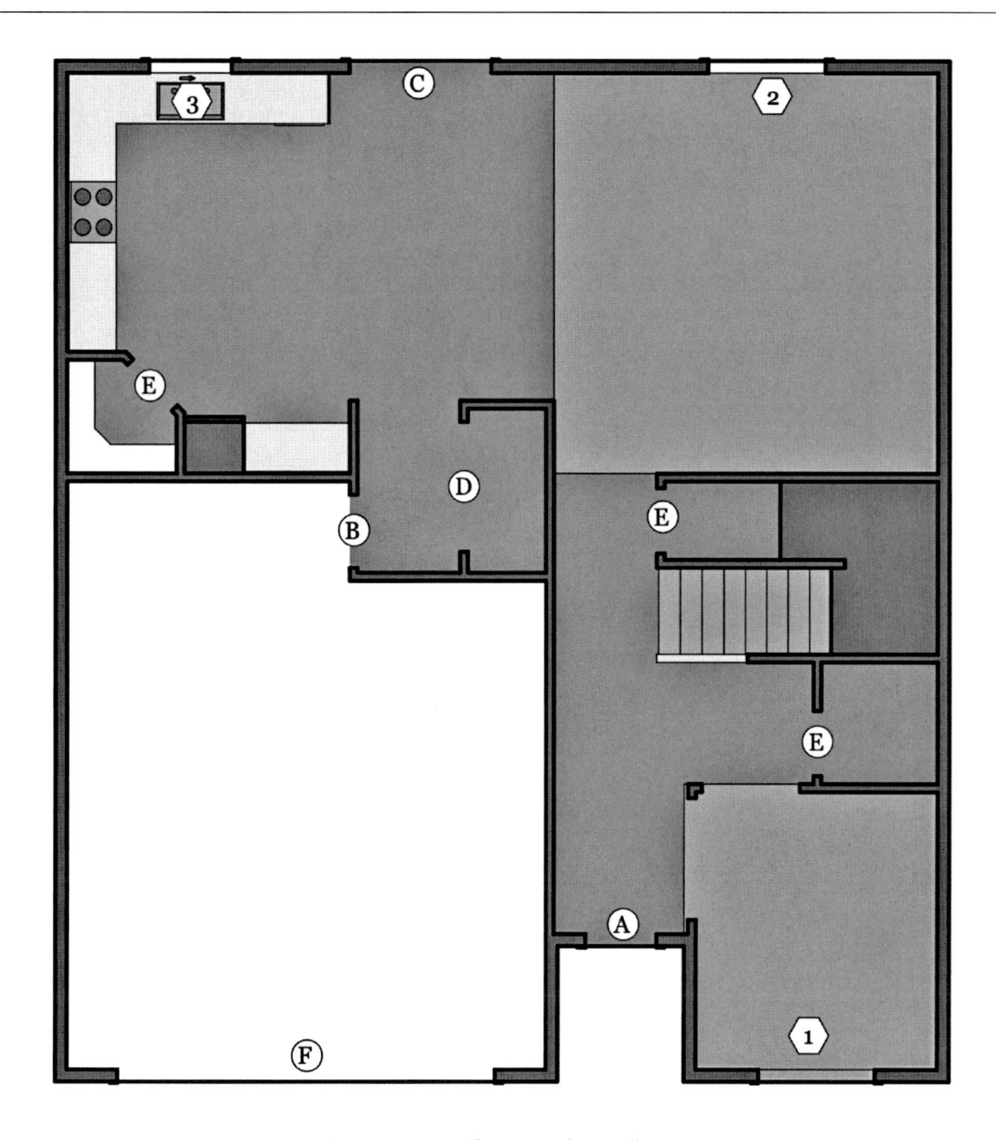

Figure 14.24 – Three window callouts

LayOut's Polygon Tool

Between using the **Circle** tool in LayOut and using the **Polygon** tool in SketchUp, you will probably figure out how to use this tool with little issue. I did want to point out that to increase or decrease the number of sides, you can tap the *Up* or *Down* arrow keys as you are drawing the shape. Also, holding down the *Shift* key will make it easier to keep the shape on the axis.

This page is looking pretty close to done, but we need to add a title to our drawing.

Adding a page title

Now, let's drop a title for this drawing onto the page along with a scale callout:

1. Expand the **Scrapbook** panel.
2. From the dropdown at the top, choose **TB-Traditional** and then **Drawing References.**

   What we want to do here is drag a copy of the drawing title at the bottom of the panel onto our page. The one we are looking for looks like this:

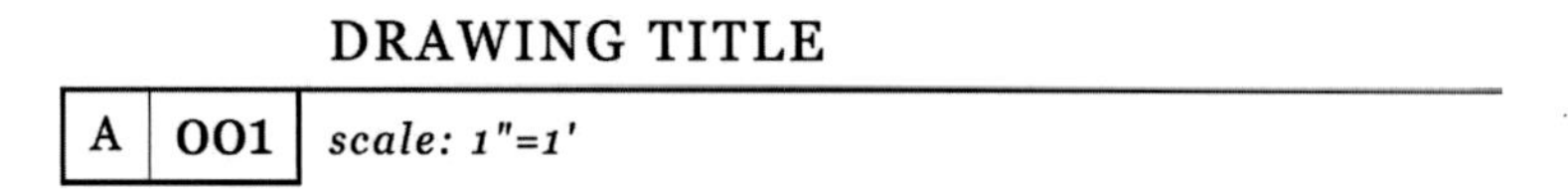

Figure 14.25 – The drawing title from the scrapbook panel

3. Click and drag the drawing title from the bottom of the **Scrapbook** panel to the page, right below our dimensioned drawing.

   When you drop an item from a scrapbook, you are simply placing LayOut items into your drawing. In this case, this is a group that contains lines, rectangles, and text. We can edit any of these items by double-clicking on them. Let's customize the information in the new group to show the appropriate drawing title and scale.

4. Use the **Select** tool to double-click on the group.
5. Double-click on the **DRAWING TITLE** text.
6. Replace the text with `First Floor Plan` and then click outside of the text box.
7. Double-click on the **scale: 1"=1'** text.
8. Change **1"** to `1/8"`.
9. Click outside of the text box twice.

   You should now have a drawing title that looks like this:

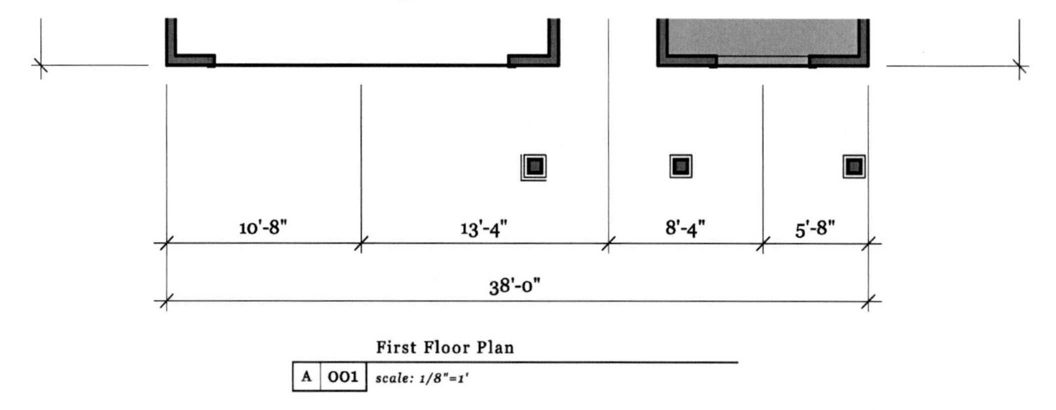

Figure 14.26 – Edited drawing title

Our plan is coming together at this point! Let's use this page to create a second floorplan.

Creating a second floorplan

While we could start a new page from scratch, as we did with the first floorplan, it is often easier to duplicate pages than make changes where appropriate to save time.

Before we get into duplicating the page we have been working on, let's make sure we give it an appropriate name. If you look in the **Pages** panel, you will see that we have two pages. The one we are currently working on (the one that is currently selected) is named **Inside Page**. This name came from the template and is different from the **Cover Page**. Let's change the name real quick:

1. In the **Pages** panel, click on the **Inside Page** text and replace it with `1st Floor Plan`. Then, press *Enter*.

   Next, let's make a duplicate of this page, rename it, and then start editing it to represent the second floor.

2. With **1st Floor Plan** still selected, click the **Duplicate** button (this is the button between the + and – buttons at the top of the **Pages** panel that shows the two pages):

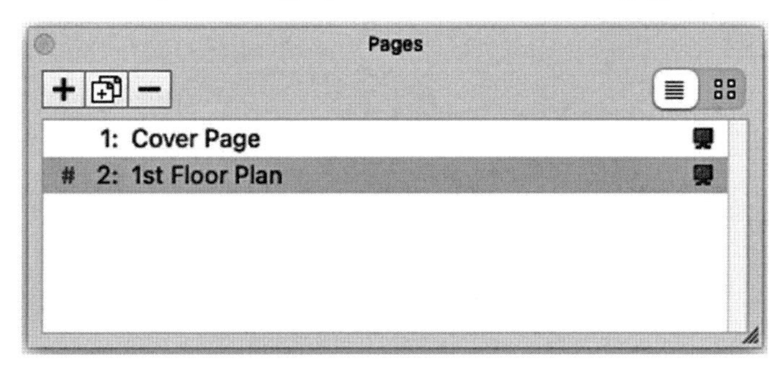

Figure 14.27 – The Pages panel

3. Click on the new page name (**Page 3**) and replace it with **2nd Floor Plan**. Then, press *Enter*.

   Now, when we look at **2nd Floor Plan**, it will look just like the **1st Floor Plan** page. Let's make some changes! First, let's change which scene is being shown in the viewport.

4. Use **Select** to highlight the SketchUp Model viewport.
5. In the **SketchUp Model** panel, change **1st Floorplan (traditional)** to **2nd Floorplan (traditional)**. Then, click the **Render** button.

   The model looks good – it is in the same location and the scale is the same, but the annotations are a bit of a mess. Since these were all placed based on the geometry of another floor, let's go ahead and wipe most of these annotations out so that we can start from scratch. I say most of the annotations, because it may make sense to keep one door callout and one window callout rather than recreate them on the new page.

6. In the **Layers** panel, click the little lock icon to the right of every layer except for **Annotations**.
7. Use the **Eraser** tool to remove the dimensions and door and window callouts (saving one of each callout if you like).
8. Unlock all of the layers and edit the drawing title so that it reads `Second Floor Plan`.
9. Finally, let's change the title bar on the side of the page so that it reads `Second Floor`.

   With that, you should have a new page that looks something like this:

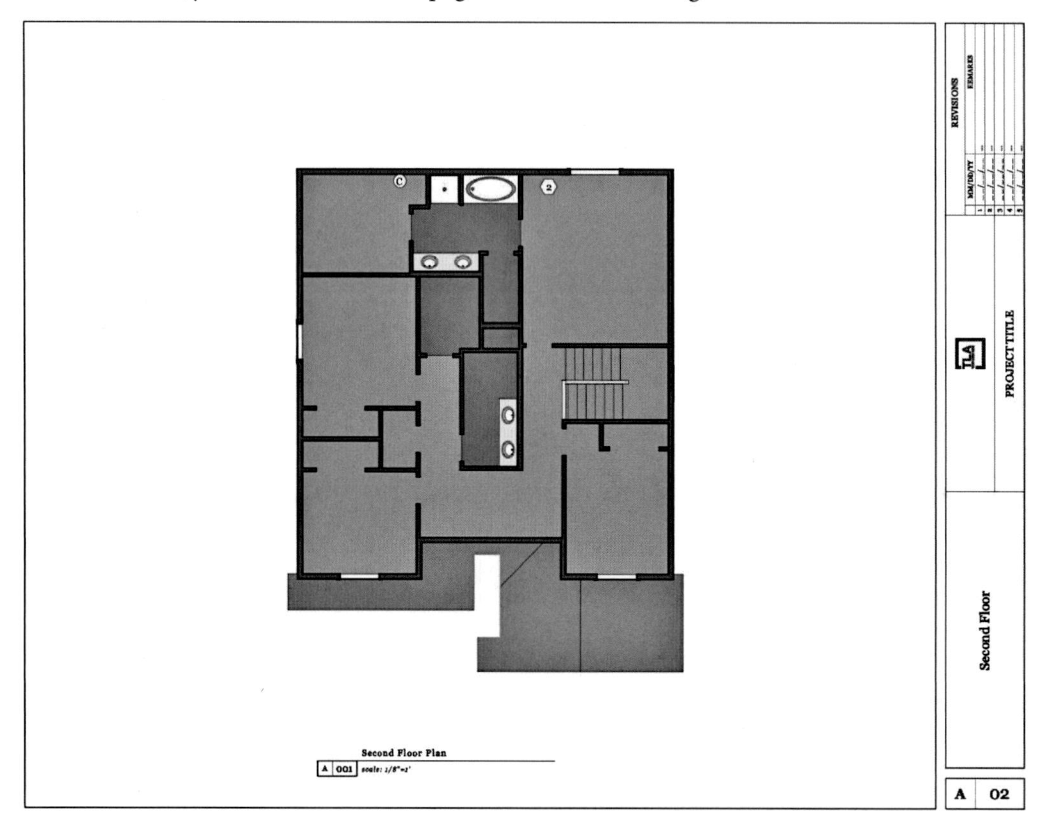

Figure 14.28 – Second floor plan page

At this point, you are ready to add new annotations and additional information specific to the second floor.

There are so many more things we could do to turn this example into a full set of plans. We could add tables referencing the door and window callouts. We could add notes from text files with information about the design or site. We could pull in an entourage from the scrapbooks. The point is that this chapter was just an introduction to what LayOut can do. If you want to keep using SketchUp, try adding more pages to the document (you can create elevation and section pages) or create a brand-new LayOut document using your own SketchUp file.

Now that we have a general idea of how we would go about creating a set of drawings from LayOut, let's spend just a little bit of time learning about where we can take these drawings.

Generating output

It sure feels good to get a nice drawing put together from a SketchUp model, but now, we need to get these pages out into the real world! There are three options when it comes to generating output from LayOut – printed drawings, digital drawings, and file exports. Let's explore why we may want any one of these types of output and touch briefly on how to generate them.

Generating printed drawings

For years and years, hard copy drawings have been a staple of almost every fabrication or building industry. In the world of building design and fabrication, these started as hand-drawn images that were eventually replaced by computer-generated 2D imagery. As much as we all talk about "going paperless," the fact is that many workflows end with someone somewhere printing out a stack of paper that represents the process that needs to be performed to transform a 3D model into reality. If you find yourself needing that stack of paper, LayOut has you covered.

Once you have finished creating your document in LayOut, printing is as simple as clicking **File**, and then **Print…**. Depending on your operating system and system setup, you will be prompted to choose the printer you want to send your drawings to. LayOut uses your computer setup to find a valid printer, so make sure that you are connected to a printer before trying to output your document.

If you are not connected to a printer or plan to use a third-party printing service, you may want to generate digital drawings.

Generating digital drawings

In addition to being perfect for sending to your neighborhood print shop to generate large format hard copies, digital drawings are great for sending to co-workers, clients, or other stakeholders of your project. Digital documents are also a nice way to generate output when you are going through design iterations and will be creating multiple versions of the output. Plus, since you do not need to wait for pages to be physically printed, digital output tends to be much faster to create.

Regardless of the reason for needing digital plans, you will likely want to generate a **PDF file**. A PDF file is pretty well accepted as the standard for digital drawings and can be opened and viewed on most devices. Additionally, once generated, there is a slew of ways to edit, annotate, or even lock your files before sharing them with others. How you generate a PDF will depend on your operating system.

Creating a PFD in Windows

To export a PDF file of your document, choose **Export** from the **File** menu, and then select **PDF...** From there, you can specify the file location from the **Export PDF** window and choose **Save**. There are no options in this dialog because the PDF is generated using the settings that were used to create your document. Things such as page size are taken from the template you used to start your LayOut document.

Creating a PDF file in macOS

There are two ways to generate a PDF in LayOut for macOS. The easiest is to use the **Print** dialog. When you choose the **Print...** command from the **File** menu, you can click the **PDF** dropdown at the bottom of the window. This will give you options for exporting a PDF file, and even creating and emailing the file directly from the **Print** dialog.

While it is great to be able to generate a PDF file that almost anyone can view, there may be cases in which you are working with someone who needs a file that they can use in their design software. This is where exporting a file comes into play.

Generating file exports

Exported files are different from the PDF file we just walked through. This process is all about generating a **DWG** or **DXF** file for use in another drawing or design software. Where PDF files can be viewed by a large group of people on most devices, DWG and DXF files can be imported and used as reference or linework in software that's specific to most design industries. Like exporting a PDF, the process of exporting a DWG/DXF file is different, depending on your operating system, and includes more options than a PDF export.

Exporting a DWG/DXF file in Windows

Exporting a DWG/DXF file from LayOut for Windows is as simple as choosing **Export** from the **File** menu, then clicking **DWG/DXF...**. In the dialog that appears, specify the location and name of the file you wish to export. As soon as you click the **Save** button, you will be prompted with the **DWG/DXF Export options** window:

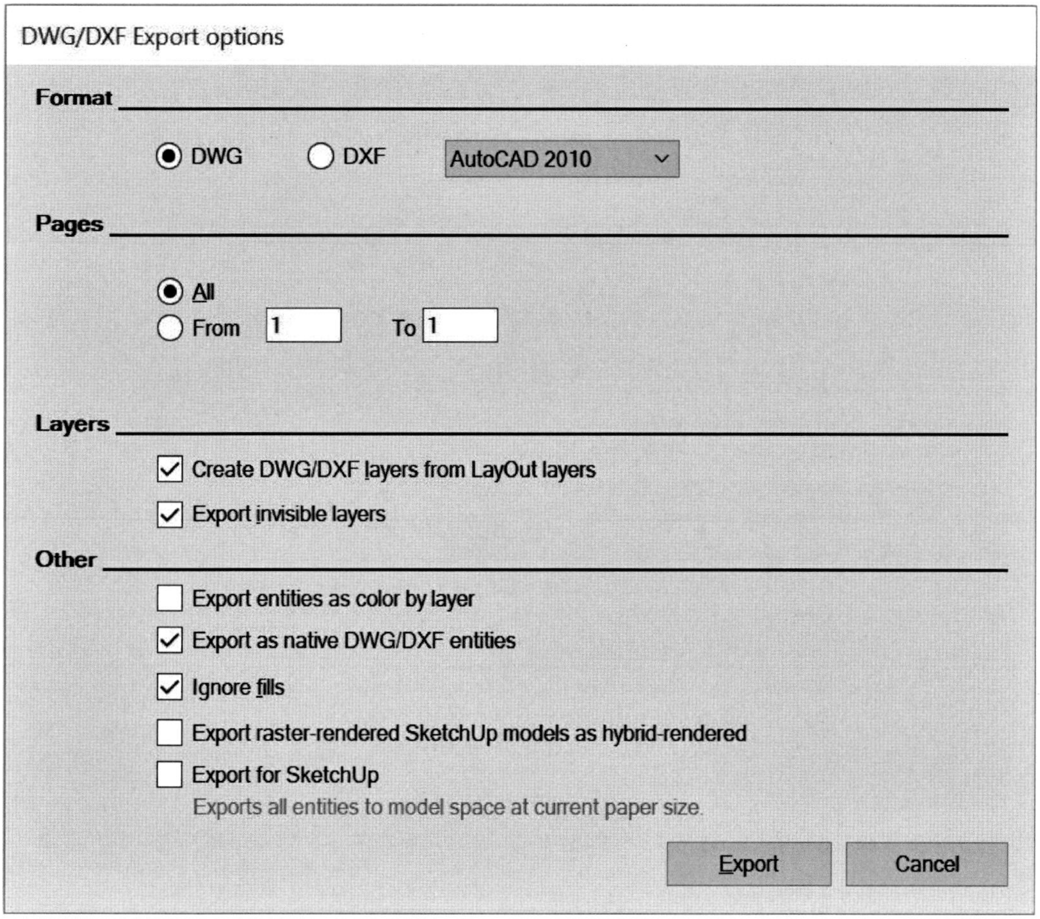

DWG/DXF Export options

Format

DWG DXF AutoCAD 2010

Pages

All
From 1 To 1

Layers

Create DWG/DXF layers from LayOut layers
Export invisible layers

Other

Export entities as color by layer
Export as native DWG/DXF entities
Ignore fills
Export raster-rendered SketchUp models as hybrid-rendered
Export for SketchUp
Exports all entities to model space at current paper size.

Export Cancel

Figure 14.29 – The DWG/DXF Export options window on Windows

Use this window to specify the exact file format and drawing options you need for the file you are exporting. These options may change based on how the file will be used once created.

Exporting a DWG/DXF in macOS

To export a DWG/DXF file while in LayOut for macOS, click the **Export** option in the **File** menu. In the dialog, specify the location and filename and choose **DWG/DXF** from the **Format** field at the bottom of the window. Clicking the **Save** button will immediately export the file. If you want to modify the settings that will be used to generate the file, click the **Options...** button to bring up the **DWG/DXF Export options** window:

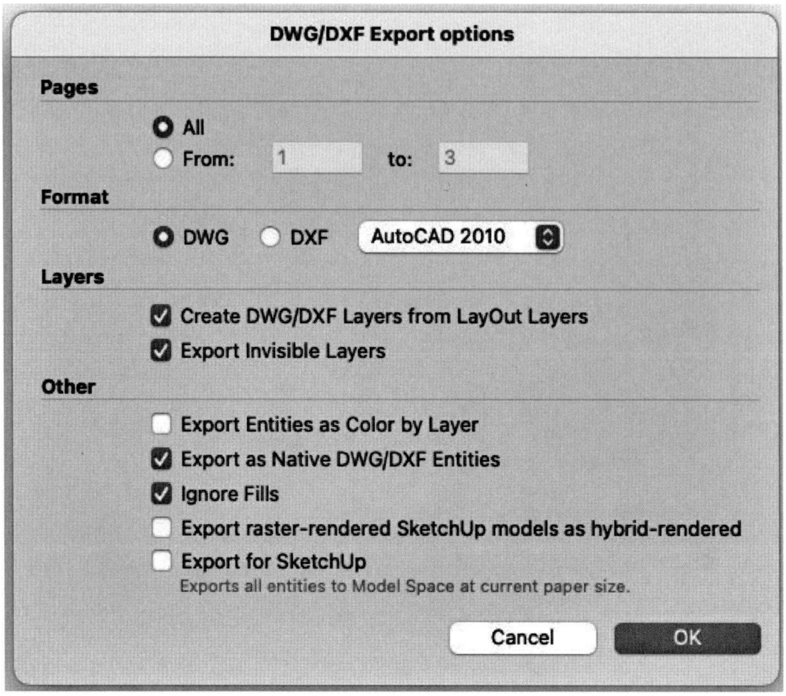

Figure 14.30 – The DWG/DXF Export options window on macOS

Use this window to specify the exact file format and drawing options you need for the file you are exporting. These options may change based on how the file will be used once created.

With that, we have covered the basics of exporting printed or digital drawings and files.

Summary

LayOut is a great resource and a wonderful tool for generating documentation from your SketchUp models. This chapter was a very high-level overview of its capabilities and intended to introduce you to what is possible. I highly recommend looking into additional training on LayOut if you plan to use LayOut in a production setting. It is a great tool but does require more instruction than I was able to present here, in this chapter.

In this chapter, you learned how you should think about setting up your SketchUp file for use in LayOut, how to import and add annotations to a SketchUp model, and finally what you need to do to generate output.

In the next and final chapter, *Chapter 15, Leveraging the SketchUp Ecosystem*, you will learn how to best leverage all of the tools that are included with your SketchUp subscription and the other resources that are available to you right now.

15
Leveraging the SketchUp Ecosystem

As the title of this book is *Taking SketchUp Pro to the Next Level*, we have, understandably, spent a lot of time talking about how to best use SketchUp Pro. Yeah, we dabbled a little bit in LayOut, but most of this book has been about leveling up your modeling ability using the desktop version of SketchUp.

In this final chapter, we will learn how to take full advantage of everything available to you, as a SketchUp user. This chapter will assume that, since you have access to SketchUp Pro, you have a SketchUp Pro subscription. We will spend much of this chapter looking at exactly what comes with that subscription, as well as some great tools that are available to continue your quest to reach the next level (and the level after that).

The goal of this chapter is to expose you to some of the software and tools that are already available to you as a SketchUp Pro subscriber. The intention here is not to try to sell you a subscription (the assumption is that you already have everything listed) and we won't be doing full step-by-step walkthroughs. By the end of this chapter, I am hoping that you realize how many tools you already have in your toolbox in the hope you will start thinking about how you may use them to take your workflows to an even higher level.

In this chapter, we will cover the following topics:

- Using SketchUp viewers
- Trying Trimble Connect
- Exploring SketchUp online
- Using SketchUp on an iPad
- Finding additional resources

Technical requirements

You do not need anything to complete this chapter. If you want to check out any of the software included with the SketchUp Pro subscription, you will need an active SketchUp Pro subscription and, for some features, internet access.

> **A Note about SketchUp Subscriptions**
>
> At the time of writing this book, there are three levels of SketchUp subscription: SketchUp Go, SketchUp Pro, and SketchUp Studio. SketchUp Go does not include the desktop software and is sort of irrelevant to almost everything we have covered in this book. SketchUp Pro is the subscription most often used by SketchUp desktop users and is the target of this chapter. SketchUp Studio includes everything in SketchUp Pro, plus some additional software that we will not specifically cover in this chapter. While the subscription offerings may change after the printing of this book, I will assume that something similar to the SketchUp Pro subscription will be available, as well as what is being used as you work through this book.

Using SketchUp viewers

SketchUp is a great way to explore ideas and bring them to life in the form of 3D models. It is a great way for you, the originator of the idea, to realize your vision, and a great way for you to share it with others. This is great if the person you want to share it with is already a SketchUp user! You can simply email them an SKP file and they will see what you are seeing. However, sharing your vision with someone who is not a SketchUp user can be a little more difficult. This is where the SketchUp Viewer family comes into play! With a SketchUp Pro subscription, you have access to three different SketchUp viewers!

SketchUp Viewer for Windows or macOS

A quick and easy way to view a SketchUp model is by using the SketchUp Viewer on a PC or Mac computer. SketchUp Viewer for Windows or macOS can be downloaded and run by anyone, even those without a SketchUp subscription. This means that if you need a client or collaborator to see your model, you can send them a download link and a copy of your model. They will install the app on their computer, which means they can open your SketchUp file.

This version of Viewer acts like a stripped-down version of SketchUp. In this version of SketchUp Viewer, you can move through the model with the same tools that are available in SketchUp, but you cannot make any changes or create any new models. As shown in the following screenshot, SketchUp Viewer looks a lot like SketchUp Pro, but with fewer options:

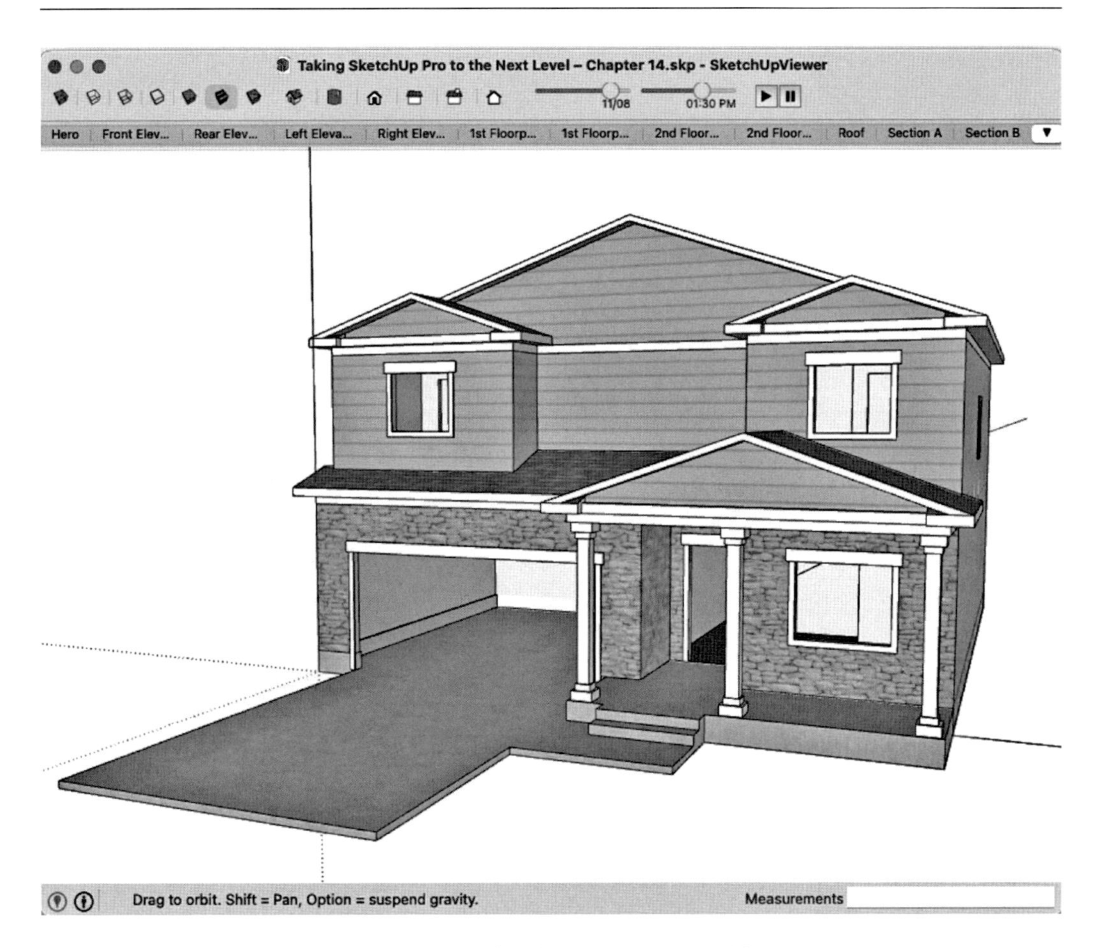

Figure 15.1 – SketchUp Viewer on macOS

As you can see, it looks a lot like SketchUp and has controls at the top to change the face style, jump to standard views, or animate the scenes. Since the scenes are listed, a user can simply click through the scenes to see them. This is a great way to help a non-SketchUp user navigate the model. If you were to click into the model, you would be able to orbit, zoom, and pan using your three-button mouse, just like in SketchUp Pro, but users of Viewer will not always understand that. If you, the SketchUp Pro pro, present them with a SketchUp model with scenes created and tell them, "*Just click through the scenes at the top of the window,*" you can help guide their experience and ensure that they are seeing the model as you intended.

Since anyone can download and run this viewer, this is a great way to allow someone to see your model. It is easier than having a client subscribe to SketchUp Pro and try to learn how to navigate in a 3D model. Of course, this means they can still orbit and zoom and get themselves "stuck" inside the model or orbit away from what they should be seeing. Fortunately, clicking back to an existing scene will make it easy to get back to where they need to be.

SketchUp Viewer for iOS and Android

Even easier than looking through a 3D model on your computer is looking through a 3D model on your phone or tablet. I know – there are those out there who will try to argue with this, reminding me that phone screens are much smaller than any computer monitor or laptop display. I agree that using SketchUp Viewer on your computer will give you a larger view of your model. However, as we touched on previously, learning to move through a model using the SketchUp Viewer tools can take a little bit of time. Looking at a 3M model on a tablet or phone is far more intuitive and is a great way for someone with no SketchUp experience to look through your creation.

When viewing a model in the SketchUp Viewer app on iOS or Android, users can intuitively move their model around with their finger and zoom in and out using pinch gestures that they are used to using on their device. Additionally, the app offers them basic control over the model, such as access to **Scenes**, **Tags**, **Styles**, and **Shadows**. Due to the limited screen space on your phone, the user interface is severely slimmed down, as shown here:

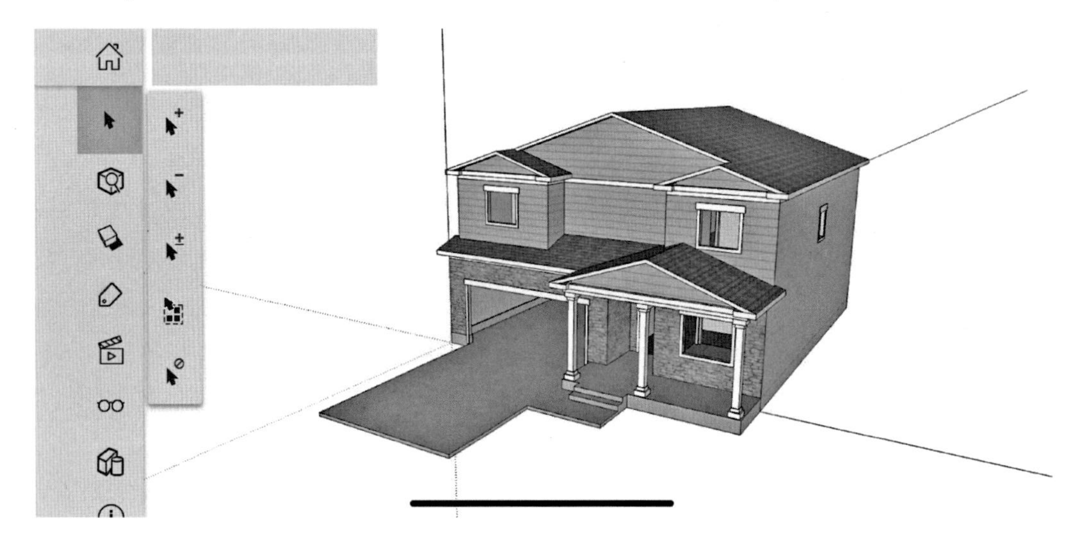

Figure 15.2 – SketchUp Viewer as seen on an iPhone

Additionally, unlike SketchUp Viewer for Windows or macOS, the mobile version allows users to select entities and view basic information about them. While you cannot do any editing, this makes it easy to pull up a wall, for example, and see its total square footage. If additional information has been added to groups or components, such as a part number of a window, **Entity Info** can be displayed when the item is selected.

With this additional level of information available, SketchUp Viewer for mobile devices is not only an easy way for your customer, co-worker, or stakeholder to view and experience your work, but it is also a good way for you to have information about your project with you when you are away from SketchUp Pro.

SketchUp Viewer for VR/AR headsets

The final viewer is SketchUp Viewer for **Augmented Reality** (**AR**). This viewer is currently available for Oculus Rift, Oculus Quest, HTC Vive, SteamVR, and Hololens. If you have any of these devices and a SketchUp Pro subscription, you can download the app and experience your SketchUp models in scaled or full-sized virtual or augmented reality:

Figure 15.3 – A SketchUp model as viewed in augmented reality

As we have discussed, viewing a 3D model means understanding the interface and tools available to move through 3D space. The great thing about viewing a model in a headset is that there is very little to learn. If you want to walk around a scaled model to see its side, you simply walk around the model and look at its side. If you want to look up at the ceiling of a full-size model, you just look up.

Not only is this an easy way for non-SketchUp users to experience your 3D models with no training, but it is also a great way for non-designers to understand things, such as how large a space is or how an item will look if it were to exist in reality. As a designer, you can look at what you have created in SketchUp and understand how it will translate to the real world. Your client/spouse/stakeholder may not have that ability. Giving them access to your design in a headset that allows them to experience it in full size, immersive 3D will make it easier for them to catch your vision.

Viewing your SketchUp models is a great way to share your work with non-designers, but what about when you want to work collaboratively with others? That is where Trimble Connect comes in!

Trying Trimble Connect

There are more than a few reasons that you may want to share your SketchUp model. You may be working collaboratively and need to pass it off to a co-worker. You may be the one generating the initial geometry, but at some point, need to pass the model to a subcontractor for material information or pricing. You may even need to share a model with yourself! In some cases, you may be moving between computers or devices and need a way to make sure that you are always working with the most recent version of your design. **Trimble Connect** can help you with all these scenarios and more.

Trimble Connect is an online tool that allows you to store and collaborate on designs with coworkers, subcontractors, or even clients. While Trimble Connect is a very robust tool with many capabilities, we will be looking at a few pieces that make it relative to us, as SketchUp designers, on our path to leveling up our skills. We will look at using Trimble Connect to share files and collaborate on designs, as well as how to leverage Trimble Connect's ability to read multiple file formats to import data from software that is not supported through SketchUp Pro's native import options.

While I hate not being able to go into everything Trimble Connect has to offer, let's spend a few minutes looking at these major features.

File sharing

Working with others is not a rarity nowadays. It is pretty common to have multiple people involved in your design process and to make that work, you need an easy way to pass files back and forth.

For years, designers working in groups have come up with solutions to pass files back and forth (with mixed success). I regularly hear about companies storing their SketchUp files on Google Drive or Microsoft Cloud. Going back to the ancient times before shared web-based storage solutions, companies would stand up dedicated servers to share their working files! While keeping files on a website such as Drive can work very well, there are some downsides.

First off, when storing on a shared site, you will have to manually track revisions and updates by creating folders or renaming files before they are updated. With Trimble Connect, revisions to files are tracked as you upload new versions. This means you can jump back to an older version of a SketchUp file by simply selecting it from a list of revisions. Additionally, Trimble Connect will not only keep track of versions of a SketchUp file as it is uploaded but also note who uploaded each revision!

Another reason to use Trimble Connect to store your files is the native integration into SketchUp itself. Regardless of the version of SketchUp that you use or what device you are using (more on SketchUp on an iPad and SketchUp on the web will be covered later in this chapter), the option to save or open a model directly from inside any version of SketchUp is a huge time saver.

One of the issues that will cause a problem in SketchUp for you is opening a file from anything other than your local storage. If you do try to open a file from a USB drive, Google Drive, a server, or any other cloud-based storage, you may run into corruption issues upon opening or saving the file. The ideal way to open a file from a remote drive is to download it locally, and then open it. When it is time to save, save it locally, then exit SketchUp and upload the finished file back to the cloud/external storage. With Trimble Connect, you cut this process in half. From inside SketchUp, you can download your file and save it locally and send it right back up to Trimble Connect when you finish without ever leaving SketchUp.

Finally, this is a great solution, even if you are not sharing files with anyone else. Even if you are doing all the work yourself, you can use Trimble Connect just to pass versions of design files from one device to another. How cool would it be to create a custom house design in SketchUp on your desktop PC, then walk out the door and show it to someone on your iPhone in SketchUp Viewer? The answer: very cool!

Collaborative design

If you are designing with a group, however small or large, Trimble Connect allows you to do much more than just share files. When you use Trimble Connect, you can create groups of people to include in your project. These could be designers who will be working in SketchUp on the actual models, subcontractors who will be responsible for supplying information about labor or materials, or even clients who will want to look at the 3D model as it is being developed. Having multiple members in a project is a great way to keep everyone involved in the process up to date on what is happening, and make sure that they are working with the most up-to-date version of the design information possible.

Design teams can be created and saved so that you can share projects with the same people on projects. Each member of a team can have a specific role and right within the project. This means not every member of a team can see all the information about a project, only the information that is pertinent to the role they play.

Plus, when you set up a project, you can assign tasks to the members of the project in the form of ToDos. With a ToDo, you can let a collaborator know that you or someone else on the project is expecting work from them. If you set up notifications, these ToDos will be emailed to them as soon as they are created. This is a great way to make sure that everyone knows what is expected of them and when they need to finish their tasks.

File conversion

Finally, Trimble Connect is a great solution for working with different file types. You have probably seen by now that, with SketchUp Pro, you can import and export several file types. While it is great to be able to work with these file types, there are plenty more out there that cannot be directly imported with SketchUp Pro. With Trimble Connect, you can import even more file types.

While different platforms support different files (Trimble Connect can be run from a web browser, on your mobile device with an app, or through a Windows desktop version), the number of file formats supported is impressive:

2D files	3D Files	Point Cloud Files	Geospatial Files
BMP	DGN (Version 8)	E57	GDB1
DOC	DWG 2018 (AutoCAD 2022 and below)	LAS	Geospatial .zip files
DOCX	DXF	LAZ	JXL
DWG	IFC (2x3, 4)	Potree	KML
GIF	IFC ZIP	URL	KMZ
JPEG	IGS, IGES	TDX	SHP2
JPG	LandXML (Version 1.2)	TZF	VCA
PDF	Navisworks (.NWD & .NWC)	XYZ	VCE
PNG	Revit		
PPT	SKP (2021 and below)		
PPTX	STEP (AP203 and AP214)		
RTF	TC ZIP		
TIF	TEKLA		
TIFF	TRB		
TXT			
XLS			
XLSX			

Figure 15.4 – Supported Trimble Connect file types

This is a great tool to have if you need to import and view files that you do not have software for! Once the files have been uploaded, Trimble Connect gives you the option to view the files and, in some cases, download alternative versions.

The simple fact is that most building or design processes do not happen in a single piece of software. With Trimble Connect, you are more likely to have a solution for working with other files without having to multiply the number of applications you must have up and running!

Trimble Connect is a great way to keep connected and leverage cloud-based applications as part of your design workflow. Another great piece of web-based software is SketchUp for the web!

Exploring SketchUp online

While it may not happen very often, there may be a point in the future where you need to get into SketchUp to model something or make a change to a model while you are away from your computer. Maybe you are working from a customer's office or using a loner laptop because yours is in the shop. Whatever the reason, there may be a point where you need to get some modeling done and do not have access to your copy of SketchUp Pro.

In some cases, such as this, SketchUp Viewer may be enough for you. If all you need to do is look at or show a model, you may have everything you need right on your phone. If you need to get in and modify or start up a new model, however, you may want to take advantage of SketchUp for the web.

There are multiple versions of SketchUp available online. If you have a SketchUp Pro subscription, then you have access to all the functionality of SketchUp for the web that's available to you right now. All you need do to access it is to go to `https://app.sketchup.com` and log in with your Trimble ID and password.

Once logged in, you can open any SketchUp files you have saved on Trimble Connect or start a new model from scratch. Even though SketchUp is running in a browser window, you still have the option to open a model that is saved on your computer!

Although SketchUp for the web does have a slightly different look and feel from SketchUp Pro on your computer, it is fairly easy to adapt and get modeling done when you need to.

This is what SketchUp for the web looks like:

Figure 15.5 – SketchUp for the web

The key to learning SketchUp for the web (which we will only get to review at a high level in this section) is to learn about the main differences between it and SketchUp Pro.

SketchUp for the web toolbar

Unlike the toolbars in SketchUp Pro, the toolbar in SketchUp for the web is not customizable. You cannot move it, you cannot change what commands it presents, and you cannot disable it. Since the user interface must all reside inside the browser window, the toolbar was designed to always bind to the left-hand side of the window. However, almost every native command that is in SketchUp Pro is available through the SketchUp for the web toolbar.

This seems a little crazy, right? There are far more commands in SketchUp Pro than are displayed on the toolbar in the preceding screenshot! How is that possible? The secret is in grouping commands. In the toolbar of SketchUp for the web, many of the icons have a small arrow to the right of them. Clicking on one of these icons will fly out a series of related icons that can then be clicked so that you can start using that tool. In this example, I have moved the cursor over the **Rectangle** icon and clicked to bring up icons for **Rectangle**, **Rotated Rectangle**, **Circle**, **Polygon**, and **3D Text**:

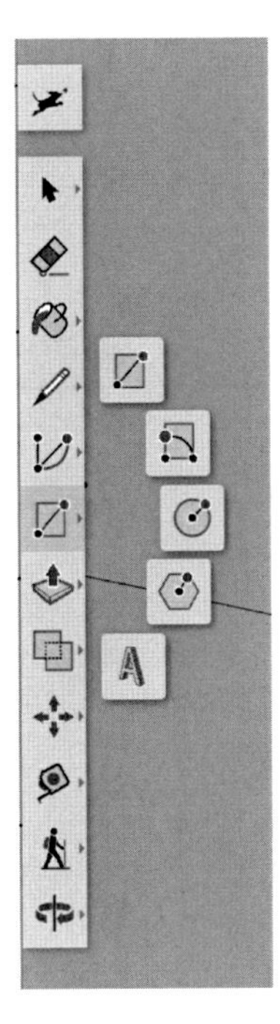

Figure 15.6 – Example of grouped tools

While this does take a little bit of getting used to (especially if you are someone who uses a lot of toolbars), it can become a habit fairly quickly. All you have to do is practice!

SketchUp for the web menu

One of the biggest difficulties I had when teaching myself SketchUp for the web was getting used to the lack of menus at the top of the screen. In SketchUp Pro, if I ever forgot a shortcut or could not find a floating toolbar, I knew that I could always head up to the menus and track down the tools I was looking for. Not so in SketchUp for the web.

Like the toolbar, the need to keep things compact and anchored meant reimagining what was needed in the menu. Any command that has an icon on the toolbar does not need to take up space in the menu. The only things in the menu are the commands that did not fit anywhere else. In addition to **Undo** and **Redo**, the menu in the top left includes the filename, save status (as a cloud-based app, SketchUp for the web is constantly saving your model, so you don't have to worry about clicking a **Save** button), and a small drop-down menu with the standard **File** menu commands, as well as **Add Location**.

Again, this is a bit of a change from what you may be used to in SketchUp Pro, but (unless you are like me and run to the menu anytime you cannot find the command you are looking for) this is a pretty easy change to adapt to.

SketchUp for the web panels

The vertical stack of icons on the right-hand side of the screen house all of the panels that would show up in the tab bars (on Windows) or floating panels (on macOS) in SketchUp Pro. Clicking any one of these icons will fly out a stack of panels. Each panel can be toggled on and off by clicking its associated icon.

While all the functionality you expect from panels is there (things such as **Entity Info**, **Tags**, **Styles**, and **Scenes**), there are also a few familiar SketchUp Pro commands that have moved into the panels, such as **3D Warehouse** and **Model Info**. There are also a few new icons, such as the ones for **Display** and **Solid Inspector**.

While the UI for these tools is easy to get the hang of, you may find yourself occasionally hunting for functionality in these panels. However, since there are only a dozen options, it is not too difficult to remember what options are there and the tools that each contains.

No extensions

OK, this is a big one. We spent a lot of time in this book talking about extensions and the role that they play in taking your SketchUp skills to the next level. Since there is no API or process to download and install third-party software into SketchUp for the web (since it is software running off a server and not your computer), there are no extensions. While this should not be discouraging to you (we have spent a lot of time talking about how to get the most out of the native commands), it may be harder on some workflows than on others.

To get over this potential hurdle you just have to... get over it! There is no softening the blow here. If you model in SketchUp for the web, you are only going to be modeling with native tools.

How is SketchUp for the web similar to SketchUp Pro?

That was a lot of explaining what makes SketchUp for the web different from SketchUp Pro, and it may have felt like everything is different! Fortunately, despite the big change to the user interface, there is a whole lot about SketchUp for the web that is just like what you are used to in SketchUp Pro.

First off, you are working in the same files that you would if you were modeling on your desktop. Even though you may not choose to download the file you are creating in SketchUp for the web, be assured that the file that is being created and auto-saved to your Trimble Connect account is a `.SKP` file, just like the one you would be saving in SketchUp Pro.

Second, the process of modeling is the same in SketchUp for the web as it is in SketchUp Pro. Your workflows (except for those that depend on extensions) will remain the same as they ever were. SketchUp for the web uses the same tools to create the same geometry as SketchUp Pro.

Navigation is another thing that works the same on either version of SketchUp. This is great because you have probably spent quite a bit of time getting used to how to use your three-button mouse to zoom, pan, and orbit! If you can connect that mouse to the computer you are using, you can navigate in SketchUp for the web just like you are used to.

Finally, default shortcuts are the same. OK, this one is a bit of a double-edged sword. Yes, the default commands come from SketchUp Pro and work for many of the tools as they do by default. If you have left these default shortcuts as they were in SketchUp Pro, then you will be able to use them in SketchUp for the web! However, if you have remapped them and rely on a completely custom version of shortcut keys to model, then you may feel the sting of these hardcoded shortcuts when you model from the web-based version.

Should you use SketchUp for the web?

In summary, SketchUp for the web is a version of SketchUp that is available to you as a part of your SketchUp Pro subscription. If you are in a situation where you do not have access to SketchUp Pro, it is a great tool to get you by. There are limitations to the size of a file that will run smoothly in SketchUp for the web and you cannot take advantage of peripherals such as a 3D mouse when you use it, but if you need SketchUp, it is a great solution to have.

One of the best things about SketchUp for the web is that you can use it on any computer, which makes it a great option if you are away from your home computer. However, if you want the full power of SketchUp with you at all times, you may want to try out SketchUp for iPad.

Using SketchUp on an iPad

If you are an iPad user, SketchUp for iPad is something you may not know that you need. SketchUp for iPad reimagines the functionality of SketchUp Pro while taking advantage of the input methods only available with the iPad. This means you can take your modeling projects with you, wherever you want, and you can work on them in SketchUp whenever you want. Boasting a custom user interface designed specifically for iPad, this version of SketchUp looks different, but feels familiar at the same time:

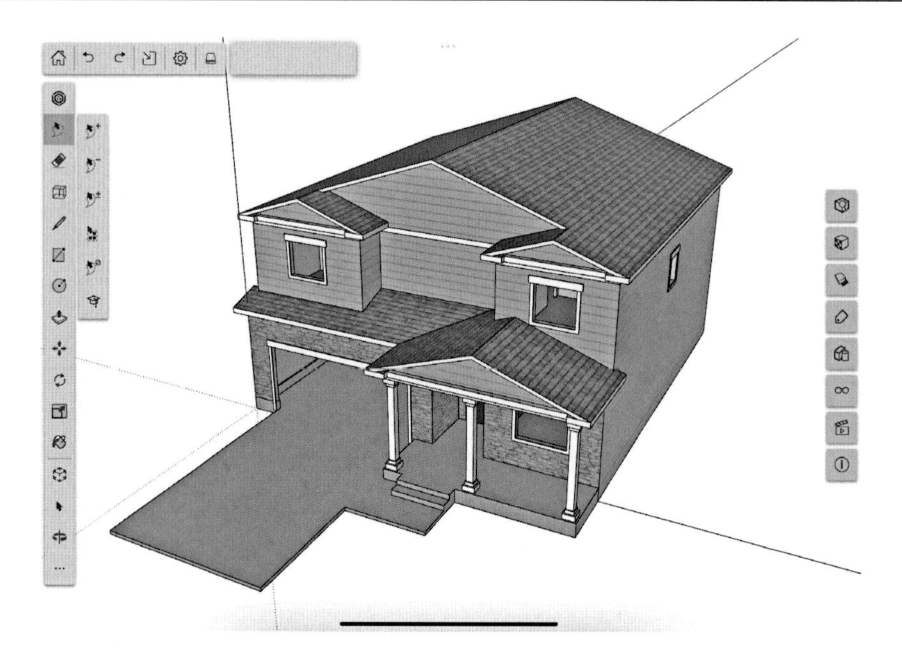

Figure 15.7 – SketchUp for iPad

I know, I already hear the pushback, "*I can already take SketchUp with me because I model on a laptop!*" Yes, having SketchUp Pro installed on a laptop is great (it is the main way I use SketchUp daily), but modeling on an iPad is different. We will get into the specific differences between SketchUp Pro and SketchUp for iPad in this section, but before that, I want to mention the advantages of working on an iPad versus a laptop.

First off, an iPad is smaller and lighter. There are some small laptops out there, sure, but few that I have held in my hand are as easy to carry around as my iPad. It is also smaller than my laptop. I am a larger human (over 6 feet tall) and I travel fairly regularly. I have spent more than a few hours sitting on trains, buses, and airplanes with my laptop open, modeling with the trackpad. Not to brag, but I have gotten to be pretty good at modeling with this setup, but it has never felt good or been nearly as effective as sitting at a desk with my three-button mouse. With an iPad, however, it is so much easier to model with the much smaller device and Apple Pencil. I can rotate the screen to whatever angle works best; I can cradle it in my arm or stand it up on my lap. It just makes for a far more pleasurable experience using SketchUp on the go.

I Am Not Here to Sell You an iPad

Before I go any further into this section, I want to point out that I honestly do not care whether you have an iPad or not. I do not want you to buy an iPad just because I enjoy using one. All I am going to show you is what using SketchUp for iPad is like. If you own an iPad and a SketchUp subscription, then you already have access to the software! If you have a SketchUp subscription and have been considering getting an iPad, then this may just be the app that pushed you over the edge!

Just as with SketchUp for the web, there are a few things that are different about using SketchUp for iPad as opposed to SketchUp Pro.

SketchUp for iPad toolbar

Just as with SketchUp for the web, SketchUp for iPad cannot rely on standard UI options that are available in SketchUp Pro. This means the toolbar must function differently. In addition to having slightly larger icons (which are easier to tap on), the toolbar is always on the left-hand side of your modeling screen. Rather than using grouped tools like SketchUp for the web does, SketchUp for iPad limits the number of visible icons. You have the option to lock your favorite commands in at the top of the toolbar, but can bring up the less commonly used commands from a complete list of commands at any point, as shown here:

Figure 15.8 – A complete list of tools on SketchUp for iPad

Since you have the option to customize your toolbar by reordering and choosing which tools are permanent, this ends up being a great way to minimize the UI, since screen space is limited.

Additionally, options for the selected tools (things such as modifiers or inference locks) are shown to the side of the selected tool in a smaller fly-out toolbar:

Figure 15.9 – The Line tool's flyout toolbar

This brings up another slight change when using SketchUp for iPad. While SketchUp for iPad can be used with a keyboard, it is designed to be fully functional, even if you are working without one. This means doing certain things in a slightly different order. For example, if you are drawing a line in SketchUp Pro and want to have it snap to the red axis, you would just tap the *Right Arrow* key after selecting a start point. In SketchUp for iPad, options such as modifiers or inferencing snaps must be activated before you can start using the tool, in many cases. This means clicking the **Line** command, then clicking the **Lock X** icon, and then picking the start point of the line.

The good news about this different toolbar is that you should recognize almost every single icon on the toolbar since they are based on the icons in SketchUp Pro!

SketchUp for iPad menu

Just like in SketchUp for the web, there is not a proper set of menus in SketchUp for iPad. The menu bar at the top right of the screen is very similar to SketchUp for the web. **Undo**, **Redo**, **Download**, **Options**, and **Save** have icons in the menu, and you can also go to the home screen (where you can open other models or start something new). Since all the tools and information are available through the toolbar or panels, there is no need for additional menus.

SketchUp for iPad panels

Also, very similar to SketchUp for the web, the panels on the right-hand side contain most of the information and control for working with your model. Just like SketchUp for the web, you can choose to expand or collapse them as needed.

SketchUp for iPad input methods

The input options available on SketchUp for iPad are one of the things that set this version of SketchUp apart from all the rest. Unlike SketchUp Pro or SketchUp for the web, where you can provide input with a mouse or trackpad, SketchUp for iPad will allow you to input using an Apple Pencil, your finger, or a Bluetooth mouse/trackpad.

While it is great to have options, there is usually a best way to use a tool. In my own humble opinion, if you are going to model in SketchUp for iPad, you will want to use an Apple Pencil. The Apple Pencil makes it easy to perform precision selections and makes it easy to snap to points and use inferencing. Plus, there are two different modes for using the Apple Pencil (Click-Move-Click or Just Draw) that give you options for modeling the way that feels most comfortable to you.

If you are ever caught without your Apple Pencil, you can use every tool with your finger (or Touch mode, as the iPad crowd likes to say). While touch will get the job done, the simple fact that your fingertip will cover up more of the screen than the tip of your Apple Pencil does make this a less than ideal input method, and more of an "in a pinch" sort of solution.

Finally, you can input using a Bluetooth mouse or trackpad. When you do this, SketchUp for iPad functions very much like SketchUp Pro. While this works perfectly well, it feels a little weird to be hunching over the smaller screen of my iPad to do input and editing with a mouse, when I could be using the same device and taking advantage of the screen and power of my laptop. Using a mouse is nice, but again, it does not take advantage of the form factor of the iPad as an Apple Pencil does.

Features only available on SketchUp for iPad

While the bulk of the tools available in SketchUp for iPad is the same across the other versions of SketchUp, there are a few that can only be used when you are modeling on iPad. Let's take a look.

Auto Shape

SketchUp for iPad has a special mode that allows you to draw gestures that are instantly transformed into SketchUp geometry. Simple shapes such as squares, circles, or solids such as boxes or spheres can be created with a quick gesture of your Apple Pencil. More complex geometry such as doors or windows can even be drawn in with the correct gesture.

Markup Mode

With SketchUp for iPad, you can draw an overlay onto any model. Using the Apple Pencil, you can sketch lines or write over the top of a model and have it saved as a scene in your model (which means you can open it and see your notes in any version of SketchUp)!

AR viewing

When you have a model open in SketchUp for iPad, you can tap the **View in AR** icon to turn on the camera of your iPad and experience your model in scaled or full-size augmented reality.

Is there a downside to SketchUp for iPad?

Ok, it may be a little extreme to say that there is a downside, but in the interest of full honesty, there are a few factors that send me back to SketchUp Pro when I need to do any serious modeling. For one thing, the hardware in my iPad is less powerful than what is in my laptop. I love my iPad, but when it comes time to work on a huge model, the RAM and processor on my laptop make working with large SketchUp models a much easier process. Additionally, working on a larger screen with my three-button mouse is more comfortable for me than holding my iPad and using the Apple Pencil when I have hours of modeling work to do.

Hopefully, at this point, you understand how much more there is to your SketchUp Pro subscription than just SketchUp and LayOut. While reviewing these additional applications is one of the big parts of what we needed to review in this chapter, there are a few resources out there that are not directly tied to your subscription.

Finding additional resources

At this point in this book (this point being the end), you have refreshed yourself on the basics of SketchUp, have come up with a game plan to take your skills and workflows to the next level, and have reviewed all of the tools that you have available to you as a SketchUp Pro subscriber. Now, what's next?

In this final section of this final chapter, I would like to attempt to equip you with as many tools as possible so that you can continue your journey of growth as a SketchUp modeler. Everything covered in this chapter is available free of charge and available to you right now. In some cases, a service or resource may be available at an additional charge. In this case, I will mention that there is a charge, but I will not dive into the specifics of pricing (primarily in the interest of making this list as timeless as possible).

Forums

I am going to start with one of my favorite learning resources available: online forums. I know that forums do not offer the structured learning that a lot of people are looking for (if that's you, feel free to skip to the *Online training* section), but for me, they offer a wealth of knowledge brought by the diverse members and the opportunity to join a community. I have said this previously but it bears repeating: not all software forums are the same. There are many forums out there for other software packages and applications that are lackluster, at best. SketchUp users are lucky in that they have access to not one but two vibrant user forums where they can get questions answered and learn about all the new things that they are not aware of!

SketchUp Community (the official SketchUp forum)

This is the forum (`https://forums.sketchup.com`) that is hosted by Trimble and a place where many of the SketchUp team staff members hang out to answer questions and hear from the SketchUp user community:

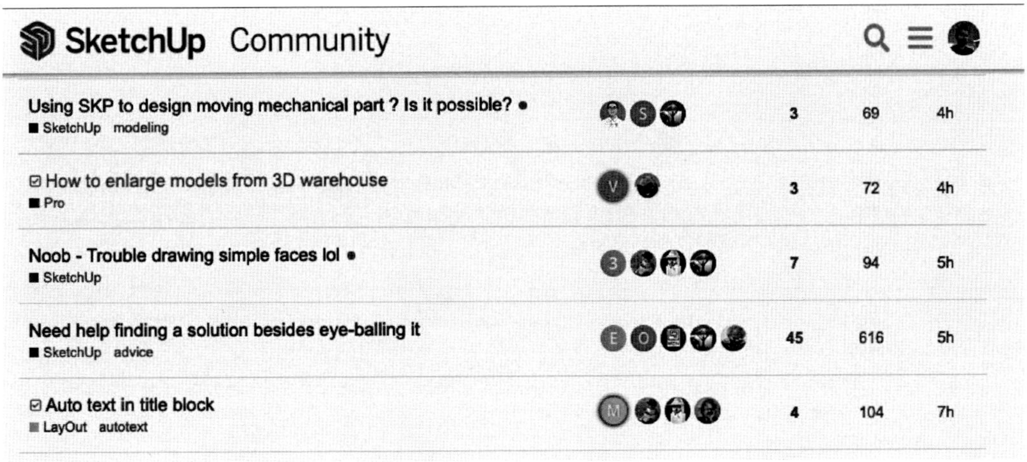

Figure 15.10 – SketchUp Community

The forum is moderated by a combination of Trimble staff and community members (experts known as SketchUp Sages). While you do need a free account to post, you can freely browse the topics and look for answers to your questions in the years of posts. Additionally, once you sign up, you can direct message any forum user, upload images or SketchUp files, and even bookmark topics or follow specific users. Not only is it a great place to get questions you have answered (and possibly answer someone else's), but it is a very active community with a **Gallery** where you can share your work and a **Corner Bar** where you can just share thoughts about anything going on that is not SketchUp-related.

SketchUcation

The SketchUcation forum (`https://sketchucation.com`) has been around longer and boasts a larger user list than SketchUp Community. Started and run completely by SketchUp users, it is a great place to find answers and talk about SketchUp:

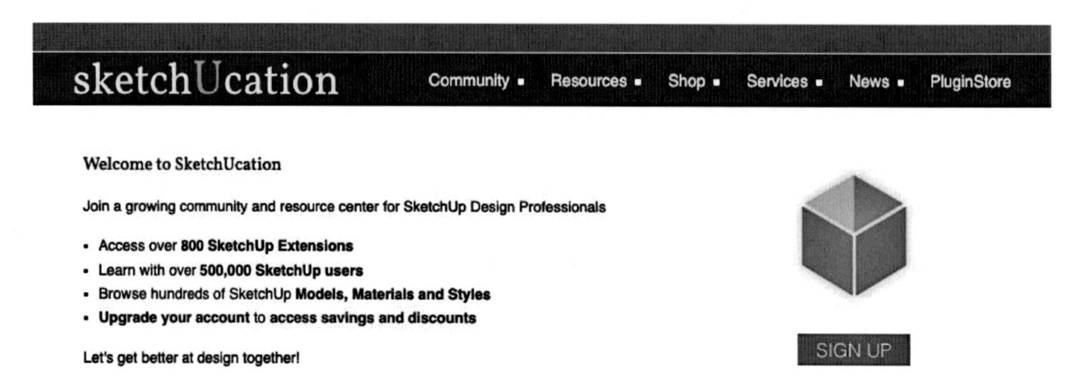

Figure 15.11 – SketchUcation

Similar to SketchUp Community, you will have to create an account to post at SketchUcation. Unlike SketchUp Community, when you sign up, you have the option to pick a paid membership. Since SketchUcation is run and supported completely by community members, paid memberships help them keep the lights on. SketchUcation offers many of the same abilities as SketchUp Community, but it also includes access to exclusive resources and services. Additionally, there are quite a few extension developers who run their beta testing and development through the forum, so you may have the opportunity to interact with the creator of your favorite extensions!

SketchUp User Group on Facebook

While not technically a forum, the SketchUp User Group on Facebook contains a lot of information on SketchUp and is a great place to show off your work, find resources, and get answers to your SketchUp questions. If you are a regular Facebook user, this is worth joining. You will have to join the group and have your application reviewed before you can get in, but it is worth it!

YouTube channels

When it comes to the best way to learn more about SketchUp, it is hard to think of a better platform than video. Don't get me wrong (he said as he wrote a 15-chapter book), I truly love books. I think that there is something great about having teaching information on a page that you can refer to time and time again or something that you can open and read while lying in bed or on the train. If you need a quick run-through of a specific SketchUp tool or wish to review a specific extension, YouTube probably has a channel that can help you out!

SketchUp

The official SketchUp channel (`https://www.youtube.com/c/SketchUp`) on YouTube hosts over 1,000 videos about SketchUp. These range from short, simple weekly training videos (called Skill Builders) to user case studies, to weekly live streams:

Figure 15.12 – The official SketchUp YouTube channel

Now, I do have to admit to a little bit of bias (I am the guy responsible for the majority of the content on this channel), but I will say that this is a YouTube channel worth following if you have any interest in SketchUp. The content is always fresh (at least two videos are uploaded each week and usually a video podcast and a 2-hour live modeling session), and the videos are high quality. Even if you do not subscribe for the recurring content, it is worth bookmarking so that you can run back there next time you have a question about a specific tool or workflow.

The SketchUp Essentials

My friend Justin Geis runs a channel called *The SketchUp Essentials* (`https://www.youtube.com/c/Thesketchupessentials`) and I love it. Justin is always upbeat and excited to talk about SketchUp and has some content that you just cannot find anywhere else:

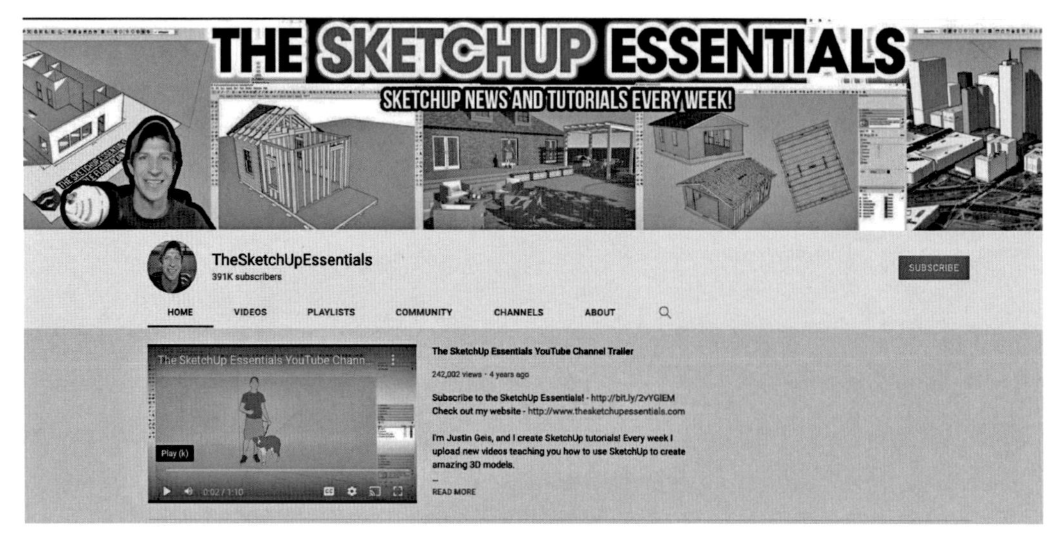

Figure 15.13 – The SketchUp Essentials YouTube channel

Justin posts videos on how to use SketchUp and the process he goes through to model certain items and shapes. The true gold on this channel, in my opinion, is the extension reviews. Justin knows how to use SketchUp and can model like a master with native tools, but the content he creates that I appreciate the most is when he installs an extension and goes through every facet to show you exactly how it works. Once again, this channel is more than worth a subscription as a SketchUp user.

Online training

If you are the type who wants a more structured approach to learning about SketchUp, here are a few resources that are worth checking out.

SketchUp Campus

SketchUp Campus (`https://learn.sketchup.com/`) is the official learning resource from Trimble. SketchUp Campus offers many multi-hour courses that can be taken all at once, or parted out, allowing you to learn over time. Since SketchUp Campus is a part of Trimble, you can use your Trimble ID and password to log in and access every course for free.

SketchUp School

Alex Olivier at SketchUp School (`https://www.sketchupschool.com/`) does an amazing job creating online learning materials for SketchUp. SketchUp School offers a huge backlog of training courses that cover all sorts of workflows. While you do have to create a paid subscription to take the courses, you can check out the quality of their work on their YouTube channel (`https://www.youtube.com/c/sketchupschool`).

Master SketchUp

Matt Donley at Master SketchUp (`https://mastersketchup.com/`) has created a site that can cover you no matter how you prefer to learn. Master SketchUp offers books, video tutorials, and online training courses. You will need to create an account to get access to the content at Master SketchUp, but you will find it well worth the effort!

I should wrap this up by stating that there are even more resources out there than I could list here. The SketchUp community is an amazing place with people constantly churning out new learning content daily. If you are a creator and I have missed you, I apologize for missing you!

Summary

Having a SketchUp Pro subscription does not just mean you can install and run SketchUp Pro on your computer. The subscription offers many tools, such as the viewers and Trimble Connect, as well as the ability to run SketchUp in a browser window or on an iPad. Outside of the subscription, remember that, as a SketchUp user, you are a part of a vibrant community of like-minded 3D modelers that are ready to share their thoughts and experiences with you whenever you are ready.

Well, that brings us to the end of this book. I truly hope that reading this book was worth your time. I did my best to try to hand you the tools that you need to take your SketchUp journey to the next level. Remember, not everyone uses SketchUp the same way. Your process and your workflows are specific to you. While this means you will not find a document that tells you step by step how to model exactly what you need to model, it does mean that you get to create your own path. In the end, like the models you create, your SketchUp workflow has been created by you.

Now, go do some 3D modeling!

Index

Symbols

A

B

C

D

E

N

O

T

U

V

W

Y

Z

Hi!

I am Aaron Dietzen, author of Taking SketchUp Pro to the Next Level. I really hope you enjoyed reading this book and found it useful for increasing your productivity and efficiency in SketchUp Pro.

It would really help me (and other potential readers!) if you could leave a review on Amazon sharing your thoughts on Taking SketchUp Pro to the Next Level here.

Go to the link below or scan the QR code to leave your review:

`https://packt.link/r/1803242698`

Your review will help me to understand what's worked well in this book, and what could be improved upon for future editions, so it really is appreciated.

Aaron Dietzen aka 'The SketchUp Guy'

Packt.com

Subscribe to our online digital library for full access to over 7,000 books and videos, as well as industry leading tools to help you plan your personal development and advance your career. For more information, please visit our website.

Why subscribe?

- Spend less time learning and more time coding with practical eBooks and Videos from over 4,000 industry professionals
- Improve your learning with Skill Plans built especially for you
- Get a free eBook or video every month
- Fully searchable for easy access to vital information
- Copy and paste, print, and bookmark content

Did you know that Packt offers eBook versions of every book published, with PDF and ePub files available? You can upgrade to the eBook version at packt.com and as a print book customer, you are entitled to a discount on the eBook copy. Get in touch with us at customercare@packtpub.com for more details.

At www.packt.com, you can also read a collection of free technical articles, sign up for a range of free newsletters, and receive exclusive discounts and offers on Packt books and eBooks.

Other Books You May Enjoy

If you enjoyed this book, you may be interested in these other books by Packt:

Learn SOLIDWORKS – Second Edition

Tayseer Almattar

ISBN: 978-1-80107-309-7

- Understand the fundamentals of SOLIDWORKS and parametric modeling
- Create professional 2D sketches as bases for 3D models using simple and advanced modeling techniques
- Use SOLIDWORKS drawing tools to generate standard engineering drawings
- Evaluate mass properties and materials for designing parts and assemblies
- Join different parts together to form static and dynamic assemblies

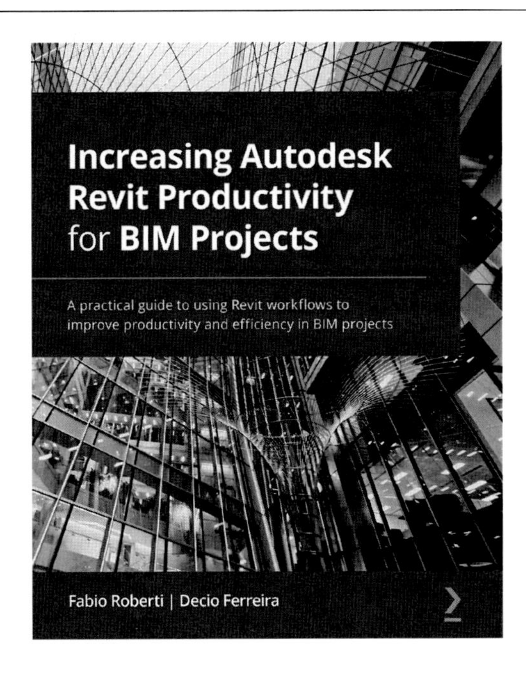

Increasing Autodesk Revit Productivity for BIM Projects

Fabio Roberti , Decio Ferreira

ISBN: 978-1-80056-680-4

- Explore the primary BIM documentation to start a BIM project
- Set up a Revit project and apply the correct coordinate system to ensure long-term productivity
- Improve the efficiency of Revit core functionalities that apply to daily activities
- Use visual programming with Dynamo to boost productivity and manage data in BIM projects
- Import data from Revit to Power BI and create project dashboards to analyze data

Packt is searching for authors like you

If you're interested in becoming an author for Packt, please visit `authors.packtpub.com` and apply today. We have worked with thousands of developers and tech professionals, just like you, to help them share their insight with the global tech community. You can make a general application, apply for a specific hot topic that we are recruiting an author for, or submit your own idea.

Made in United States
Orlando, FL
27 June 2023

34578277R00248